THE CONCISE HANDBOOK OF

ANALYTICAL SPECTROSCOPY

Theory, Applications, and Reference Materials

Volume 4

Infrared Spectroscopy

THE CONCISE HANDBOOK OF ANALYTICAL SPECTROSCOPY

Theory, Applications, and Reference Materials

Volume 4

Infrared Spectroscopy

Jerry Workman

Unity Scientific, USA and National University, USA

World Scientific

NEW JERSEY • LONDON • SINGAPORE • BEIJING • SHANGHAI • HONG KONG • TAIPEI • CHENNAI • TOKYO

Published by

World Scientific Publishing Co. Pte. Ltd.
5 Toh Tuck Link, Singapore 596224
USA office: 27 Warren Street, Suite 401-402, Hackensack, NJ 07601
UK office: 57 Shelton Street, Covent Garden, London WC2H 9HE

Library of Congress Cataloging-in-Publication Data
Workman, Jerry.
The concise handbook of analytical spectroscopy : theory, applications, and reference materials / by Jerry Workman (Unity Scientific, USA & National University, USA).
volumes cm
Includes bibliographical references and index.
ISBN 978-9814508056 (Set)
ISBN 978-9814508087 (v. 1 : hardcover : alk. paper)
ISBN 978-9814508094 (v. 2 : hardcover : alk. paper)
ISBN 978-9814508100 (v. 3 : hardcover : alk. paper)
ISBN 978-9814508117 (v. 4 : hardcover : alk. paper)
ISBN 978-9814508124 (v. 5 : hardcover : alk. paper)
1. Spectrum analysis. I. Title.
QD95.W67 2015
543'.5--dc23
2014042606

British Library Cataloguing-in-Publication Data
A catalogue record for this book is available from the British Library.

Typeset by Stallion Press
Email: enquiries@stallionpress.com

Printed in Singapore

Preface

The Concise Handbook of Analytical Spectroscopy: Theory, Applications, and Reference Materials (In 5 Volumes)

Overview and Description

The concept of improving the use of electromagnetic energy to achieve a variety of qualitative and quantitative spectroscopic measurements on solid and liquid materials has been proliferating at a rapid rate. The use of such technologies to measure chemical composition and appearance; for classification; and to achieve a detailed understanding of material interactions has prompted a dramatic expansion in the use and development of spectroscopic techniques over a variety of academic and commercial fields. This multi-volume handbook is designed specifically as a reference tool for students, commercial development and quality scientists, and researchers or technologists in a variety of measurement endeavors. Each volume covers the theory, instrumentation, sampling methods, experimental design, and data analysis techniques, as well as essential reference tables, figures, and spectra for each spectroscopic region. The detailed practical aspects of applying spectroscopic tools for many of the most exciting and current applications are covered. Featured applications are included as well as common commercial analysis methods (i.e., color, appearance), spectroscopic quantitative and qualitative techniques, and advanced methods.

Outline

Each volume covers a specific region of the electromagnetic spectrum used for electronic and vibrational spectroscopic measurements. The volumes are not a series of essays, or a compilation of isolated papers on specific topics, but represent an integrated whole. The Handbook is compiled as a highly useful reference set for those directly involved in making and understanding spectroscopic measurement techniques. Each volume contains a logical progression of the material as follows.

Organization of Contents for Each Volume

Each volume commences with Front Reference Material for General Spectroscopy, preceding the following chapters.

1. General Introduction and History of Technique (includes useful basic concepts)
2. Theory of Technique (including delineation of common equations and references)
3. Types of Instrumentation (Types and design aspects)
4. Sampling Considerations (Cookbook approach to preparing samples)
5. Spectrum Data Acquisition (Experimental design and methods for data acquisition)
6. Data Analysis and Presentation Methods (The chemometrics and data analysis techniques applicable for each spectral region)
7. Applications and Spectral Examples (Cookbook approaches, results, and extensive literature reviews)
8. Reference Materials, Tables, Figures, and Spectra (The key look-up aspects for each technique)
9. Glossary of Specific Terms: Detailed Index for Reference Material and Text (Each volume includes a detailed table of contents and index that delineates the minutia of using the text and reference material in each volume.)

Each volume contains specifics for the spectral region covered, such as: common and unusual sampling devices; sampling methods (i.e., reagents, solvents, and sampling cells); and reference standard materials. Optics include: filters, etalons, prisms, and lenses; and fiber optics (principles, equipment, and connectors). In addition, reference tables and spectra, glossaries of terms, common equations and calculation methods, cookbook directions, and solved problems are included.

Reference Materials

Fundamental reference sources and early publications are referred to whenever possible as a primary source of data and information. Reference material common to each method is summarized within each volume to provide a more practical set of volumes for bench use. The intention is that each volume will stand somewhat alone for the working analyst using a specific spectroscopic method, such that required referral to other volumes is kept to a minimum and that key information is available regarding the individual spectroscopic technique within each volume.

Terminology and Symbols

Terminology use is included within the chapters and glossary sections for each volume. Note that *spectrophotometer* is the precise term used for each instrument described within the five volumes. However for brevity, the term *spectrometer* is sometimes used less precisely. Also, the symbols used are from different disciplines and thus there are some symbols used for more than one variable or parameter, depending on the common usage and the specific academic discipline where the symbology was formalized. A list of symbols used in this set of volumes may be located using the index and front material of each chapter. The terms used for each equation within the volumes are defined where the equation is used.

Dedication

This book is dedicated to my fabulous wife Rebecca and to my family who have patiently sacrificed many hours of their lives for the cause of spectroscopy; and to everyone who spends many hours in laboratories analyzing samples for new knowledge, applied research, quality improvement, and the overall betterment of the lives of people… And to the One Great Designer who made all this marvelous measurement with light possible, discoverable, and such a pleasure to work with…

Author Biography

Jerome (Jerry) J. Workman, Jr. is Executive Vice President of Research & Engineering for Unity Scientific; Certified Core Adjunct Professor at National University, CA; and Principal at Biotechnology Business Associates. He was formerly Vice President of Technology Research for Masimo Corporation; Director of Research, Technology & Applications Development for Molecular Spectroscopy & Microanalysis for ThermoFisher Scientific; Senior Research Fellow at Kimberly-Clark Analytical Science & Technology; and Principal Scientist at Perkin-Elmer.

Dr. Workman has played a major role in defining and developing over twenty scientific instrument advancements and applications of novel vibrational spectroscopy technology for laboratory, process and remote monitoring of commercial synthetic and natural products and processes; and medical applications of spectroscopy. Over 55 U.S. and International Patent Applications (since 1998); 17 U.S. and International Patents issued, and multiple trade secrets. He has a total of 475 Technical Publications; and 18 reference books on a broad range of spectroscopy, chemometrics, and data processing techniques. He has received awards from the Eastern Analytical Symposium, ASTM International, and the Coblentz Society, as well as multiple fellowships, technical, and government appointments. He has taught annual courses in spectroscopy, chemometrics, and statistics for the Association of Official Analytical Chemists, the American Chemical Society, the Instrument Society of America, and the Federation of Analytical Chemists and Spectroscopy Societies; and at several universities, government agencies, and corporations.

Patented inventions include: multi-spectral monitoring (data fusion), web-based information technology in spectroscopy, scanning systems for positioning materials in manufacturing technology, invention of small molecule metabolite reporters for monitoring multiple parameters in human metabolism, advanced technology and algorithm development for instrument automation, standardization, and calibration, and for manufacturing monitoring methods for consumer products.

He is a Fellow of the American Institute of Chemists, the American Society for Testing and Materials International, and the Royal Society of Chemistry (U.K.). He is also a Chartered Chemist (CChem) of the Royal Society of Chemistry (England) and a Chartered Scientist of the Science Council (UK). He has served as Chair of the Industrial Advisory Board for the Center for Process Analytical Chemistry (CPAC) at the Univ. of Washington; and The Council for Near Infrared Spectroscopy; and is past Chairman of ASTM Main Committee E13 on Molecular Spectroscopy and Separation Science. He has served on the U.S. National Academies NRC Panel for assessment of NIST programs.

With a B.A. in natural sciences and an M.A. in biological sciences from Saint Mary's University of Minnesota, Dr. Workman's Ph.D. is in biological chemistry from Columbia Pacific University. He is a graduate of the Columbia Senior Executive Program and also holds Columbia Business School Certificates in Executive Development (CIED) and in Business Excellence (CIBE). He also holds a Certificate in Strategy and Innovation from the M.I.T. Sloan School. He is listed in Who's Who in the World, Who's Who in America, and Who's Who in Science and Engineering.

Contents

Reference Tables: Comparison of Spectroscopic Techniques, the Electromagnetic Spectrum, Beer's Law, and Useful Spectral Ranges for Sources, Detectors and Optical Materials

0.1 Physical Constants and Conversion Factors for SI and non-SI Units

Table 0.1. Spectroscopy constants.[1]

Physical constant	Value	Standard uncertainty	Concise form
Avogadro constant. N_A, L	$6.022\ 141\ 29 \times 10^{23}$ mol^{-1}	$0.000\ 000\ 27 \times 10^{23}$ mol^{-1}	$6.022\ 141\ 29(27) \times 10^{23}$ mol^{-1}
Boltzmann constant, k	$1.380\ 6488 \times 10^{-23}$ J K^{-1}	$0.000\ 0013 \times 10^{-23}$ J K^{-1}	$1.380\ 6488(13) \times 10^{-23}$ J K^{-1}
Faraday constant, F	96 485.3365 C mol^{-1}	0.0021 C mol^{-1}	96 485.3365(21) C mol^{-1}
Molar gas constant, R	8.314 4621 J mol^{-1} K^{-1}	0.000 0075 J mol^{-1} K^{-1}	8.314 4621(75) J mol^{-1} K^{-1}
Planck constant, h	$6.626\ 069\ 57 \times 10^{-34}$ J s	$0.000\ 000\ 29 \times 10^{-34}$ J s	$6.626\ 069\ 57(29) \times 10^{-34}$ J s
Speed of light (in vacuum), c	299 792 458 m s^{-1}	(exact)	299 792 458 m s^{-1}
Stefan-Boltzmann constant, ö	$5.670\ 373 \times 10^{-8}$ W m^{-2} K^{-4}	$0.000\ 021 \times 10^{-8}$ W m^{-2} K^{-4}	$5.670\ 373(21) \times 10^{-8}$ W m^{-2} K^{-4}

Table 0.2. Wavelength conversions.[2]

To convert Column 1 into Column 2 multiply by	Column 1 SI Units	Column 2 non-SI Units	To convert Column 2 into Column 1 multiply by
2.54	Centimeter, cm (10^{-2} m)	Inch, in	0.394
25.4	Millimeter, mm (10^{-3} m)	Inch, in	3.94×10^{-2}
1.0	Micrometer, μm (10^{-6} m)	Micron, μ	1.0
0.1	Nanometer, nm (10^{-9} m)	Angstrom, A (10^{-10} m)	10
$1/(\text{nm} \times 10^7)$	Nanometer, nm (10^{-9} m)	Wavenumber (cm^{-1})	$1/(\text{cm}^{-1} \times 10^7)$
$1/(\text{nm} \times 2.99792458 \times 10^{17})$	Nanometer, nm (10^{-9} m)	Frequency (Hertz, Hz)	$1/(\text{Hz} \times 2.99792458 \times 10^{17})$

Example 1: To convert Nanometer, nm (10^{-9} m) in Column 1 to Angstrom, A (10^{-10} m) in Column 2 to multiply by 0.1; therefore 0.1 nm = 1 Angstrom.

Example 2: To convert Angstrom, A (10^{-10} m) in Column 2 to Nanometer, nm (10^{-9} m) in Column 1 multiply by 10; therefore 10 Angstroms = 1 nm.

Table 0.3. Length conversions.

To convert Column 1 into Column 2 multiply by	Column 1 SI Units	Column 2 non-SI Units	To convert Column 2 into Column 1 multiply by
1.609	Kilometer, km (10^3 m)	Mile, mi	0.6213
0.914	Meter, m	Yard, yd	1.094
0.304	Meter, m	Foot, ft	3.28
1.0	Micrometer, μm (10^{-6} m)	Micron, μ	1.0
1000	Micrometer, μm (10^{-6} m)	Millimeter, mm (10^{-3} m)	0.001
25.4	Millimeter, mm (10^{-3} m)	Inch, in	3.94×10^{-2}
0.1	Nanometer, nm (10^{-9} m)	Angstrom, A	10

Table 0.4. Area conversions.

To convert Column 1 into Column 2 multiply by	Column 1 SI Units	Column 2 non-SI Units	To convert Column 2 into Column 1 multiply by
0.405	Hectare, ha	Acre	2.47
4.05×10^{-3}	Square kilometer, km^2 $(10^3\ m)^2$	Acre	2.47
2.590	Square kilometer, km^2 $(10^3\ m)^2$	Square mile, mi^2	0.386
4.05×10^{3}	Square meter, m^2	Acre	2.47×10^{-4}
9.29×10^{-2}	Square meter, m^2	Square foot, ft^2	10.76
645	Square millimeter, mm^2 $(10^{-3}\ m)^2$	Square inch, in^2	1.55×10^{-3}

Table 0.5. Volume conversions

To convert Column 1 into Column 2 multiply by	Column 1 SI Units	Column 2 non-SI Units	To convert Column 2 into Column 1 multiply by
102.8	Cubic meter, m^3	Acre-inch	9.73×10^{-3}
2.83×10^{-2}	Cubic meter, m^3	Cubic foot. ft^3	35.3
1.64×10^{-5}	Cubic meter, m^3	Cubic inch, in^3	6.10×10^{4}
35.24	Liter, L $(10^{-3}\ m^3)$	Bushel, bu	2.84×10^{-2}
0.946	Liter, L $(10^{-3}\ m^3)$	Quart (liquid), qt	1.057
28.3	Liter, L $(10^{-3}\ m^3)$	Cubic foot, ft^3	3.53×10^{-2}
3.78	Liter, L $(10^{-3}\ m^3)$	Gallon	0.265
2.96×10^{-2}	Liter, L $(10^{-3}\ m^3)$	Ounce (fluid), oz	33.78
0.473	Liter, L $(10^{-3}\ m^3)$	Pint (fluid), pt	2.11

Table 0.6. Mass conversions.

To convert Column 1 into Column 2 multiply by	Column 1 SI Units	Column 2 non-SI Units	To convert Column 2 into Column 1 multiply by
453.6	Gram, g (10^{-3} kg)	Pound, lb	2.2×10^{-3}
28.4	Gram, g (10^{-3} kg)	Ounce (Avoirdupois, avdp), oz	3.52×10^{-2}
0.454	Kilogram, kg	Pound, lb	2.205
100	Kilogram, kg	Quintal (metric), q	0.01
907	Kilogram, kg	Ton (2000 lb), U.S. ton	1.10×10^{-3}
0.907	Megagram, Mg (Tonne)	Ton (U.S.), ton	1.102
0.907	Tonne, t	Ton (U.S.), ton	1.102

Table 0.7. Rate conversions.

To convert Column 1 into Column 2 multiply by	Column 1 SI Units	Column 2 non-SI Units	To convert Column 2 into Column 1 multiply by
0.447	Meter per second, m s^{-1}	Mile per hour	2.237

Table 0.8. Specific surface conversions.

To convert Column 1 into Column 2 multiply by	Column 1 SI Units	Column 2 non-SI Units	To convert Column 2 into Column 1 multiply by
0.1	Square meter per kilogram, $m^2 kg^{-1}$	Square centimeter per gram, $cm^2 g^{-1}$	10
0.001	Square meter per kilogram, $m^2 kg^{-1}$	Square millimeter per gram, $mm^2 g^{-1}$	1000

Table 0.9. Density conversions.

To convert Column 1 into Column 2 multiply by	Column 1 SI Units	Column 2 non-SI Units	To convert Column 2 into Column 1 multiply by
1.00	Megagram per cubic meter, Mg m^{-3}	Gram per cubic centimeter, g cm^{-3}	1.00

Table 0.10. Pressure conversions.

To convert Column 1 into Column 2 multiply by	Column 1 SI Units	Column 2 non-SI Units	To convert Column 2 into Column 1 multiply by
0.101	Megapascal, MPa (10^6 Pa)	Atmosphere	9.869
0.1	Megapascal, MPa (10^6 Pa)	Bar	10
47.9	Pascal, Pa	Pounds per square foot, lb ft^{-2}	2.09×10^{-2}
6.90×10^3	Pascal, Pa	Pounds per square inch, lb in^{-2}	1.45×10^{-4}

Table 0.11. Temperature conversions.

To convert Column 1 into Column 2 multiply by	Column 1 SI Units	Column 2 non-SI Units	To convert Column 2 into Column 1 multiply by
100(K-273)	Kelvin, K	Celsius, °C	1.00 (°C + 273)
(°R − 491.67) × 5/9	Rankine, °R	Celsius, °C	(°C + 273.15) × 9/5
(9/5 °C) + 32	Celsius, °C	Fahrenheit, °F	5/9 (°F − 32)
°R − 459.67	Rankine, °R	Fahrenheit, °F	°F + 459.67
[K] × 9/5 − 459.67	Kelvin, K	Fahrenheit, °F	(°F + 459.67) × 5/9
[K] × 9/5	Kelvin, K	Rankine, °R	[°R] × $^5/_9$

Table 0.12. Energy, work, quantity of heat conversions.

To convert Column 1 into Column 2 multiply by	Column 1 SI Units	Column 2 non-SI Units	To convert Column 2 into Column 1 multiply by
1.05×10^{3}	Joule, J	British thermal unit, Btu	9.48×10^{-4}
4.19	Joule, J	Calorie, cal	0.239
10^{-7}	Joule, J	erg	107
1.36	Joule, J	Foot-pound	0.735
4.19×10^{4}	Joule per square meter, $J\,m^{-2}$	Calorie per square centimeter (Langley)	2.387×10^{-5}
10^{-5}	Newton, N	Dyne	10^{5}
698	Watt per square meter, $W\,m^{-2}$	Calorie per square centimeter minute (irradiance), $cal\,cm^{-2}\,m^{-1}$	1.43×10^{-3}

Table 0.13. Plane angle conversions.

To convert Column 1 into Column 2 multiply by	Column 1 SI Units	Column 2 non-SI Units	To convert Column 2 into Column 1 multiply by
$1.75 \times 10^{-2} = \pi/180$	Radian, rad	Degrees (angle), °	57.3
60	Minutes (of arc), arcminute	Degrees (angle), °	$1/60 = 0.01667$
3600	Seconds (of arc), arcsecond	Degrees (angle), °	$1/3600 = 0.0002778$

Table 0.14. Electricity and magnetism conversions.

To convert Column 1 into Column 2 multiply by	Column 1 SI Units	Column 2 non-SI Units	To convert Column 2 into Column 1 multiply by
10	Siemen per meter, S m-1	Millimho per centimeter, mmho cm^{-1}	0.1
10^{-4}	Tesla, T	Gauss, G	10^{4}

Table 0.15. Fluid flow conversions.

To convert Column 1 into Column 2 multiply by	Column 1 SI Units	Column 2 non-SI Units	To convert Column 2 into Column 1 multiply by
102.8	Cubic meter, m^3	Acre-inch, acre-in	9.73×10^{-3}
101.9	Cubic meter per hour, $m^3 h^{-1}$	Cubic foot per second, $ft^3 s^{-1}$	9.81×10^{-3}
0.227	Cubic meter per hour, $m^3 h^{-1}$	U.S. gallon per minute, gal min^{-1}	4.40
0.123	Hectare meter, ha m	Acre-foot, acre-ft	8.11
1.03×10^{-2}	Hectare meter, ha m	Acre-inch, acre-in	97.28
12.33	Hectare centimeter, ha cm	Acre-foot, acre-ft	8.1×10^{-2}

Table 0.16. Concentration conversions.

To convert Column 1 into Column 2 multiply by	Column 1 SI Units	Column 2 non-SI Units	To convert Column 2 into Column 1 multiply by
1	Centimole per kilogram, cmol kg^{-1}	Milliequivalent per 100 grams, meq 100 g^{-1}	1
0.1	Gram per kilogram, g, kg^{-1}	Percent, %	10
1	Milligrams per kilogram, mg kg^{-1}	Parts per million, ppm	1

Table 0.17. Radioactivity conversions.

To convert Column 1 into Column 2 multiply by	Column 1 SI Units	Column 2 non-SI Units	To convert Column 2 into Column 1 multiply by
3.7×10^{10}	Becquerel, Bq	Curie, Ci	2.7×10^{-11}
37	Becquerel per kilogram, Bq kg^{-1}	Picocurie per gram, pCi g^{-1}	2.7×10^{-2}
0.01	Gray, Gy (Absorbed dose)	Rad, rd	100
0.01	Sievert, Sv (equivalent dose)	Rem (Roentgen equivalent man)	100

0.2 Introduction and Comparison of Spectroscopic Methods[3]

0.2.1 *Ultraviolet (UV) Spectroscopy*

The Ultraviolet region is specified as 190 to 360 nanometers (nm) or 10^{-9} meters. The types of electrons that can be excited by UV/Vis light are few in number: nonbonding electrons, electrons in single bonds, and electrons involved in double bonds. These may be excited to several excited states. The distinction between molecules is that the ability to "jump" to higher states is affected by attached moieties — for example: double bonds, conjugations, and elements such as: oxygen, bromine, and others with pairs of nonbonding electrons. As a consequence, most elements with UV/Vis absorbances have specific wavelengths at which their peak absorbances occur. These peaks may be used to identify a particular molecule. The UV/Vis region is not as "rich" in information as, say, the infrared spectrum, but it reveals enough detail to enable a comparison of a material with a previously identified substance. A common use of this capability is in the Pharmaceutical industry, where UV/Vis detectors are frequently used with HPLC instruments as a final check before a drug product is released for sale to consumers. The various moieties or chromophores associated with ultraviolet absorption include: Nitriles (R-C≡N), 160 nm; Acetylenes (-C≡C-), 170 nm; Alkenes (>C=C<), 175 nm; Alcohols (R-OH), 180 nm and 175–200 nm; Ethers (R-O-R), 180 nm; Ketones (R-C=O -R′), 180 nm and 280 nm; Amines, primary ($R\text{-}NH_2$), 190 nm and 200–220 nm; Aldehydes (R-C=O-H), 190 nm and 290 nm; Carboxylic acids (R-C=O-OH), 205 nm; Esters (R-C=O-OR), 205 nm; Amides, primary ($R\text{-}C{=}O\text{-}NH_2$), 210 nm; Thiols (R-SH), 210 nm; Nitrites ($R\text{-}NO_2$), 271 nm; and Azo-group (R-N=N-R′), 340 nm.

0.2.2 *Visible (Vis) Spectroscopy*

Visible light (360 to 780 nm) is the light or electromagnetic energy seen by the human eye. The actual light demonstrated by different colors is visible light and spectra in the visible region are subjected to specific mathematical algorithms to determine color coordinates or color space coordinates to enable specific colors and brightness or darkness to be precisely described in mathematical terms. The aspects of color measurement and computation are contained in the text volume on visible spectroscopy and may be located in the index for this set of volumes. The terms "color measurement" and "visible spectroscopy" are the main index topic headings to find detailed information in this set of volumes. Colors are produced by electrons in a pigment moving from one orbital transition to another around the atoms within the molecules of the colored substance. Visible reflectance or transmittance spectra demonstrate the various colors. The color appearance of the human eye to a particular visible spectrum is dependent on: the light source color temperature and emission spectrum, the observer angle of observation, and the object background brightness and color differences. The size, direction, and scattering/absorption characteristics of the object also have an effect on its color appearance. The basic seven colors first described by Isaac Newton, with their corresponding wavelength scale in nanometers, consist of: violet (360–415 nm); indigo (415–444 nm), blue (444–487 nm), green (487–540 nm), yellow (540–590 nm), orange (590–690 nm), and red (690–830 nm). A twelve color system, with corresponding wavelength information is more common in modern terminology, although technically color is defined by a series of color measurement coordinate systems. These color measurement coordinate systems include: XYZ tristimulus values, Yxy color space, L*a*b* color space, L*C*h color space, and Hunter Lab color space. These are described in detail within sections of these volumes (see index and coordinate system names for more information).

There are also molecular absorptions that occur within the visible region and these are described in various chapters within this text and include the notations for the harmonic band or overtone involved: O-H Alkyl alcohol (6ν, no hydrogen bonding), Aromatic C-H Stretch (5ν), O-H Alkyl alcohol (4ν, no hydrogen bonding), Methyl C-H Stretch (5ν), O-H Phenols (4ν, no hydrogen bonding), Methylene C-H Stretch (5ν), O-H Primary Alcohols (4ν), O-H Water (4ν), O-H Secondary Alcohols (4ν), O-H Tertiary Alcohols (4ν), and Alkenes, conjugated RC=C-C=C-R′. The visible spectra include information related to both electronic and molecular interactions of energy with atoms and molecules. Volume 2, Chapter 7 gives more details.

0.2.3 *Near Infrared (NIR) Spectroscopy*

Near infrared spectroscopy (780–2500 nm) is used where multicomponent molecular vibrational analysis is required in the presence of interfering substances. The near infrared spectra consist of overtones and combination bands of the fundamental molecular absorptions found in the mid infrared region. Near infrared spectra consist of generally overlapping vibrational bands that are non-specific and poorly resolved. The use of chemometric mathematical data processing can be used to calibrate for qualitative or quantitative analysis despite these apparent spectroscopic limitations. Traditional near infrared spectroscopy was used in agricultural product analysis for lignin polymers (2270 nm), paraffins and long alkane chain polymers (2310 nm), glucose based polymers such as cellulose (2336 nm), amino acid polymers as proteins (2180 nm), carbohydrates (2100 nm), and moisture (1440 and 1940 nm). The dominant near infrared spectral features include the methyl C-H stretching vibrations, methylene C-H stretching vibrations, aromatic C-H stretching vibrations, O-H stretching vibrations, methoxy C-H stretching, and carbonyl associated C-H stretching. In addition, N-H from primary amides, secondary amides (both alkyl, and aryl group associations), N-H from primary, secondary, and tertiary amines, and N-H from amine salts predominate near infrared spectral features of polymers and organic compounds.

0.2.4 *Infrared (IR or Mid-IR) Spectroscopy*

Mid infrared spectroscopy (2500–25,000 nm or 4000-400 cm^{-1}) provides a measurement technique for intense, isolated and reliable absorption bands of fundamental molecular vibrations from polymers and other organic compounds. The spectrometric methodology allows for univariate calibration with higher signal strength (absorptivity) required for solid, liquids or gas phase measurements. Relatively small pathlengths of 0.1 to 1.0 mm are required for hydrocarbon liquids and solids. The technique is generally incompatible with the use of fiber optics, but specialized fiber materials exist. Instrumentation is higher cost than near infrared spectrophotometers for the most part.

Dominant mid infrared spectral features include the C-H (methyl, methylene, aromatic, methoxy, and carbonyl) fundamental stretching and bending molecular vibrations, O-H (hydroxyl) stretch fundamental vibrations; N-H (amine) stretching, C-F (fluorocarbon) stretching, -C≡N (nitrile) stretching, -C=O (carbonyl) stretch from esters, acetates, and amides; C-Cl stretch from chlorinated hydrocarbons, and -NO_2 from nitro- containing compounds.

0.2.5 *Raman Spectroscopy*

Raman spectroscopy (2500–25,000 nm or 4000-400 cm^{-1}) provides a variety of measurements on samples that are aqueous in nature or where glass sample holders are present. Carbon dioxide, water, and glass (silica) are weak scatterers and thus there is generally no problems in analyzing samples having these properties. There is typically no sample preparation involved in samples measured using Raman. Raman spectroscopy is complementary to mid infrared spectroscopy in the measurement of fundamental molecular vibrations. Raman measurements are compatible with fiber optics. Raman measurements exhibit high signal-to-noise and a reasonable cost for instrumentation. The dominant Raman spectral features are acetylenic C≡C stretching, olefinic C=C stretch at 1680–1630 cm^{-1}, N=N (azo-) stretching, S-H (thio-) stretching, C=S stretching, C-S stretching, and S-S stretching bands.

Raman spectra also contain such molecular vibrational information as CH_2 twist and wagging, carbonyl C=O stretch associated with esters, acetates, and amides; C-Cl (halogenated hydrocarbons) stretching, and -NO_2 (nitro-/nitrite) stretching. In addition, Raman yields information content of phenyl- containing compounds at $1000\,cm^{-1}$.

0.3 The Electromagnetic Spectrum

The electromagnetic spectrum consists of many different types of radiation (energy): gamma, x-ray, ultraviolet (UV), visible (Vis), infrared (IR), microwave, and radio waves (Figures 0.1 through 0.3). Each type of radiation occupies its own region of the electromagnetic spectrum, and the major difference between the individual spectral regions is merely the energy involved. This energy is expressed in units of: photon energy (eV), Frequency (Hz), wavenumber (cm^{-1}), or wavelength (nm) and the manner in which these energies interact with matter.

With the possible exception of gamma radiation, most people are familiar with these types of radiation through their experiences in daily living. X-rays penetrate our bodies, allowing physicians to visualize our internal anatomy. UV light is associated with sunburn and tanning. We see colors and objects in the visible spectrum. In the kitchen, we toast our bread with IR radiation, and "zap" our meals with microwaves. We use microwave transmission for cell phone technology and we use radio waves to broadcast sound and images through AM/FM radio and television signals. The spectral region most useful in day-to-day analytical chemistry is the range of wavelengths from just below our visual perception (i.e., Ultraviolet or UV) through the colors we see (i.e., visible or Vis), known collectively as the UV/Vis spectral region. The wavelengths covered in the UV/Vis are measured in nanometers (nm), a unit of length representing one billionth of a meter. The generally accepted ranges for the UV/Vis region are ultraviolet (190–380 nm) and visible (380–750 nm). Some UV/Vis work involves longer wavelengths in the neighboring shortwave near-IR spectral region, from about 750–950 nm.

0.3.1 *Molecular and Electronic Spectroscopy Unit Conversions*

The units used to describe spectra for molecular spectroscopy vary with the academic or engineering discipline describing the spectral regions. For the most part the spectra are described in terms of an optical response specific to energy from a particular spectral region. The responses are in terms of inelastic or elastic collisions of the energy with various materials and are referred to as scattering absorption, absorbance, reflection, transmission, and so on. The various spectral energy for molecular and electronic spectroscopy is described in terms of units of wavelength (e.g., microns (μm) as 10^{-6} meters, or nanometers (nm) as 10^{-9} m, or Angstroms (Å) as 10^{-10} m). Energy is also expressed as frequency (cycles per second or sec.$^{-1}$) termed Hertz (Hz), as Wavenumbers (cm^{-1}), and as electron Volts (eV).

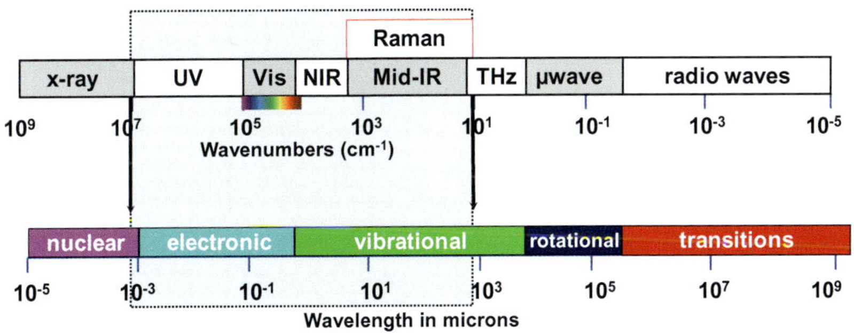

Fig. 0.1 The electromagnetic spectrum highlighting the ultraviolet to infrared/Raman spectral regions. Ultraviolet through Infrared/Raman range is shown in the rectangle. This region is from 190 nanometers to 25,000 nanometers, or 0.19 to 25 microns and represents electronic transitions through vibrational absorptions.

The various formulas used for the inter-conversion of units for spectroscopy include the following.

Molecular Electromagnetic Spectrum Regions

Region	Wavelength (nanometers)	Wavenumbers (cm^{-1})	Frequency (Hz)	Energy (eV)
Radio Waves	$1e^9$-$1e^{17}$	$1e^{-2}$-$1e^{-10}$	$3.0e^8$-3.0	$1.2e^{-6}$-$1.2e^{-14}$
Microwaves	$1e^6$-$1e^9$	10-0.01	$3.0e^{11}$-$3.0e^8$	$1.2e^{-3}$-$1.2e^{-6}$
Terahertz	2.5×10^4-10^6	400-10	$1.2e^{13}$-$3.0e^{11}$	0.05-0.001
Infrared	2500-25,000	4000-400	$1.2e^{14}$-$1.2e^{13}$	0.50-0.05
Raman	2500-25,000	4000-400	$1.2e^{14}$-$1.2e^{13}$	0.50-0.05
Near Infrared	780-2500	14,493-3,333	$3.8e^{14}$-$1.2e^{14}$	1.6-0.50
Visible	360-780	27,778-12,821	$8.3e^{14}$-$3.8e^{14}$	3.4-1.6
Ultraviolet	190-360	52,632-27,778	$1.6e^{15}$-$8.3e^{14}$	6.5-3.4
Vacuum UV	10-190	$1e^6$-$5.26e^4$	$3.0e^{16}$-$1.6e^{15}$	124-6.5
X-Rays	0.01-10	$1e^9$-$1e^6$	$3.0e^{19}$-$3.0e^{16}$	$124e^3$-124
Gamma Rays	<0.01	>$1e^9$	>$3e^{19}$	>$124e^3$

Fig. 0.2 Comparative Regions of the Electromagnetic Spectrum in wavelength (nanometers), wavenumbers, Frequency (Hertz), and Energy (electron Volts).

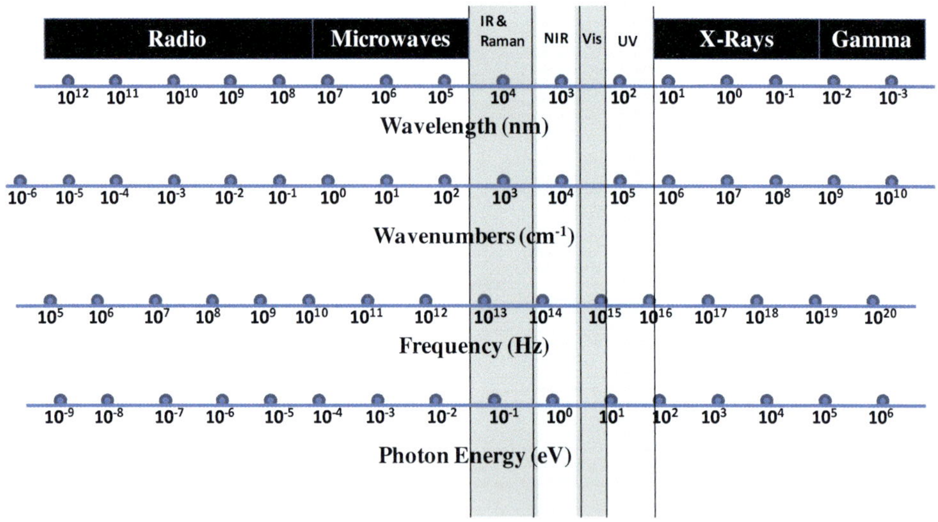

Fig. 0.3 Graphical Illustration of the Comparative Regions of the Electromagnetic Spectrum in wavelength (nanometers), wavenumbers, Frequency (Hertz), and Energy (electron Volts).

Light has both particle and wave properties, quantum theory tells us that the energy of a light "particle" or photon E_p is given by:

$$E_p = h\nu \tag{0.1}$$
$$E_p = hc\bar{\nu} \tag{0.2}$$

where: h = Planck's constant (or 6.6256×10^{-27} erg-sec.); and ν (nu) is the frequency of light, also known as Hertz (Hz) or the number of vibrations per second, in units of sec.$^{-1}$; $\bar{\nu}$ is the wavenumber units as cm^{-1} (i.e., the number of waves per centimeter), c is the velocity of light in a vacuum (or 2.9979×10^{10} cm·sec^{-1}).

$$\bar{\nu} = \frac{\nu}{c \cdot n^{-1}} = \frac{1}{\lambda} \tag{0.3}$$
$$\nu = \frac{c \cdot n^{-1}}{\lambda} \tag{0.4}$$

where: $\bar{\nu}$ is the wavenumber units as cm^{-1} (i.e., the number of waves per centimeter), c is the velocity of light in a vacuum (or 2.9979×10^{10} cm·sec^{-1}), ν (Greek: nu) is the frequency of light, also known as Hertz (Hz) or the number of vibrations per second, in units of (sec.$^{-1}$), and n is the refractive index of the medium the light is passing through (e.g., air = 1.0003), λ (Greek: lambda) is the wavelength in units of cm. Combining the above equations we also note:

$$\bar{\nu} = \frac{hc}{E_p} \Rightarrow \frac{1}{\lambda} = \frac{hc}{E_p} \Rightarrow \lambda = \frac{E_p}{hc} \tag{0.5}$$

0.3.2 *Direct Unit Conversions*

Wavenumbers (cm^{-1}) to wavelength (in nanometers):

$$\lambda_{nm} = \frac{10^7}{\bar{\nu}_{cm^{-1}}} \tag{0.6}$$

Wavelength (in nanometers) to wavenumbers (cm^{-1}):

$$\bar{\nu}_{cm^{-1}} = \frac{10^7}{\lambda_{nm}} \tag{0.7}$$

Wavelength (in nanometers) to frequency (in Hz):

$$\nu_{HZ} = \frac{2.9979 \times 10^{17} nm \cdot \sec^{-1}}{\lambda_{nm}} \tag{0.8}$$

Wavelength (in nanometers) convert to energy (in eV):

$$E_{p_{eV}} = \frac{1240}{\lambda_{nm}} \tag{0.9}$$

0.4 Description of Basic Quantitative Spectroscopic Measurements using Beer's Law[4]

Spectroscopic measurements depend upon the principle that light energy interacting with a material will cause absorption at a specific frequency depending upon the chemical characteristics of that material. The amplitude of the absorption at any particular frequency (or wavelength or wavenumber) is determined by the absorptivity of the molecule being measured and the number of molecules encountered by the beam path of the measuring instrument. It is assumed that a change in spectral response is related to a concentration as described by the Bouguer, Lambert, and Beer relationship, most often termed Beer's law. The Beer's law relationship is described as the absorbance (A, Au, or signal strength) of an analyte being measured using a spectrophotometer is equivalent to the product of the absorptivity (ε) of a specific type of molecular vibration at a given frequency; the concentration (c) of the molecules in the measurement beam; and the pathlength (l) of the sample holder within the measurement beam. This relationship between measured spectral signal and concentration of a molecule is most often expressed as:

$$A = \varepsilon cl \tag{0.10}$$

where ε is the molar absorptivity (referred to as molar extinction coefficient by earlier physicists) in units of Liter·Mole^{-1}·cm^{-1}; c is the concentration of molecules in the spectrometer beam in units of Mole·Liter^{-1} (Note: this is a scaled volume fraction unit); and pathlength (l) is the thickness of the sample in units of cm of the measured sample at a specific concentration. The absorptivity for any specific molecule type is calculated by careful measurements of the absorbance of a compound, generally diluted in a suitable organic solvent, and by applying the relationship:

$$\varepsilon = \frac{A}{cl} \tag{0.11}$$

Note that for transmittance (where T = 0.0 to 1.0) and percent transmittance (where %T = 0 to 100.0) spectroscopy, a more complete delineation of the relationships between the various terms is contained in an expression such as:

$$T = \frac{I}{I_0} = 10^{-\varepsilon cl} \Rightarrow \text{Abs.} = A = -\log_{10}\left(\frac{I}{I_0}\right) = -\log_{10} T = \varepsilon cl \tag{0.12}$$

Here the symbols I and I_0 represent the attenuated energy detected after sample interaction and the initial energy incident to the sample, respectively. For reflectance (where R = 0.0 to 1.0) and percent reflectance (Where %R = 0.0 to 100.0) spectroscopy the various relationships are expressed as:

$$R = \frac{I}{I_0} = 10^{-\varepsilon cl} \Rightarrow Abs. = A = -\log_{10}\left(\frac{I}{I_0}\right) = -\log_{10} R = \varepsilon cl \tag{0.13}$$

Noting the relationship exists where the change in intensity (I) of the transmitted or reflected light from a sample is a function of the change in pathlength (l) of the sample as expressed by the absorptivity (ε) of a specific analyte (or molecular substance) and its concentration (c) by:

$$-\frac{\partial I}{\partial l} = \varepsilon cl(\ln 10) \tag{0.14}$$

Modern spectrophotometers utilize these assumptions for making spectroscopic measurement and generally display spectroscopic data as transmission/transmittance (T), reflection/reflectance (R), and absorbance A (y-axis or ordinate axis) versus wavelength (nm, microns) or wavenumber (cm^{-1}) (as x-axis, or abscissa axis).

0.5 Spectral Ranges for Sources, Detectors and Optical Materials

Table 0.18 Source emissivity ranges (Adapted with permission from reference 3).

Source emissivity ranges (μm)	Start	End
Quartz tungsten halogen	0.22	2.7
Glass tungsten halogen	0.25	2.25
DC deuterium lamp	0.185	3.75
Pulsed xenon arc lamp	0.18	2.5
DC arc lamp	0.20	2.5
Globar (Silicon Carbide)	1.0	100
Nernst glower	0.30	35
Carbon arc	0.50	100
Mercury lamp	0.30	100
Helium-neon laser (He:Ne)	0.6327	0.6328
Neodymium yttrium aluminum garnet (Nd:YAG) laser	1.0639	1.0640

Table 0.19 Detector sensitivity ranges.[3]

Detector sensitivity ranges (μm)	Start	End
Silicon (Si)	0.30	1.1
Lead Sulfide (PbS)	1.1	3.0
Indium Arsenide (InAs)	1.7	5.7
Indium Gallium Arsenide (InGaAs), typical	0.90	1.7
Indium Gallium Arsenide (InGaAs), extended	0.65	2.4
Indium Gallium Arsenide (InGaAs), IR extended	0.65	2.70
Germanium (Ge:X)	2	40
Germanium Gold (Ge:Au)	2	9
Germanium Cadmium (Ge:Cd)	2	24
Lead Selenide (PbSe)	1.7	5.5
Germanium Zinc (Ge:Zn)	2	40
Indium Antimonide (InSb)	1.8	6.8
Lead Telluride (PbTe)	1.5	4.5
Deuterated Triglycine Sulfate (DTGS/KBr)	0.83	25
Deuterated Triglycine Sulfate (DTGS/PE)	10	120
Mercury Cadmium Telluride (MCT)	1	17
Triglycine Sulfate (TGS)	10	120
Pyroelectric Lithium Tantalate ($LiTaO_3$) (PLT)	1.5	30
Silicon Photodiode Detectors	0.19	1.1
Photomultipler Tube (PMT), S1 Type Photocathode	0.30	1.2
Photomultipler Tube, Cesium Gallium Arsenide (CsGaAs) Photocathode	0.30	0.93
Photomultipler Tube, Cesium Indium Gallium Arsenide (CsInGaAs) Photocathode	0.30	1.0
Photomultipler Tube, Cesium Antiminide CsSb Photocathode	0.25	0.75
Photomultipler Tube, Multialkali (Na-K-Sb-Cs) Photocathode	0.30	0.93
Photomultipler Tube, CsI Photocathode	0.20	0.90
Photomultipler Tube, S1 Type Photocathode	0.30	1.2
Photomultipler Tube, Cesium Gallium Arsenide (CsGaAs) Photocathode	0.30	0.93
Photomultipler Tube, Cesium Indium Gallium Arsenide (CsInGaAs) Photocathode	0.30	1.0
Silicon (Si) CCD	0.20	1.1
Indium Gallium Arsenide (InGaAs) CCD	0.80	1.7
Ultraviolet range	0.19	0.36
Visible Range	0.36	0.78
Near Infrared Range	0.78	2.5
Infrared	2.5	25
Raman	2.5	25

Table 0.20 Optical material spectral ranges.[3]

Optical material spectral ranges	Start	End
Methacrylate, Polymer	0.25	1.1
Silica, UV Grade Fused (SiO_2)	0.20	2.5
Silica, Synthetic Fused (SiO_2)	0.23	2.6
Quartz, Crystalline (SiO_2)	0.24	2.6
Quartz, extremely low O-H	0.19	2.7
Glass, Flint (SF 10)	0.38	2.35
Glass, Flint (SF 8)	0.355	2.35
Glass (BK 7)	0.315	2.35
Glass, "Optical Crown"	0.32	2.3
Glass, "Borosilicate Crown"	0.36	2.35
Glass, Pyrex®	0.31	2.5
Glass, Zerodur®Schott	0.40	2.85
Glass, Tempax®	0.36	2.35
Sapphire (Aluminum Oxide) (Al_2O_3) 5 mm is 50 %T at 5.1 microns	0.15	5.1
Sodium Chloride (NaCl)	0.25	16.0
Suprasil 300	0.19	3.6
Diamond (C)	0.22	4.0
Silica, Fused as pure SiO_2 (IR Quality)	0.20	4.0
Quartz, Anhydrous	0.20	2.75
Zirconium Fluoride (ZrF)	0.90	4.76
Chalcogenide (NIR-IR measurements)	2.22	11.1
Cesium Iodide (CsI)	0.3	50.0
Lead Sulfide (PbS)	1.1	3.0
Potassium Bromide (KBr)	0.25	26.0
Potassium Chloride (KCl)	0.25	20.0
Thallium Bromide-Iodide (KRS-5)	0.60	40.0
Germanium (Ge)	1.1	30.0
Arsenic Sulfide (As_2S_3)	0.6	15.0
Magnesium Fluoride (MgF_2)	0.6	9.5
Zinc Selenide (ZnSe) (IRTRAN-4)	0.6	26.0
Barium Fluoride (BaF_2)	0.2	13.0
Zinc Sulfide (ZnS) (Cleartran or IRTRAN-2)	0.6	15.0
Calcium Fluoride (CaF_2) (IRTRAN-3)	0.2	9.0
Silver Bromide (AgBr)	0.5	35.0
Polyethylene, Polymer (High Density)	16	333
Glass, Silica, Fused "Water Free" (SiO_2); 10 mm is 82%T at 3.15 microns	0.15	3.15
Glass, Corning Optical Glasses as HPFS®7978 and 7979 50%T/cm	0.15	4.35

(Continued)

Table 0.20 (*Continued*)

Optical material spectral ranges	Start	End
Glass, Corning Optical Glasses as HPFS® 7980; 50%T/cm	0.2	2.6
Glass, Corning Optical Glasses as HPFS® 865X and 797X 50%T/cm	0.15	2.6
Glass, ESCO Optical Glasses as S1-UVA and S1-UVB Working Range (Band 1)	0.17	2.2
Glass, ESCO Optical Glasses as S1-UVA and S1-UVB Working Range (Band 2)	2.9	3.6
Glass, ESCO Optical Glasses as I2-IR Working Range	0.25	3.6
Glass, ESCO Optical Glasses as G1 and A1 Working Range	0.27	2.5
Glass, ESCO Optical Glasses as BK7 Working Range	0.35	2.0
Ultraviolet range	0.19	0.36
Visible Range	0.36	0.78
Near Infrared Range	0.78	2.5
Infrared	2.5	25
Raman	2.5	25

Table 0.21a UV-Vis gas emission sources and peak positions (in nanometers).[5]

Position	Source	Position	Source	Position	Source
253.652	Hg	452.186	Xe	612.884	Ne
296.728	Hg	462.420	Xe	614.306	Ne
302.150	Hg	466.849	Xe	616.359	Ne
313.155	Hg	469.097	Xe	621.728	Ne
334.148	Hg	469.804	Xe	626.649	Ne
341.790	Ne	473.415	Xe	630.479	Ne
342.391	Ne	479.262	Xe	633.442	Ne
344.770	Ne	480.702	Xe	638.299	Ne
345.076	Ne	482.971	Xe	640.225	Ne
345.419	Ne	484.329	Xe	650.653	Ne
346.052	Ne	491.651	Xe	653.288	Ne
346.658	Ne	492.315	Xe	659.895	Ne
347.257	Ne	503.135	Ne	667.828	Ne
349.806	Ne	503.775	Ne	671.704	Ne
350.121	Ne	508.038	Ne	692.947	Ne
351.519	Ne	511.367	Ne	703.241	Ne
352.047	Ne	511.650	Ne	717.394	Ne
359.353	Ne	540.056	Ne	724.512	Ne
360.017	Ne	546.074	Hg	733.930	Xe
363.366	Ne	556.222	Kr	738.600	Xe
365.015	Hg	557.029	Kr	739.379	Xe
368.573	Ne	576.441	Ne	740.040	Xe
370.122	Ne	576.960	Hg	743.890	Ne
404.656	Hg	579.066	Hg	747.244	Ne
407.783	Hg	582.015	Ne	748.887	Ne
431.958	Kr	585.249	Ne	753.577	Ne
435.833	Hg	587.096	Kr	754.404	Ne
436.264	Kr	588.189	Ne	755.979	Xe
437.612	Kr	594.483	Ne	758.468	Xe
439.997	Kr	597.553	Ne	758.741	Kr
445.392	Kr	602.000	Ne	760.155	Kr
446.369	Kr	607.433	Ne	764.391	Xe
450.235	Kr	609.616	Ne	768.525	Kr
				769.454	Kr

Table 0.21b NIR gas emission sources and peak positions (in nanometers).[5]

Position	Source	Position	Source	Position	Source
780.265	Xe	1107.887	Ar	1390.748	Ar
785.482	Kr	1117.752	Ne	1409.364	Ar
788.132	Xe	1144.183	Ar	1414.244	Xe
791.343	Kr	1148.811	Ar	1442.679	Kr
796.734	Xe	1152.275	Ne	1473.281	Xe
805.726	Xe	1166.871	Ar	1473.444	Kr
805.95	Kr	1171.949	Ar	1504.65	Ar
806.134	Xe	1181.938	Kr	1517.269	Ar
810.436	Kr	1211.233	Ar	1520.31	Kr
819.006	Kr	1213.974	Ar	1532.934	Ar
823.163	Xe	1220.353	Kr	1537.204	Kr
826.324	Kr	1234.339	Ar	1541.839	Xe
826.652	Xe	1243.932	Ar	1598.949	Ar
829.811	Kr	1248.766	Ar	1605.328	Xe
837.761	Ne	1262.339	Xe	1620.872	Kr
849.536	Ne	1270.228	Ar	1647.29	Xe
866.794	Ar	1273.342	Ar	1656.023	Xe
877.675	Kr	1280.274	Ar	1672.815	Xe
878.375	Ne	1295.666	Ar	1689.676	Kr
881.941	Xe	1300.826	Ar	1694.058	Ar
892.869	Kr	1317.741	Kr	1704.288	Ar
904.545	Xe	1322.811	Ar	1755.35	Kr
912.297	Ar	1322.811	Ar	1763.882	Xe
916.265	Xe	1327.264	Ar	1785.738	Kr
922.45	Ar	1331.321	Ar	1790.45	Xe
935.422	Ar	1336.711	Ar	1800.223	Kr
965.779	Ar	1350.419	Ar	1809.09	Xe
978.45	Ar	1362.266	Ar	1816.733	Kr
979.97	Xe	1363.422	Kr	1832.53	Xe
992.319	Xe	1365.706	Xe	1959.94	Xe
1047.005	Ar	1367.855	Ar	1984.638	Xe
1067.357	Ar	1371.858	Ar	2190.851	Kr
1083.837	Xe	1382.572	Ar		

References

1. NIST: http://physics.nist.gov/cuu/Constants/index.html. (Used with Permission).
2. Adapted from Near Infrared Spectroscopy in Agriculture xx-xxiii (Used with Permission). Tables have been changed with different values and conversion factors.
3. J. Workman, The Academic Press Handbook of Organic Compounds: NIR, IR, Raman, and UV-VIS Spectra Featuring Polymers, and Surfactants, (3 Volume Set: Vol. 1, Methods and Interpretation; Vol. 2, UV-Vis and NIR Spectra; Vol. 3, IR and Raman Spectra), Academic Press, Boston (2000).
4. J. Workman and L. Weyer, Practical Guide and Spectral Atlas to Interpretive Near Infrared Spectroscopy, Second Edition CRC — Taylor and Francis, Boca Raton, FL (2012).
5. Ocean Optics Wavelength Calibration Sources, www.oceanoptics.com. (Used with Permission).

Chapter 1

Introduction to Infrared Spectroscopy

Contents

1.1 Introduction

As modern instrumentation for infrared (IR) spectroscopy advances, it becomes increasingly automated and easy to use; laboratories are now able to assign the routine analysis of samples to workers just getting started in their laboratory career. This section is intended to describe the basic tenets of infrared spectroscopy. This chapter includes an explanation of the electromagnetic spectrum and how radiation interacts with matter. Methods of describing what the spectrum tells us about a sample's composition are explained. On a more practical note, the components of infrared spectrometers are discussed in detail and useful information about sample preparation, sampling device selection, and routine maintenance is provided. Finally, a glossary of key terms and concepts and a selective bibliography are provided to guide the reader to further discovery on the subject of infrared spectroscopy.

Mid infrared (MIR) spectroscopy provides a measurement technique for intense, isolated and reliable absorption bands of fundamental molecular vibrations from polymers and other organic compounds. The spectrometric methodology allows for univariate calibration with higher signal strength (absorptivity) required for solid, liquids or gas phase measurements. Relatively small pathlengths of 0.1 to 1.0 mm are required for hydrocarbon liquids and solids. The technique is generally incompatible with the use of fiber optics, but specialized fiber materials exist. Instrumentation is higher cost than near infrared spectrophotometers for the most part.

Dominant mid infrared spectral features include the C-H (methyl, methylene, aromatic, methoxy, and carbonyl) fundamental stretching and bending molecular vibrations, O-H (hydroxyl) stretch fundamental vibrations, N-H (amine) stretching, C-F (fluorocarbon) stretching, -C≡N (nitrile) stretching, -C=O (carbonyl) stretch from esters, acetates, and amides, C-Cl stretch from chlorinated hydrocarbons, and $-NO_2$ from nitro- containing compounds.

1.1.1 *Basic Theory*

1.1.1.1 *The Electromagnetic Spectrum*

The electromagnetic spectrum consists of many different types of radiation (energy): gamma, x-ray, ultraviolet (UV), visible (Vis), infrared (IR), microwave, and radio waves (Figures 1.1 and 1.2). Each type of radiation occupies its own region of the electromagnetic spectrum, and the major difference between the individual spectral regions is merely the energy involved. This energy is expressed in units of: photon energy (eV), Frequency (Hz), wavenumber (cm^{-1}), or wavelength (nm) and the manner in which these energies interact with matter.

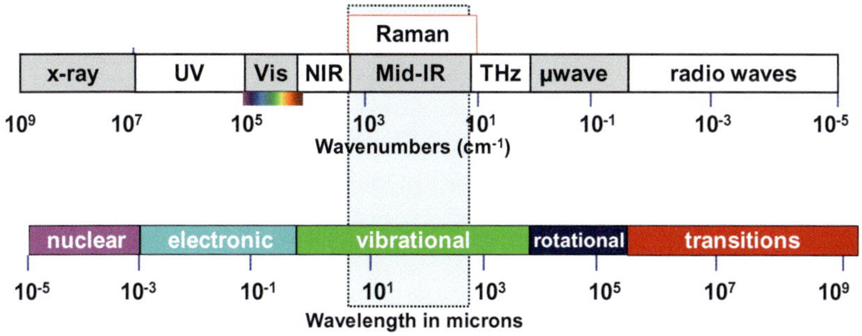

Fig. 1.1 The electromagnetic spectrum, highlighting the IR region. The infrared spectral region is defined as 4000 through 400 cm^{-1}.

Molecular Electromagnetic Spectrum Regions

Region	Wavelength (nanometers)	Wavenumbers (cm^{-1})	Frequency (Hz)	Energy (eV)
Terahertz	$2.5x10^{4}$-10^{6}	400-10	$1.2e^{13}$-$3.0e^{11}$	0.05-0.001
Infrared	2500-25,000	4000-400	$1.2e^{14}$-$1.2e^{13}$	0.50-0.05
Raman	2500-25,000	4000-400	$1.2e^{14}$-$1.2e^{13}$	0.50-0.05
Near Infrared	780-2500	14,493-3,333	$3.8e^{14}$-$1.2e^{14}$	1.6-0.50

Fig. 1.2 IR spectral region of the electromagnetic spectrum in wavelength (nanometers), wavenumbers (cm^{-1}), frequency (Hertz), and energy (electron volts).

Infrared spectroscopy measures the specific vibrations occurring in molecules providing unique interpretation of the types of molecular bonds present in a sample; the quantity of molecules under measurement; and the basic physical form of molecules under measurement. These properties of a material are measured using transmittance or reflectance techniques and by recording the intensity of such measurements with respect to energy, wavelength, frequency, or wavenumbers.

1.1.1.2 *Molecular and Electronic Spectroscopy Unit Conversion*

The units used to describe spectra for molecular spectroscopy vary with the academic or engineering discipline describing the spectral regions. For the most part the spectra are described in terms of an optical response specific to energy from a particular spectral region. The responses are in terms of inelastic or elastic collisions of the energy with various materials and are referred to as scattering absorption, absorbance, reflection, transmission, and so on. The various spectral energy for molecular and electronic spectroscopy is described in terms of units of wavelength (e.g., microns (μm) as 10^{-6} meters, or nanometers (nm) as 10^{-9} m, or Angstroms (Å) as 10^{-10} m). Energy is also expressed as frequency (cycles per second or sec.$^{-1}$) termed Hertz (Hz), as Wavenumbers (cm^{-1}), and as electron Volts (eV).

The various formulas used for the inter-conversion of units for spectroscopy include the following. Light has both particle and wave properties, quantum theory tells us that the energy of a light "particle" or photon E_p is given by:

$$E_p = h\nu \tag{1.1}$$

$$E_p = hc\overline{\nu} \tag{1.2}$$

where: h = Planck's constant (or 6.6256×10^{-27} erg-sec.); and ν (nu) is the frequency of light, also known as Hertz (Hz) or the number of vibrations per second, in units of sec.$^{-1}$; $\overline{\nu}$ is the wavenumber units as cm^{-1} (i.e., the number of waves per centimeter); and c is the velocity of light in a vacuum (or 2.9979×10^{10} cm·sec^{-1}).

$$\overline{\nu} = \frac{\nu}{c \cdot n^{-1}} = \frac{1}{\lambda} \tag{1.3}$$

$$\nu = \frac{c \cdot n^{-1}}{\lambda} \tag{1.4}$$

Where: $\overline{\nu}$ is the wavenumber units as cm^{-1} (i.e., the number of waves per centimeter); c is the velocity of light in a vacuum (or 2.9979×10^{10} cm·sec^{-1}); ν (Greek: nu) is the frequency of light, also known as Hertz (Hz) or the number of vibrations per second, in units of sec.$^{-1}$; and n is the refractive index of the medium the light is passing through (e.g., air = 1.0003); λ (Greek: lambda) is the wavelength in units of cm. Combining the above equations we also note:

$$\overline{\nu} = \frac{hc}{E_p} \Rightarrow \frac{1}{\lambda} = \frac{hc}{E_p} \Rightarrow \lambda = \frac{E_p}{hc} \tag{1.5}$$

1.1.1.3 *Unit Conversions*

Wavenumbers (cm^{-1}) to wavelength (in nanometers):

$$\lambda_{nm} = \frac{10^7}{\overline{\nu}_{cm^{-1}}} \tag{1.6}$$

Wavelength (in nanometers) to wavenumbers (cm^{-1}):

$$\overline{\nu}_{cm^{-1}} = \frac{10^7}{\lambda_{nm}} \tag{1.7}$$

Wavelength (in nanometers) to frequency (in Hz):

$$\nu_{Hz} = \frac{2.9979 \times 10^{17} nm \cdot \mathrm{sec}^{-1}}{\lambda_{nm}} \tag{1.8}$$

Wavelength (in nanometers) convert to energy (in eV):

$$E_{p_{eV}} = \frac{1240}{\lambda_{nm}} \tag{1.9}$$

1.2 Applications of Infrared Spectroscopy

Infrared energy is the electromagnetic energy of molecular vibration. The energy band is defined for convenience as the mid infrared or MIR region from

(2.5 to 25.0 microns); and the far infrared (25.0 to 1000 microns). Official standards, textbooks, and the scientific literature generally state that the infrared (mid infrared) spectral region extends from 2500 to 25,000 nanometers (4000 to 400 cm^{-1}).

The predominant mid infrared spectral features include the following fundamental bands (Note: Chapter 7 includes tables and figures demonstrating this). Primary infrared fundamental absorption bands occur as C-H stretching, from alkanes, alkenes, alkynes, and aromatic groups; C-H bending, from alkanes, alkenes, alkynes, and aromatic groups; C-C stretching, from alkenes, alkynes; >C=O carbonyl stretching from ketones, aldehydes, and esters; -O-C=O – carboxylic acids stretching; R′-C=O-O-C=O-R′ anhydride groups; acyl halides; amides, primary, secondary, and tertiary types; O-H stretch from alcohols, phenols, and water; N-H stretch from amines; C=N stretch from unsaturated nitrogen containing molecules; C-X from halogenated organic compounds; S-H from sulfur containing compounds; and Si-O stretching from silicones.

Qualitative and quantitative mid infrared (MIR) spectroscopic methods are used for analysis and identification of pure materials and mixtures of various molecules. In the past generation, many classically trained spectroscopists were often quite adroit at interpreting infrared spectra simply by viewing the spectrum, noting the main transmittance bands and identifying pure compounds or the molecules present in complex mixtures. Due to the many software programs now available, both qualitative and quantitative chemometric tools are used for interpretation of spectra. Advanced applications and infrared methods require more sophisticated software algorithms. The use of multivariate calibration algorithms and statistical methods (i.e., chemometrics) are used to model MIR spectral response to chemical or physical properties of the samples used for calibration.[1–2]

The MIR method relies on the spectra-structure correlations existing between a measured spectral response caused by fundamental vibrations occurring at infrared frequencies. These harmonic vibrations occur at unique frequencies depending upon the quantity of absorber (analyte), type of absorbing molecules present within the sample, and the sample thickness (or pathlength). This chapter introduces the use of mid infrared spectroscopy as a technique for qualitative identification, quantitative analysis, and structure-correlation assessment of MIR functional group frequencies.

1.2.1 *Early Work in MIR Spectroscopy*

The history of infrared detectors and multiple applications over a very active time in modern infrared developments is described in Chapter 7 of this volume. Table 1.1

Table 1.1 Relative intensities of C-H stretch bands for infrared fundamental and the various NIR overtone regions.

Band	Wavelength region (nm)	Wavenumber region (cm^{-1})	Relative intensity
Fundamental (ν)	3380–3510 nm	2959–2849 cm^{-1}	100
1st over. (2ν)	1690–1755 nm	5917–5698 cm^{-1}	1
2nd over. (3ν)	1127–1170 nm	8873–8547 cm^{-1}	0.1

demonstrates the relative intensities of the various spectral regions as a consideration for pathlength and instrumentation requirements.

1.3 Basic MIR Spectroscopic Instruments (Optical Configurations)

One of the basic issues of spectroscopic measurement is that physical matter in each of its phases (i.e., gaseous, liquids, or solid substances) will respond in a unique way when exposed to any type of electromagnetic radiation. The atoms and molecules that make up matter will absorb and remit certain amounts of the MIR radiation as it passes through or collides with each molecule; this response is based on the frequency of the radiation source. In the earliest spectroscopic experiments, scientists discovered that each atom or molecule will react in exactly the same way every time it is exposed to a particular type of radiation, predictably the absorption or remission of radiation at the different frequencies. This behavior is unique to various types of matter and allows the science of spectroscopy to exist. It should be noted that no two different types of atoms or molecules react to radiation in exactly the same manner. However, types of molecules that are very similar produce very similar spectra and thus the spectral detail and signal-to-noise of a spectrophotometer measurement determine its ability to distinguish very similar molecules from one another.

In the MIR spectral range, light is absorbed by matter based on molecular content. When the vibrational energy of a specific molecule type corresponds to the frequency produced by the measuring spectrometer instrument, the molecule absorbs the photon, so the various frequencies are absorbed by the molecule creating a spectrum of frequency or wavenumbers (abscissa, x-axis) versus absorption recorded as transmittance, reflectance, or absorbance (for the ordinate, y-axis). Because energy absorbance in the MIR range is predictable and reproducible, measured spectra are useful for both qualitative and quantitative work. Spectral analysis is used to reveal the types of molecules in a sample, and the amount of each type

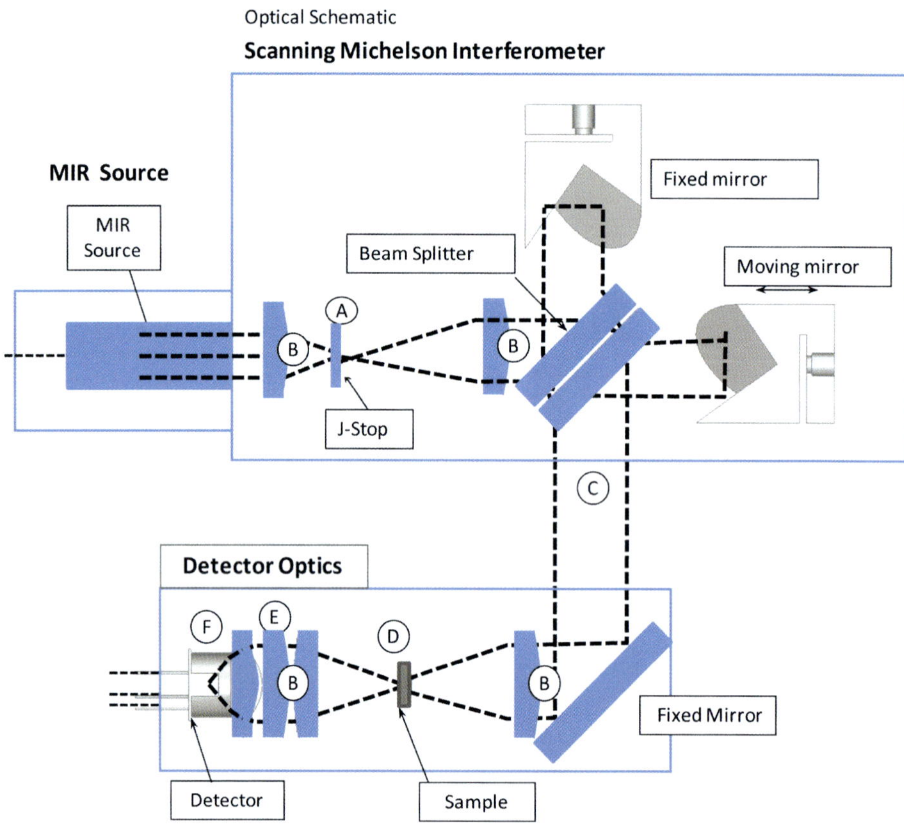

Fig. 1.3 Laboratory Interferometer based optical system (Interferometer/FT-type). A = J-stop (aperture), B = collimating optics, C = optical train, D = sample cell, E = focusing optics, F = detector.

of molecule in a sample specimen. Figures 1.3 through 1.6 show the basic optical configurations for various MIR spectrometer systems. For information on other spectrophotometer designs, such as dispersive based systems, please refer to the index and Volume 3, Chapter 3 on NIR Instrumentation. Figure 1.3 illustrates a typical laboratory FTIR instrument, Figure 1.4 is an open path FTIR system, Figure 1.5 is an FTIR Microscope, and Figure 1.6 is a filter based MIR spectrometer.

1.3.1 *Components of a Mid Infrared Spectrometer*

(a) General Components: MIR spectrometers share the same basic macro-components, which can be configured to achieve specific optical effects as shown in Figures 1.3 through 1.6. These components include: an MIR radiation source, a wavelength selector (e.g., filter(s), or interferometer, etc.),

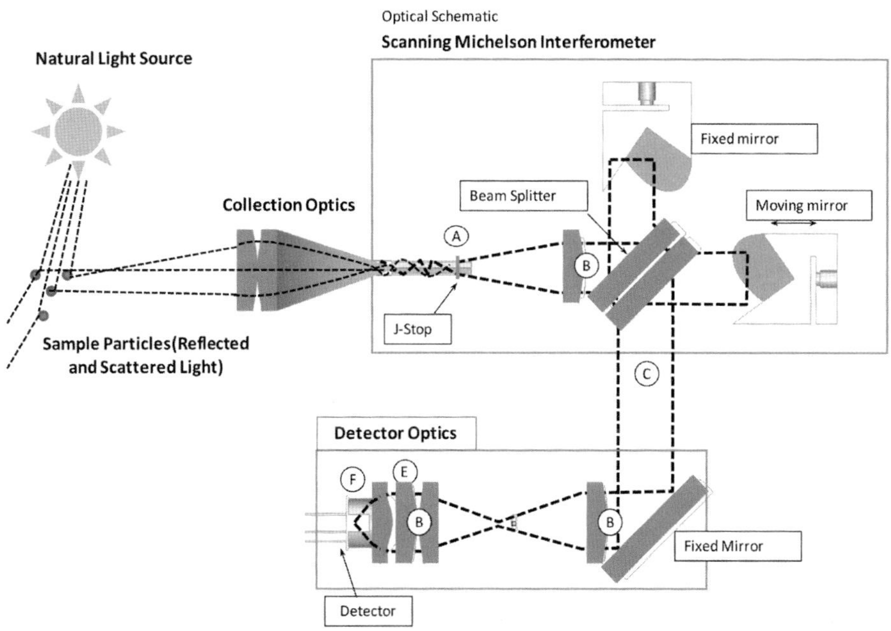

Fig. 1.4 Open-path interferometer based optical system (interferometer/FT-type). A = J-stop (aperture), B = collimating optics, C = optical train, D = sample cell, E = focusing optics, F = detector.

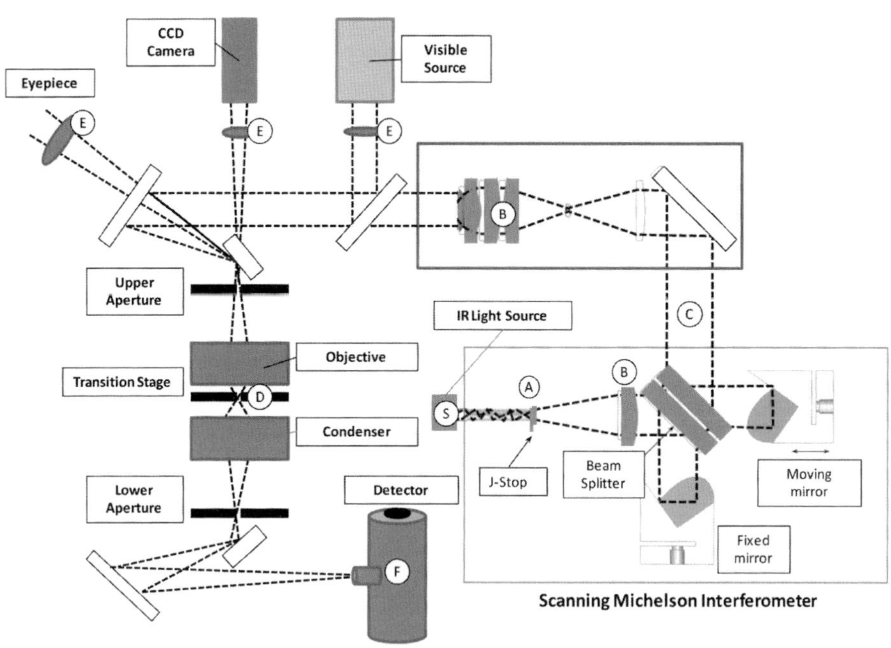

Fig. 1.5 Microscope interferometer based optical system (interferometer/FT-type). A = J-stop (aperture), B = collimating optics, C = optical train, D = sample cell, E = focusing optics, F = infrared detector and Photomultiplier Tube (PMT).

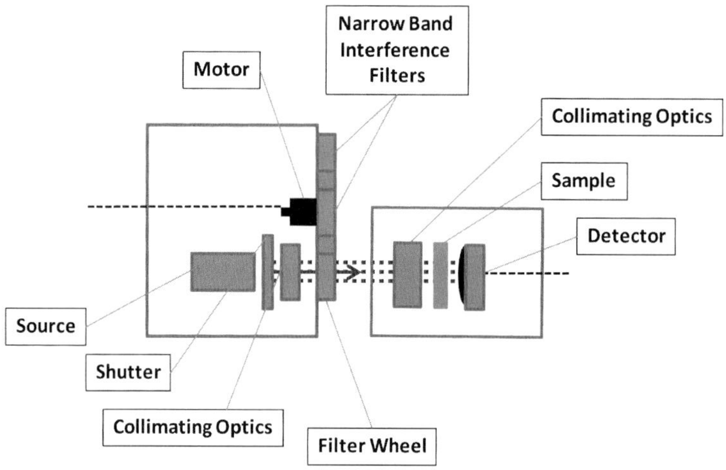

Fig. 1.6 Interference filter based MIR photometer optical system (filter type).

a sample holder, and a detector (e.g., generally single detector). The electronics include: a basic power supply to distribute power to all motors, and electronic components; a computer board for processing data, low-level instrument commands, and communications; a controller board for timed driving of motors, shutters, etc.; and a detector board for amplifying signal, and for analog to digital (A to D) conversion of basic instrument signal (as voltage) to generate the MIR frequency versus absorption spectral profiles.

(b) Energy Sources: The energy used in MIR spectrometry is usually produced by a Globar generating heat energy or the energy of infrared radiation. The following Table 1.2 shows the variety of sources for spectrometers using the MIR range.

Table 1.2 Infrared energy radiation sources (microns).

Source emissivity ranges	Start	End	Difference
Globar	1.0	30	29
Nernst glower	1.0	30	29
Carbon arc	0.5	100	100
Mercury lamp	0.3	100	100
Near Infrared Region	0.78	2.5	2
Infrared Region	2.5	25	23
Raman Region	2.5	25	23

The infrared radiation from the glowing source is transmitted through filters or an interferometer to the sample and yield spectra of various materials. The reader is referred to the index for additional information on sources and their emission properties.

(c) Sample Holders: The most common sample holders in MIR analysis are windows for sample deposition, film holders, mull crystals for solid particles, KBr pellet sample holders, specular reflectance, attenuated total reflectance (ATR), and diffuse reflectance (DRIFTS) sampling. Chapter 4 of this volume describes the details of sample preparation for these and other MIR sampling techniques.

(d) Detectors: MIR energy reflected or transmitted through a sample and reference material is captured by using a Mercury Cadmium Telluride (MCT), Deuterated Triglycine Sulfate (DTGS/KBr), or Indium Antimonide (InSb) detector (Table 1.3). The voltage signals generated by these detectors are amplified, converted from analog to digital signals, and used to quantify the amount of energy emerging from a sample. A more complete list of common detectors, common abbreviations, and their useful working wavelength ranges would include the following table. The reader is referred to the index and Chapter 3 on Instrumentation for more details on the physics of infrared detectors.

Table 1.3 Infrared detectors (most commonly used).

Detector sensitivity ranges (microns or μm)	Start (μm)	End (μm)	Difference (μm)
Germanium (Ge:X)	2	40	38.0
Germanium Cadmium (Ge:Cd)	2	24	22.0
Germanium Zinc (Ge:Zn)	2	40	38.0
Deuterated Triglycine Sulfate (DTGS/KBr)	0.83	25	24.2
Indium Antimonide (InSb)	1.8	6.8	5.0
Deuterated Triglycine Sulfate (DTGS/PE)	10	120	110.0
Mercury Cadmium Telluride (MCT)	1	17	16.0
Triglycine Sulfate (TGS)	10	120	110.0
Pyroelectric Lithium Tantalate ($LiTaO_3$) (PLT)	1.5	30	28.5
Near Infrared Range	0.78	2.5	1.7
Infrared Range	2.5	25	22.5

(e) Basic Measurement Modes: The measurement modes for mid infrared spectroscopy are given in Table 1.4. A well designed and flexible research spectrophotometer is used to make any of these measurements.

Table 1.4 Infrared measurement modes.

Instrument measurement mode	Description of measurement
Scan	Absorbance* vs. wavenumbers Reflectance or Transmittance vs. wavenumbers
Timedrive	Absorbance* vs. time at each wavenumber
Step-Scan Phase-Resolved	Amplitude Modulation, Phase Modulation, or Sample Modulation
Step-Scan Time-Resolved	Microsecond to nanosecond time resolution for timed events
Step-Scan Space-Resolved	Imaging spectra for spatial data acquisition for focal-plane imaging
Polarization Modulation	Modulation of light polarization for measuring polarization effects
Individual wavenumber step-scan	Individual Absorbance(s)* at selected wavenumbers(s)
Two-dimensional Correlation	Correlates multiple spectra for conditions of applied stress and relaxation to determine rheo-optical properties. Also for correlation of spectra from different regions such as NIR-IR.
Quantitative methods	Concentration of analyte vs. Absorbance
Qualitative methods	Correlation or spectral matching for Identification
Kinetics	Kinetic rates of reaction

*As $-\log_{10}$ (transmittance) or $-\log_{10}$ (reflectance).

1.3.2 *The Concepts Behind Spectroscopy*

(a) Beer's Law: Beer's law is a mathematical way of expressing how radiation is absorbed by matter in a spectroscopic measurement. The law states that the amount of monochromatic light emerging from a sample is diminished by three physical phenomena: (1) the amount of absorbing material in its pathlength (concentration as a scaled volume fraction, i.e., weight/volume or moles/volume); (2) the distance the light must travel through the sample (pathlength); and (3) the probability that the photon of that particular wavelength will be absorbed by the material (as the absorptivity or extinction coefficient).

This relationship may be expressed as:

$$A = \varepsilon c l \tag{1.10}$$

where A = absorbance (unitless); ε = molar extinction coefficient (i.e., absorptivity); c = molar concentration as weight/volume or moles/volume; and l = pathlength (in cm). (See index and section following for more details on units of measure.)

In practice, the absorbance is measured by comparing the light emerging from a sample with the light emerging from a reference material or reference cell. The intensity of the reference radiation, I_0, is ratioed to the light emerging from a sample, I. The ratio I/I_0 is referred to as *reflectance* (*R*) or *transmittance* (*T*), depending upon the sample measurement geometry, and may be expressed in units of 0 to 1 or 0% to 100%. This can be plotted against the concentration, but the relationship is not linear. The negative log base 10 of the reflectance or transmittance, however, is mostly linear with concentration up to the limits of dynamic range for the measuring instrument. Absorbance is thus measured as:

$$-\log_{10}(I/I_0) = \log_{10}(I_0/I) = \log_{10}(1/R) = \log_{10}(1/T) = A = \varepsilon cl \quad (1.11)$$

Harmonic vibrations follow a functional description such that the types of vibrations in a molecule determine the frequency at which the different functional groups absorb MIR energy. The amplitude of the absorption at any particular wavelength or wavenumber is determined by its absorptivity and the number of molecules encountered within the beam path (i.e., scaled volume fraction)[3] of the measuring instrument. It is assumed that a change in spectral response is related to a concentration change as described by the Bouguer, Lambert, and Beer relationship most often termed Beer's law. The Beer's law relationship is described as the absorbance (A, Au, or signal strength) of an analyte being measured using a spectrophotometer and is equivalent to the product of the absorptivity (ε) of a specific type of molecular vibration; the concentration (c) of the molecules in the measurement beam; and the pathlength (l) of the sample holder within the measurement beam. This relationship between measured spectral signal and concentration of a molecule is most often expressed as:

$$A = \varepsilon cl \quad (1.12)$$

where ε is the molar absorptivity (referred to as molar extinction coefficient by earlier physicists) in units of $\text{Liter} \cdot \text{Mole}^{-1} \cdot \text{cm}^{-1}$; c is the concentration of molecules in the spectrometer beam in units of $\text{Mole} \cdot \text{Liter}^{-1}$ (Note: this is a volume fraction unit not weight fraction unit)[19]; and pathlength (i.e., the thickness of the sample in units of cm). The absorptivity for any specific molecule type is calculated by careful measurements of the absorbance of a sample by applying the relationship:

$$\varepsilon = \frac{A}{cl} \quad (1.13)$$

Note that for transmittance (where $T = 0.0$ to 1.0) and percent transmittance (where $\%T = 0\%$ to 100.0%) spectroscopy, a more complete delineation of the relationships between the various terms is contained in an expression such as:

$$T = \frac{I}{I_0} = 10^{-\varepsilon cl} \Rightarrow Abs. = A = -\log_{10}\left(\frac{I}{I_0}\right) = -\log_{10} T = \varepsilon cl \qquad (1.14)$$

Here the symbols I and I_0 represent the attenuated energy detected after sample interaction and the initial energy incident to the sample, respectively. For reflectance (where R = 0.0 to 1.0) and percent reflectance (where %R = 0.0% to 100.0%) spectroscopy the various relationships are expressed as:

$$R = \frac{I}{I_0} = 10^{-\varepsilon cl} \Rightarrow Abs. = A = -\log_{10}\left(\frac{I}{I_0}\right) = -\log_{10} R = \varepsilon cl \qquad (1.15)$$

Modern spectrophotometers utilize these assumptions for making spectroscopic measurement and generally display spectroscopic data as transmission (T), reflectance (R), and absorbance A (y-axis or ordinate axis) versus wavelength (nm, microns) or wavenumber (cm^{-1}) (as x-axis, or abscissa axis). Note that spectroscopy measures in mass per volume not weight percent.[3]

(b) Bandpass and Resolution: The terms *bandpass* and *resolution* are used to express the capability of a spectrometer to separate spectral bands or lines that are separated by some finite spectral distance. The bandpass of a spectrometer is used to describe the separation distance (in wavelength or wavenumbers) at which closely positioned absorption bands can be isolated (or resolved) by the spectrometer into distinct peaks. See Chapter 3 in this volume for more detailed information.

(c) Signal-to-Noise: Simply stated, the *signal-to-noise ratio* (S/N) in a spectrometer is the numerical ratio of the amount of electrical current generated by the light from the sample striking the detector at 100%T or 100%R (i.e., equivalent to 1.0 T or 1.0 R) relative to the standard deviation of replicate measurements at 100%T or 100%R under various typical measurement conditions. For MIR measurement conditions include differences in resolution, detector area, absorptivity, measurement time, and throughput. A high signal-to-noise is optimum and decreases as throughput decreases and resolution increases for any given instrument. See Chapter 3 for details of signal-to-noise calculations for an interferometer based spectrophotometer.

(d) Dynamic Range: The range of a specified analyte concentration over which a sensor response is directly proportional (or theoretically linear) to a change in concentration is termed the *dynamic range* of a spectrophotometer. Dynamic range is noise limited for an interferometric system. Calculating the maximum theoretical dynamic range for an instrument, the relationship of detector non-linearity, and the maximum observable absorbance value requires knowledge of Beer's law and a few simple mathematical relationships. There is detailed information and computations shown in the various sections of these volumes, particularly the Instrumentation chapters (e.g., Chapter 3).

1.3.3 *Practical Aspects for MIR Spectral Measurements*

A variety of sample presentation methods are available to the infrared spectroscopist; Chapter 4 covers many aspects of sampling for MIR Applications. Basic sampling methods include: transmittance/transmission (straight), reflectance/reflection (specular and diffuse) — combined with interferometer measurement modes described in Table 1.4. Pathlength selection for optimum mid infrared measurements involves the following: gases should be measured with a pathlength of at least 10 meters for STP (Standard Temperature and Pressure) cells; and liquid cells should have a pathlength of 0.01 mm (10 microns) to 1.0 mm (1000 microns) depending upon bands of interest that are to be measured, e.g., low absorbing bands require a greater pathlength for accurate measurement. Chapter 4 in this volume is on sampling and provides much more detail. There are many sample preparation and sampling systems available for standard MIR measurements.

The sample presentation geometries most often used in mid infrared measurements are given in Table 1.5. For this table, the abbreviations used designate Transmission (T), Diffuse Transmission (DT), and Diffuse Reflection (DR). Transmission and Diffuse Transmission are the identical sample presentation geometry, but the sample for DT is a matrix of particles allowing light to penetrate without significant backscatter. DR is used for samples of infinite optical thickness that will not transmit light if T or DT is used. The ideal surface for DR is a Lambertian or isotropic surface where light is reflected diffusely or evenly throughout a hemisphere above the surface.

Samples exhibit different optical characteristics which must be considered in order to optimize spectroscopic measurement. Standard clear samples or solutions are measured using transmission spectroscopy. Highly colored samples are generally measured using transmission spectroscopy unless the optical density exceeds the linear range of the measuring instrument. At this point, either dilution or reducing the pathlength is preferred.

Fine scattering particulates from 5 to 25 times the measuring wavelength for mid infrared (i.e., particle diameters from 25 to 625 microns, 0.025 to 0.625 mm) or larger are measured using diffuse transmission or diffuse reflection methods. The scattering produced by some of the reflected light creates a pseudo-pathlength effect. This effect is compensated for by using scatter correction data processing methods for quantitative measurements, sieving the particles, or grinding to improved particle size uniformity.

Table 1.5 Sample presentation methods for mid infrared sampling.

Sample type	Pathlength	Sample presentation	Comments
Liquids (lower viscosity)	0.01 to 1.0 mm	T	Maintain Instrument within linear measurement range, that is generally less than 1.5 Au
Slurries (high viscosity)	0.01 to 0.05 mm	T and DT	Maintain Instrument within linear measurement range, that is generally less than 1.5 Au
Solids and smaller particles	Infinite optical thickness (∞)	DR	For DR use infinite optical thickness
Gases	10 meters	T	High Pressure cells may have shorter pathlengths
Polymer films	0.03 to 0.04 mm is optimum; up to 0.05 mm	T	Maintain Instrument within linear measurement range, that is generally less than 1.5 Au
Webs	Infinite optical thickness (∞)	DR	For DR use infinite optical thickness
Pellets (large particles)	1 mg or less of fine ground sample powder in 100 mg KBr (fine ground)	DT	DT thickness should allow discrimination of high and low absorbing bands. Peak troughs should be less than 0.3 Au.

Large scattering particulates on the order of 100 times the measuring wavelength in diameter (i.e., >2.5 mm for mid infrared) present a challenge for measurements as the particles intercept the optical path at random intervals. Signal averaging can be employed to compensate for random signal fluctuations. Reflection spectroscopy can be used to measure the size, velocity, and concentration of scattering particulates within a flowing stream.

Very high absorptivity (optically dense) materials with absorbances above 4.0 to 6.0 Au are difficult to measure accurately without the use of high precision photometric instruments, such as high performance Fourier Transform interferometers. Measurements of such samples can be made using higher numbers of repetitive scans, in the case of interferometric measurements often with FTIR photoacoustic spectroscopy (FTIR-PAS).

1.3.3.1 *Basic Applications of MIR Spectroscopy*

MIR spectrometers are useful in analytical situations for identification of compounds in pure or mixture forms. Also oxidation or degradation are readily observed in MIR spectra. MIR is useful for reverse engineering various compounds and molecular structures, for qualitative analysis, and for quantitative analysis. MIR is used for quality control, manufacturing, material and drug development and identification of chemical active groups, and in the role as a tool for pure and applied research. The following descriptions of basic analytical MIR applications are helpful for clarification of uses and measurement techniques.

(a) Quantitative Applications: The ability of light to be absorbed in proportion to its concentration has found use in many industries, just a few of which are described here.

- *Pharmaceutical companies* use MIR spectroscopy in numerous applications including identification of raw materials and quality control of finished dosage forms. It is used for reverse engineering of molecules; for molecular structure verification of new compounds; for production monitoring; for determination of contaminants and production problems.
- *Chemical companies* use the MIR spectroscopy to follow the progress of chemical reactions, identify raw materials, and monitor the kinetics and completion of chemical reactions.
- *Biotechnology companies* use MIR to monitor bio-fermentation during batch processing and to identify raw materials for fermentation reactions.
- *Universities*, technical schools, and other teaching environments use MIR spectroscopy to instruct students on vibrational spectroscopy, spectra-structure correlation, and qualitative and quantitative spectroscopy.
- *Food and Agricultural Producers* use MIR to measure raw materials for identification, and for natural and commercial food product production quality and nutrition.
- *Consumer products companies* use MIR spectroscopy for identification of raw materials and quality control of finished materials. It is used for reverse engineering of products; for production monitoring, and for determination of contaminants, oxidation, and production problems.

(b) Quantitative and Qualitative Applications: Mid infrared spectroscopy is used to measure the fundamental molecular absorptions found in molecular stretching, bending, and wagging. The use of chemometric mathematical data processing can be used to calibrate for qualitative or quantitative analysis. See Chapter 6 in each Volume, and the index, for more detailed information on chemometric methods.

(c) Selection of Measurement Technique and Pathlength: Samples are analyzed by transmitting or reflecting light from a sample contained in a sample holder. See Table 1.5, and Chapter 4 for details. Spectroscopists have the choice of measuring the MIR radiation just as it is transmitted through the sample, *transmission* (straight and diffuse); as it is reflected off the sample, *reflectance* (specular and diffuse); or as it emits from a sample.

The physical nature and appearance of the sample should determine which measurement technique is used. Samples can possess any of several different physical or optical properties that affect the type of analysis and sample presentation used. The following guidelines may be used for selecting a specific sample presentation geometry based on the optical properties of a sample.

- *Gases (long pathlength cells)*: Transmittance
- *Clear solids (optical materials)*: Transmittance
- *Translucent or opaque solids*: Diffuse reflectance or diffuse transmittance (for turbid samples)
- *Reflecting optical surfaces*: Specular reflectance
- *Clear liquids*: Transmittance
- *Translucent or opaque liquids*: Reflectance or diffuse transmittance
- *High optical density (highly absorbing)*: Photoacoustic Spectroscopy (PAS)
- *Mirrored surfaces*: Specular or diffuse reflectance
- *Colored Solid Materials*: Diffuse reflectance

Chapter 4 on sampling devices and accessories both illustrates and describes the various sampling methods used in mid infrared spectroscopy.

1.3.3.2 *Calibration and Basic Maintenance*

(a) General Concepts of Calibration:

Photometric Accuracy: Materials are available for reference measurements to verify photometric accuracy of MIR spectrophotometers. The materials are referred to as Standard Reference Materials (SRMs). ASTM E1421 is a consensus standard specifying the methods for determining the measuring performance of an FTIR instrument.[4]

Linearity Checks: The linearity of the instrument may be checked by using variable transmittance samples, such as attenuation screens for transmittance, or variable reflectance standards for reflectance measurements. These photometric materials are nonspecific (i.e., they cover all wavelengths) that reduce the amount of reflected light by specified amounts, for example, by 10% T or R, 20% T or R, and so forth. If there is a problem with the linearity, the instrument manufacturer should be contacted. These checks may be made anytime, but should be performed at least once per month. The correlation between the reference values for photometric materials and the measured values for those materials should be very close to 1.0 or better than 0.99 up to as high as 1.5 absorbance units (Au).

Wavenumber Accuracy Checks: If an instrument is not indicating the correct wavenumber registration during an analysis, a measurement error will be generated and all libraries collected during a poor wavenumber registration state will have to be later discarded. Reference materials are measured with known peak positions and these are compared to the measured values and should be within specific tolerances as pre-specified as instrument performance specifications. Sections and data describing the details of wavelength alignment for spectrometers are included in this text and may be found in the index for these volumes. For MIR a NIST standard SRM 1921b polystyrene film is recommended for testing the wavelength scale for infrared spectrophotometers.[5–8.]

(b) Instrument Care: The proper procedure for using an MIR spectrometer is prescribed by the manufacturer.

(c) Sample Preparation and Handling: *Solvents and Operating Conditions*: Choosing the proper solvent is important for several reasons: (1) because the solvent must interact with the analyte to dissolve it, it has an effect on the spectrum, and (2) the solvent itself might have an absorbance at the wavelength used for analysis. Good solvents have their spectral properties delineated on the label: "Spectro-Grade" or "UV-Grade", and so forth. HPLC solvents may not be sufficient for sensitive spectroscopic work. Select a solvent that dissolves the sample but does not react with it. Use the purest solvents possible when performing spectroscopic work and make sure there is no residual materials left on sample holding devices when making measurements.

(d) Troubleshooting: Because modern instruments are complex, usually only limited maintenance can be performed directly by the user. When a problem does arise, though, its cause must be isolated to determine if the problem is from the instrument or the sample. Isolating and testing each variable individually allows diagnostics of a measurement problem.

Sample/Solvent Problem: If you suspect a problem with your sample or the solvent you are using, first try running the sample on a second instrument if available. It is always quicker to change instruments than to solve a complex instrument failure problem. If the problem persists on a different instrument, then the chemistry and sample must be checked. Remeasure one or two standards or samples in a different lot of solvent, using freshly cleaned sample cells or holders. Because most modern spectrometers are quite stable, the fault is often in sample handling or changes in sample chemistry.

Instrument Problems: If it has been determined that the samples are correct, then the instrument should be checked. Other than any special steps recommended by the manufacturer, the linearity, photometric accuracy, and wavelength accuracy checks should be completed. If the problem is not easily solved or performance seems to be not within published specifications, the best thing to do is call the instrument company for assistance. See Chapters 3 and 6 for instrument performance details.

Avoiding Problems: Good laboratory techniques and reproducible sample work-up are good habits to develop. Each step of an analysis must be clearly stated and adhered to. Beyond that, routine maintenance is probably the best option for avoiding problems. Keeping a log record for each instrument allows the analyst to determine whether a spectrometer source needs to be replaced or whether the detector may be getting old. Cleaning is also quite important. Fumes, dust, and spills in and around the instrument will eventually destroy or cloud optical surfaces.

1.4 Interpretive Infrared Spectroscopy

Sections of this text are used by permission from J. Workman, *Handbook of Organic Compounds* (Academic Press-Elsevier, Boston 2000).

1.4.1 *Group Frequencies*

A comparison of the intensity of MIR and NIR bands for C-H Stretching is presented in Table 1.6. Groups of atoms (termed functional groups) with common group names for mid infrared vibrational spectroscopy are shown in Table 1.8. Mid infrared fundamental bands have high absorptivity values. Note that a group of figures and tables illustrating band positions for infrared bands are given in detail within Chapter 7 of this text.

Table 1.6 Relative band intensities: MIR vs. NIR for C-H stretching bands.

Band	Wavenumber (wavelength) region	Relative intensity	Recommended sample cell pathlength (for liquid hydrocarbons)
Fundamental (ν)	2959–2849 cm^{-1} (3380–3510 nm)	1	0.1–4 mm
First overtone (2ν)	5917–5698 cm^{-1} (1690–1755 nm)	0.01	0.1–2 cm
Second overtone (3ν)	8873–8547 cm^{-1} (1127–1170 nm)	0.001	0.5–5 cm

Table 1.7 Spectroscopic regions of interest for chemical analysis.

Region	Wavenumbers/ (Wavelength)	Characteristic measured
Ultraviolet	52,632–27,778 cm^{-1} (190–360 nm)	Electronic transitions: Delocalized Pi electrons
Visible	27,778–12,821 cm^{-1} (360–780 nm)	Electronic transitions: Color measurements
Near Infrared (NIR)	14,493–3,333 cm^{-1} (690–3000 nm) or 12,821–4,000 cm^{-1} (780–2500 nm)*	Overtone and combination bands of fundamental molecular vibrations, especially stretching and bending (some deformation as well)
Infrared (MIR)	4,000–400 cm^{-1} (2500–25,000 nm)	Fundamental molecular vibrations: Stretching, bending, wagging, scissoring
Far-Infrared (FIR or Terahertz)	400–10 cm^{-1} (2.5×10^4 to 10^6 nm)	Molecular rotation

*Official ASTM International Definition.

The broad band of infrared energy is the electromagnetic energy of molecular vibration. The energy band is defined for convenience as the near infrared covering 12,821 to 4,000 cm^{-1} (780 to 2500 nanometers); the infrared (or mid infrared) as 4,000 to 400 cm^{-1} (2500 to 25,000 nanometers; and the far infrared (or Terahertz) from 400 to 10 cm^{-1} (25,000 to 1,000,000 nanometers). Table 1.7 illustrates the region of the EMR (electromagnetic radiation) spectrum referred to as the MIR region. The table shows the molecular interactions associated with the energy frequencies (or corresponding wavelengths) of the various regions.

As mentioned, specific molecular bonds most active in the MIR are listed in Table 1.8.

Table 1.8 MIR active molecular bonds.

-OH, hydoxyl- [O-H]
R-OH, alcohols [O-H]
-CH -OH, primary alcohols
R′R″CH-OH, secondary alcohols
R′R″R‴C-OH, tertiary alcohols
Ar-OH, phenolics (or phenols) [O-H]
-OCH, methoxy (or ether group)
C-O-C, ethers
R-C=O -H, aldehydes
R-C=O -R′, ketones
C=C=O -R′, ketenes

(*Continued*)

Table 1.8 (*Continued*)

R-C=O-OR′, esters
S-C=O -O-R′, thiol esters
R-C=O -C=O -R, anhydrides
-C-O-O-C-, peroxides
-O-C=O -C=O -O-, oxalates
-O-C=O t, carboxy-
-C=O -, carbonyl group
-C=O -NH, primary amides
-C=O -NHR, secondary amides
-C=O -NR′R″, tertiary amides
-C≡N, nitriles
C-C, paraffinics (or alkanes)
CH -(CH) - CH, normal paraffins
R-CH, methyl C-H
R-CH R′, methylene C-H
R-CHR′R″, methine C-H
C=C, olefinic group (or alkenes)
-CH=CH, vinyls
R′R″C=CH, vinylidenes
>C=C=CH, allenes
-C≡C-, acetylinics
R-C≡N, nitriles
C H, aromatics
C H, naphthalenes (or cycloalkanes)
C-NO, nitroso- group
CONO, nitrites
R′R″NNO, nitrosamines
R-NO, nitro- group
R-NH, primary amines
R′R″NH, secondary amines
N=N, azo- group
R-N=N =N, R-N -N ≡N; azides
NNO, azoxy group
R-O-N-N=O, organic nitrites
-NO, nitrates
ON=NO, nitroso- group
H N -CH-COO, amino acids

1.4.2 *Fingerprint Frequencies*

Fingerprint frequencies are due to molecular vibrations of the entire molecule rather than being specific to a particular functional group. Since the infrared spectrum is unique to an individual molecule it is termed the fingerprint of that molecule, much as the fingerprint of a person indicates the specific identity of that person. However, in identifying the unique “fingerprint vibrations” of a molecule, information is given as to the approximate molecular formula and composition of the molecule. (Incidentally William James Hershel, grandson of F. William Hershel, was one of the first to identify the uniqueness of individual human fingerprints).[9]

1.4.3 *Dividing the Infrared Spectrum*

The group frequencies are given as those bands occurring below 6667 nm (above 1500 cm^{-1}); and the fingerprint frequencies are those above 6667 nm (below 1500 cm^{-1}). The group frequencies above 1500 cm^{-1} usually demonstrate bands of reasonable intensity, with reliable interpretation based upon frequency position; whereas the fingerprint frequencies contain bands for both group and fingerprint frequencies. This mixed (or overlapping) band region below 1500 cm^{-1} may give inaccurate interpretation based only upon wavelength position. Therefore, in the fingerprint region the most useful interpretive information is based upon shape, e.g., broad, very broad, sharp, very sharp; or intensity, e.g., weak, medium, intense. The fingerprint region is also useful for interpretation relative to the absence of a particular band. In summary, the group frequency region is used to derive a first assumption about the identity of the sample based on its infrared spectrum, and then the fingerprint region is used to verify the assumptions by the presence or absence of bands as well as their shapes and intensities.

Group frequencies are used with general rules and comparative known spectra to make positive identification of a sample spectrum. The ideal group frequencies have the characteristics of: (1) always being found above 1500 cm^{-1}, (2) have intense amplitude, (3) have identifiable and consistent position in frequency, (4) are reliable and always appear when a particular group is present, (5) are isolated without interference, (6) are narrow (in the range of ± 25 cm^{-1}), and (7) the group has more than one characteristic frequency (used for further verification of the group's presence).[10–12]

1.4.4 *Coupling of Vibrations*

Coupling indicates that the oscillators or molecular vibrations of two or more molecules are interactive, so that the original vibrational energy states (if the vibrations could occur independently of one another) result in split energy states due to the interaction of the vibrations. Coupling is divided into two basic orders, first and second (Fermi resonance). First order coupling can be involved in several important infrared group frequencies. For example, CO_2 has the two separate (uncoupled oscillators) of C=O, each occurring at approx. 5405 nm (~1850 cm^{-1}). The interactive (coupled oscillators) energy states occur at 4257 nm (2349 cm^{-1}, asymmetric stretch) and 7463 nm (1340 cm^{-1}, symmetric stretch). First order coupling is involved for multiple infrared group frequencies as listed.[13–14]

a. The stretches for all cumulated double bonds, X=Y=Z, e.g., C=C=N.
b. The stretches in XY_2, including -CH_2-, and H_2O.
c. The stretches in XY_3 groups, including -CH_3.
d. The deformations of XY_3 groups, including -CH_3, CCl_3.
e. The N-H in-plane bend of secondary amides, e.g., R-CO-NH-R′.

1.4.5 *Fermi Resonance (or Second Order Coupling)*

Fermi resonance is the interaction or coupling of two vibrational energy states with resultant separation of the states where one of the states is an overtone or a sum tone.[15]

An overtone vibration occurs as the integer multiple of a fundamental vibration in frequency space (intensity falls off rapidly with the higher multiples):

a. The first overtone of a fundamental vibration (ν_i) is equal to $2 \times \nu_i$.
b. The second overtone of a fundamental vibration is equal to $3\ \nu_i$.
c. The third overtone is equal to $4\ \nu_i$.

Note: overtones are a special case of sum tones where the frequencies are identical.

A sum tone is the general case of an overtone where the frequencies are not equal and where a variety of vibrational energy states can occur:

a. A binary sum tone is equal to the sum of two fundamentals, e.g., $\nu_i + \nu_k$.
b. A ternary sum tone is equal to the sum of three fundamental vibrations, e.g., $\nu_i + \nu_k + \nu_m$.
c. Other sum tones can occur such as the sum of an overtone and a fundamental vibration, e.g., $2\nu_i + \nu_k$.

Three requirements are stated for Fermi resonance.[16,18]

a. The zero-order frequencies must be close together (typically within 30 cm^{-1}).
b. The fundamental and the overtone or sum tone must have the same symmetry.
c. There must be a mechanism for the interaction of the vibrations.

(1) The vibrations cannot be separate (or localized) in distinctly different parts of the molecule.

(2) The vibrations must be mechanically interactive in order that the interaction of one vibration affects another.

The results of Fermi resonance are important for Infrared, Raman, and Near Infrared spectroscopy. Fermi resonance causes the following effects on spectral bands:

a. The resultant bands are moved in position from their expected frequencies.
b. Overtone bands are more intense than expected.
c. There may exist doublet bands where only singlets were expected.
d. Solvent changes can bring about slight shifts in frequency location of a band and intensities can be greatly changed.[17–19]

1.4.6 *Frequencies of Alkanes (Paraffins)*

1.4.6.1 *Normal Hydrocarbons (n-Alkanes)*

For infrared absorptions, the number of normal vibrations occurring within the infrared region for saturated hydrocarbons is 3N-6, where N is the sum of carbon plus hydrogen atoms, e.g., n-hexane (C_6H_{14}) exhibits $3(20) - 6 = 20$ normal vibrational modes. Internal rotation and overlapping of bands complicate the interpretation of alkane (paraffinic) group spectra. C-H stretching vibrations can be expressed to a first approximation by the straight-forward use of Hooke's law describing the vibration between two masses connected by a vibrating spring.

The first consideration in the description of alkane group spectral interpretation would include the C-H stretching vibrations for methyl (-CH_3) and methylene (-CH_2-) groups. The *methyl C-H stretching vibrations* (Figure 1.7) are found in two basic modes:

a. The asymmetric (degenerate or out-of-phase) methyl C-H stretch at 3365 nm to 3388 nm ($2962 \pm 10\,cm^{-1}$).
b. The symmetric (in-phase) methyl C-H stretch at 3470 nm to 3494 nm methyl C-H stretch at 3470 nm to 3494 nm ($2872 \pm 10\,cm^{-1}$).
c. Three C-H stretching modes correspond to three coupled C-H stretching vibration oscillators.
d. The molecular extinction coefficient for methyl C-H stretch remains constant irrespective of increasing alkane chain length.

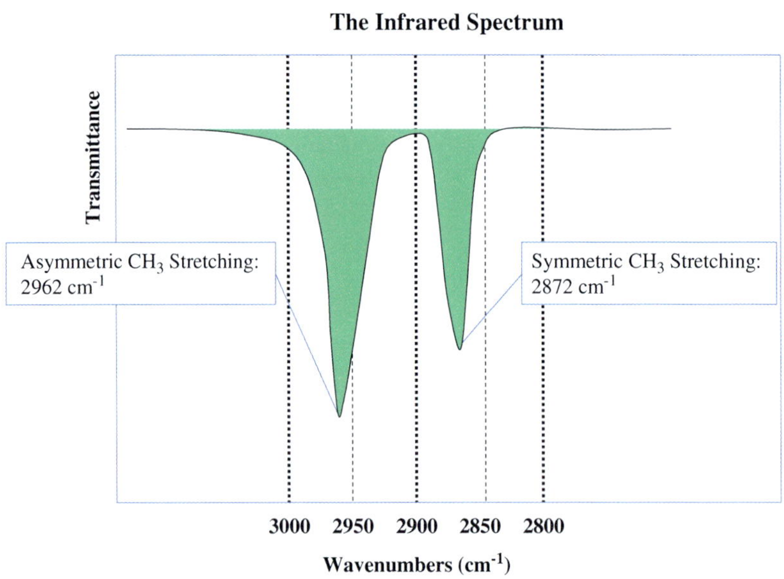

Fig. 1.7 Illustration of aliphatic methyl C-H stretching band locations as a transmittance spectrum.

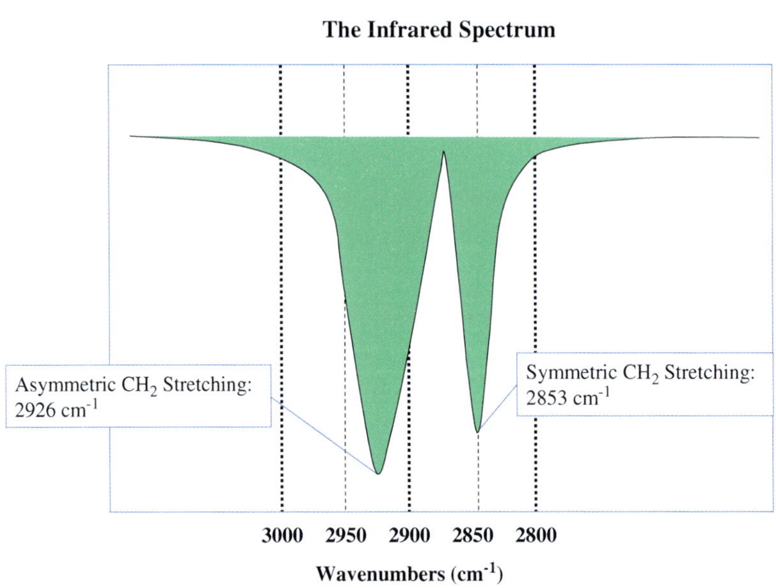

Fig. 1.8 Illustration of aliphatic methylene C-H stretching band locations as a transmittance spectrum.

The *methylene C-H stretching vibrations* (Figure 1.8) are also found in two basic modes:

a. The asymmetric (degenerate or out-of-phase) methylene C-H stretch at 3406 nm to 3429 nm ($2926 \pm 10\,cm^{-1}$).
b. The symmetric (in-phase) methylene C-H stretch at 3493 nm to 3517 nm ($2853 \pm 10\,cm^{-1}$).
c. Two methylene stretching modes result from two coupled C-H stretching vibration oscillators.
d. Increasing the chain length within alkane groups brings about a linear increase in the molecular extinction coefficient of methylene C-H stretching bands.

The carbon-hydrogen vibrational modes for alkanes also contain both methyl and methylene bending vibrations. The *methyl C-H bending vibrations* are found in two modes:

a. The asymmetric (degenerate or out-of-phase deformation) methyl C-H bending at 6803 nm to 6897 nm ($1460 \pm 10\,cm^{-1}$).
b. The symmetric (in-phase or "umbrella") methyl C-H bending at 7220 nm to 7326 nm ($1375 \pm 10\,cm^{-1}$). This functional group vibrational frequency is dependent upon the atom to which the methyl group is attached. Examples from the literature illustrating band frequency location relative to the atom attached to methyl group ($X\text{-}CH_3$) include those shown in Table 1.9.[20]
c. There is no coupling of methyl C-H bending modes to the rest of the system.

The *methylene C-H bending vibrations* are found in four basic modes:

a. The symmetric (scissoring) methylene C-H bending at 6826 nm to 6920 nm ($1455 \pm 10\ cm^{-1}$) where no coupling is observed.
b. The in-phase twisting vibration at 7634 nm to 7752 nm ($1300 \pm 10\,cm^{-1}$) with observed coupling.
c. The coupled wagging vibrations at 7692 nm to 10,000 nm (1300 to 1000 cm^{-1}).
d. The in-phase rocking vibration for alkanes with four or more carbons is located at 13,699 nm to 14,085 nm ($720 \pm 10\,cm^{-1}$) where coupling is observed. In addition, other bending modes of methyl and methylene can be observed.

The alkane carbon-to-carbon stretching, and carbon-to-carbon bending vibrations also occur as weak group frequencies.

Table 1.9 Infrared band frequency locations relative to the specific atom attached to a methyl group (X-CH_3).

Functional group	Fundamental frequency ($\pm 10\,cm^{-1}$)	Fundamental wavelength (in nm)
Pb-CH_3	$1165\,cm^{-1}$	8511–8658 nm
Sn-CH_3	1190	8333–8475
Sb-CH_3	1200	8264–8403
Hg-CH_3	1200	8264–8403
Ge-CH_3	1235	8032–8163
As-CH_3	1250	7937–8065
I-CH_3	1252	7924–8052
Se-CH_3	1280	7752–7874
Br-CH_3	1305	7605–7722
Si-CH_3	$1255 \pm 5\,cm^{-1}$	7937–8000
P-CH_3	1300	7634–7752
B-CH_3	1320	7519–7634
S-CH_3	1325	7491–7605
Cl-CH_3	1355	7326–7435
Carboxyl C- CH_3	1355	7326–7435
C-CH_3	$1378 \pm 5\,cm^{-1}$	7231–7283
N-CH_3	1425	6969–7067
O-CH_3	1455	6826–6920
F-CH_3	1475	6734–6826

1.4.6.2 *Branched Chain Hydrocarbons (Iso-, Gem-Dimethyl, Tertiary Butyl, and Cyclic Groups)*

For infrared absorptions the following spectra-structure correlations apply.

a. Methine groups (tertiary methyl, R_3C-H) are not important for the infrared region and do not exhibit unique infrared spectral characteristics.
b. Isopropyl and gem-dimethyl groups exhibit a symmetric bending vibration doublet at 7246 nm to 7353 nm ($1370 \pm 10\,cm^{-1}$) with coupling absent.
c. Tertiary butyl substitution exhibits a symmetric bending vibration doublet from 7194 nm to 7326 nm (1390 to $1365\,cm^{-1}$) that does not exhibit mechanical coupling.
d. Cyclic hydrocarbons demonstrate methylene C-H stretching modes based upon ring size. In addition, there are a number of spectral effects based upon lone electron pair interactions with carbon-hydrogen molecular orbitals. The reader is referred to a number of references describing the unique spectra of cyclic hydrocarbons with various substituted functional groups.[18–19]

1.4.7 *Group Frequencies of Alkenes (Olefins)*

1.4.7.1 *C-H Stretch of Alkenes*

For infrared, the olefins include bands for saturated hydrocarbons as well as the olefinic C=C stretching and olefinic C-H stretching and bending vibrations.

1.4.7.2 *Olefinic C-H Stretch*[21]

a. Olefinic C-H stretch occurs at 3226 nm to 3333 nm (3100 to 3000 cm^{-1}). C-H stretch for an unsaturated carbon occur below 3333 nm (above 3000 cm^{-1}), whereas the saturated carbon C-H bond stretching occurs above 3333 nm (below 3000 cm^{-1}). The sp-type C-H stretch for ≡C-H occurs at approximately 3030 nm (3300 cm^{-1}); the sp^2-type C-H stretch associated with aromatic and olefinic carbons occurs at 3226 nm to 3333 nm (3100 to 3000 cm^{-1}); and the sp^3-type stretch for saturated carbons is observed at the high frequency ranges between 3344 to 3571 nm (2990 and 2800 cm^{-1}).

b. Olefinic C-H stretch can occur as an end-group, e.g. R-R-C=C; or as an imbedded group associated with other carbons on either end, e.g., R-C=C-R. The absorption bands for these two olefinic types occur from 3268 nm to 3226 nm (3060 to 3100 cm^{-1}) for the end-group type, and from 3322 nm to 3279 nm (3010 to 3050 cm^{-1}) for the imbedded type. Note that compounds containing the vinyl group (-CH=CH_2) have weak absorption bands at approximately 3247 nm to 3300 nm (3080 cm^{-1} and 3030 cm^{-1}).

1.4.7.3 *Olefinic C=C Stretch*

a. Olefinic C=C stretch is infrared active but the intensity is quite variable depending upon associated atoms to the C=C group. The table below summarizes the appearance of absorption bands associated with olefinic C=C stretch for vinyl-type molecules, as illustrated in Table 1.10.

b. Factors affecting the frequencies for the various olefinic groups include conjugation. Conjugation is the sharing of electrons between the C=C olefinic bond and atoms or molecules attached to either of the carbons in the C=C group. The associated groups lower the frequency by 75 nm to 115 nm (20 to 30 cm^{-1}). The conjugated groups that lower the frequency of the -C=C- stretch include: nitrile (-C≡N), occurring at 6215 nm (1609 cm^{-1}); carbonyl (-CO-R), occurring at 6184 nm (1617 cm^{-1}); phenyl (-C_6H_5), at 6135 nm (1630 cm^{-1}); vinyl (-CH=CH_2), at 6169 nm (1621 cm^{-1}); and methyl (-CH_3), at 6057 nm (1651 cm^{-1}).

Table 1.10 Showing locations of C=C stretch vibrations related olefinic containing molecules.

Olefinic molecule type (R-type substituents)*	C=C stretch band location	Relative band intensity
trans, tri- and tetra substituted	5935 nm to 6006 nm (1685 to 1665 cm^{-1})	weak
cis olefins	6006 nm to 6154 nm (1665 to 1625 cm^{-1})	medium to strong
mono-substituted	6024 nm to 6135 nm (1660 to 1630 cm^{-1})	medium to strong
trans, di- substituted	5952 nm to 5935 nm (1680 to 1685 cm^{-1})	weak
cis, di-substituted	5992 nm to 6135 nm (1669 to 1630 cm^{-1})	medium to strong
di-substitution on a single carbon	5992 nm to 6135 nm (1669 to 1630 cm^{-1})	medium to strong
tri-substitution	5952 nm to 5935 nm (1680 to 1685 cm^{-1})	weak
tetra-substitution	5952 nm to 5935 nm (1680 to 1685 cm^{-1})	weak

*Substituents other than R-groups will change the frequencies of absorption.

c. The attachment of =C to a variety of ring sizes affects the frequency of the C=C olefinic stretch. A six member phenyl group attached to a (=C) group yields a 6057 nm band (1651 cm^{-1}); a five member ring yields a 6035 nm band (1657 cm^{-1}); a four membered ring exhibits a band at 5959 nm (1678 cm^{-1}); a three-membered ring at 5615 nm (1781 cm^{-1}); and the allene two carbon C=C olefinic group ($H_2C{=}C{=}CH_2$) occurs at 5136 nm (1974 cm^{-1}).

d. Carbonyl groups attached to olefinic groups yields a variety of frequency shifts depending upon the conjugation between carbonyl, the olefinic group, and the other attached groups; the section on carbonyls discusses this in greater detail.

e. Cyclic olefins exhibit unique frequencies for -C=C- stretch depending upon the number of carbons within the cyclic olefin group, and the type of attached molecule or atom to the olefinic carbons. The following table illustrates the frequency for olefinic C=C stretch dependent upon the size of the cyclic olefin and the attached group types (Table 1.11).

An increase in the electronegativity of the attached group raises the frequency of the olefinic C=C stretch; resonance lowers the frequency of the C=C stretch and there are interactions when the attached groups exhibit both electronegative and resonance features.

Table 1.11 Demonstrating C=C stretch frequency with respect to carbon size of cyclic olefin and type of attached group.

Attached atom or molecule to both sites of olefinic carbons as (-C=C-)	Number of carbons in cyclic olefin			
	6	5	4	3
-H	$1653\,cm^{-1}$ (6050 nm)	$1611\,cm^{-1}$ (6207 nm)	$1565\,cm^{-1}$ (6390 nm)	$1648\,cm^{-1}$ (6068 nm)
$-CH_3$	$1683\,cm^{-1}$ (5942 nm)	$1686\,cm^{-1}$ (5931 nm)	$1695\,cm^{-1}$ (5900 nm)	$1887\,cm^{-1}$ (5299 nm)

f. In the IR region the intensity of the fundamental olefinic carbon-carbon stretch, designated by $\tilde{\nu}$ (C=C) is highly variable. In some cases the band is completely absent; in other cases the band may be very strong. If there is little change in the dipole moment of an olefinic containing molecule, the C=C stretching mode will be infrared inactive. Thus, if the C=C group is across the center of symmetry for a molecule the C=C stretch will be forbidden in the infrared. A center of symmetry is where identical atoms or molecules are attached coaxially across the C=C olefinic group. As a general rule, the more symmetrically substituted the olefin is the weaker is the infrared band for the olefinic molecule. Bands present in the infrared spectrum for an olefin can indicate the presence of an olefinic group, but the absence of such bands does not give conclusive evidence for the absence of an olefin.

1.4.7.4 *Olefinic C-H Bend*

a. Two modes of bending occur in olefinic hydrogen bonds, in-plane and out-of-plane. In-plane bend represents a change in the C-H bond angle along the same plane, occurring at 6920 nm to 8333 nm (1455 to $1200\,cm^{-1}$) as a weak and not very useful group frequency for the presence of olefinic hydrogen bonds. Out-of-plane bending occurs as the C-H bonds vibrate outside the flat olefinic group plane. Out-of-plane bending occurs at approximately 10,000 nm to 15,385 nm (1000 to $650\,cm^{-1}$) and represent excellent bands for identifying the presence of olefinic groups. The out-of-plane C-H bend related bands are group frequencies which are distinct, not being affected by conjugation or coupling with other vibrations in the olefinic molecule.

b. There is only one infrared active C-H bend of the four possible modes for olefinic bending vibrations. The four bending modes include *twisting or rotational motion*; opposite twisting at the individual olefinic carbons producing a *torsion*

Table 1.12 Showing locations of C-H out-of plane bending vibrations for olefinic molecules.[22]

Olefinic molecule type (R-type substituents)*	C-H bend frequencies	Relative band intensity/shape
mono-substituted	995 to 985 cm^{-1} (10,050 to 10,152 nm)	very strong
	915 to 905 cm^{-1} (10,929 to 11, 050 nm)	very strong
trans, di-substituted cis olefins	970 to 960 cm^{-1} (10,309 to 10,417 nm)	very strong
cis, di-substituted	730 to 650 cm^{-1} (13,699 to 15,385 nm)	weak and broad
di-substitution on a single carbon	895 to 885 cm^{-1} (11,173 to 11,299 nm)	very strong
tri-substitution	840 to 790 cm^{-1} (11,905 to 12,658 nm)	strong
tetra-substitution	N/A	N/A

*Substituents other than R-groups will change the frequencies of absorption for the C-H out-of plane bend.

at the C=C bond; wagging, or a motion of the hydrogen bonds; and two forms of wag, one where the two groups of two each hydrogens move in opposite planes causing a rotational motion around a common center of axis; and the second wag where all four hydrogens move in or out of the plane together. The first type of wag is Raman only active and the second type of wag is the only infrared active type. Table 1.12 summarizes the frequencies for C-H out-of plane bending given the number of R groups attached to the two olefinic carbons.

1.4.8 *Molecules with Triple Bonds (-C≡)*

1.4.8.1 *Acetylenes (-C≡C-)*

a. Triple bonded carbon-to-carbon functional groups are termed acetylinic (or acetylenes). Due to symmetry about the C≡C bond the absorptions for C≡C are weak in the infrared. For acetylene (H-C≡C-H) the C≡C stretch is forbidden in the infrared and thus not observed. For R-C≡C-H (mono-substituted) the C≡C stretch occurs at 4673 nm to 4762 nm (2140 to 2100 cm^{-1}) and is weak in intensity. For R-C≡C-R′ (di-substituted acetylenes) the C≡C stretch occurs from 4425 nm to 4566 nm (2260 to 2190 cm^{-1}) with very weak, sharp bands. The ≡C-H stretch band for mono-substituted acetylenes is medium and sharp in the infrared occurring at 3021 nm to 3040 nm (3310 to 3290 cm^{-1}).

b. The bending vibrations for ≡C-H occur as medium-strong bands at approximately 15,385 nm (650 cm^{-1}).

1.4.8.2 *Nitriles (R-C≡N)*

a. The frequency for unconjugated nitriles, where -C≡N is not associated with electron withdrawing groups, occurs at 4425 nm to 4464 nm (2260 to 2240 cm^{-1}). Where conjugation occurs such as Aromatic-C≡N, or C=C-C≡N, the frequency is found from 4464 nm to 4525 nm (2240 to 2210 cm^{-1}).

b. Hydrogen cyanide produces the only differences to the above rules. In the case of HCN, frequencies occur at 4787 nm (2089 cm^{-1}) for gas phase and 4850 nm (2062 cm^{-1}) for liquid phase. In the case of acetonitrile, the frequency for C≡N stretch is found at 4444 nm (2250 cm^{-1}); for RCN the frequency of 4425 nm to 4464 nm (2260 to 2240 cm^{-1}) is observed.

c. Related groups have different frequencies depending upon electron sharing of the associated atom or group with the C≡N bond. As examples, nitrile oxide -C≡NO occurs at 4255 nm (2350 cm^{-1}, approximately 100 cm^{-1} higher than C-C≡N); and thiocyanate RS-C≡N occurs at 4651 nm (2150 cm^{-1}, approximately 100 cm^{-1} lower than C-C≡N).

1.4.9 *Cumulated Double Bond Systems (A=B=C)*

Table 1.13 illustrates the frequencies for a linear and non-linear system in the order of frequency. Note that the linear system, carbon dioxide, has a much greater frequency differential between the asymmetric and symmetric modes due to the high interaction between two co-linear oscillators sharing a common central atom. This first order interaction between two (hypothetical) fundamental vibrations is termed *coupling*.

Table 1.13 Frequencies for a linear and non-linear A=B=C systems in the order of frequency.

Frequency in cm^{-1} (wavelength in nm)	Functional group	Vibrational mode	Bond angle (degrees)
2349 cm^{-1} (4257 nm)	→O=C=O →	asymmetric	180, linear
1340 cm^{-1} (7463 nm)	←O=C=O →	symmetric	
1362 cm^{-1} (7342 nm)	→O=S=O →	asymmetric	119, non-linear
1151 cm^{-1} (8688 nm)	←O=S=O →	symmetric	

Note: Bond angle is only one consideration for frequency difference for interacting oscillators, another important consideration is the mass of the central atom.

1.4.10 *Aromatic Systems*

a. Aromatic systems produce strong and sharp infrared bands due to the relatively rigid molecular structures. There are multiple vibrational modes present in aromatic systems including: (1) carbon-hydrogen stretching, (2) in-plane bending, and (3) out-of-plane bending. Carbon-carbon interactions include (4) ring stretching, and (5) ring bending modes. Internal vibrations caused by groups associated with the aromatic system also exhibit vibrational group frequencies. The total vibrational modes for aromatic systems is equal to 3N-6 vibrations, where N is the number of carbons in the ring structure. These vibrational modes are shown in several references.[23–26]

b. Carbon-hydrogen stretching vibrations occur from approximately 3226 nm to 3333 nm (3100 to 3000 cm^{-1}) as shown in Figure 1.9. Infrared spectra of aromatics using high resolution instrumentation shows a pair of composite bands near 3257 nm (3070 cm^{-1}) and 3300 nm (3030 cm^{-1}). These composite bands arise as the combined interactions of several stretching modes discussed briefly above.

c. The carbon-carbon ring stretching occurs as a vibrational pair near 6250 nm (1600 cm^{-1}) as illustrated in Figure 1.9. This pair of frequencies is termed a degenerate pair and results from two stretching modes of the carbon-carbon ring stretching.

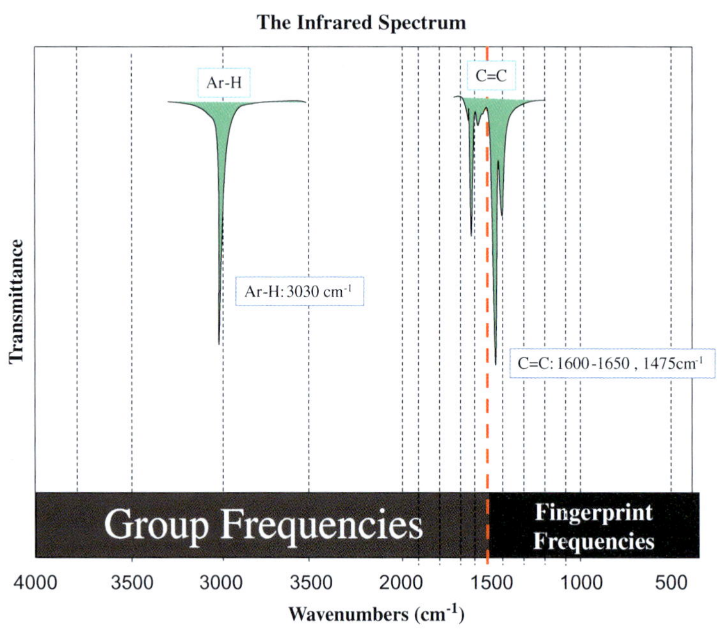

Fig. 1.9 Aromatic C-H stretching band as a transmittance spectrum.

However, due to Fermi resonance, one of the vibrations (near 6270 nm or 1595 cm^{-1}) is not observed in the infrared spectrum. When there is substitution on the aromatic ring, the degeneracy is lost and the pair of bands occur near 6250 nm (1600 cm^{-1}) and 6329 nm (1580 cm^{-1}). The 6329 nm (1580 cm^{-1}) band will be more intense if there is conjugation with the aromatic system by an electron withdrawing group, or if substitution by O or N has occurred. Para-disubstitution of a benzene ring by two identical groups yields to the center of symmetry rule, and thus the carbon-to-carbon ring stretching is infrared forbidden. It follows that if the para-disubstituted groups are not identical that some weak bands will occur near the 6250 nm (1600 cm^{-1}) region.

d. The carbon-hydrogen bending (wag) vibrations occur as in-plane bending and out-of plane bending. The in-plane bending vibrations are much less consistent than the out-of-plane bending. The in-plane bending vibrational frequency depends upon the number of C-H groups present and the substituent groups. Out-of-plane bending is much more consistent and intense in the infrared region. Out-of-plane bending is not particularly influenced by the substituents present, but the frequency for the bending vibrations depends upon the numbers of adjacent hydrogens. The typical presentation for the out-of-plane bending vibrations for benzene is given in Table 1.14.

Table 1.14 Out-of-plane bending vibrations for benzene.

Substitution on ring	Adjacent hydrogen atoms	Frequency (in cm^{-1})	Wavelength (in nm)
None	6	765–735	13072–13605
Mono-	5	770–735	12987–13605
Ortho (di-)	4	775–740	12903–13514
Tri-	3	810–750	12346–13333
Tetra-	2	860–800	11628–12500
Penta-	1	910–850	10989–11765

The carbon-hydrogen bending vibrations give rise to a series of combination bands termed *sumtones* or *summation bands*. These sumtones, summation bands, or combination bands occur from 5000 nm to 5882 nm (2000 to 1700 cm^{-1}). The patterns are used in infrared spectroscopy to determine ring substituents. The pattern for the sumtones is more important than their exact position. These composite band patterns are given in most infrared texts.[27–28]

1.4.11 *Vibrational States in Benzene*

Hayward and Henry calculate the frequencies for all eight members of the C-H stretching overtone spectrum in benzene. The paper provides a limited generalization method for calculating the anharmonicity constants for molecules where nearly identical oscillators are involved, and when these oscillators have vibrational motion which is approximately independent of the rest of a polyatomic molecule. The authors report the overtone frequencies (in cm^{-1}) of $\upsilon = 1-8$ as 3053, 5940, 8737, 11411, 14003, 16458, 18807, and 21025, respectively. The equivalent wavelength positions (in nm) for these overtone bands are 3275, 1684, 1145, 876, 714, 608, 532, and 476.[24–26,29]

Heller and Mukamel describe a general theory for estimating the overtone absorption line shapes for polyatomic molecules. The authors expand on the molecular Hamiltonian and local mode theory, and provide details for overtone line shape estimation. Numerical estimation of benzene overtone line shapes are given along with the derivation and detailed explanation of both theory and methodology.[30]

A description of the aspects related to local mode combination bands for aromatic and aliphatic hydrocarbons is discussed in a paper by Burberry and Albrecht. The authors explain the origin of side bands in the local mode (LM) overtone vibrations as resulting from LM-LM combination bands of various C-H oscillators. The local mode overtones are described by theory and observation for methyl groups in tetramethylsilane (TMS) as well as benzene.[31]

Child discusses issues on the use of local modes for spectroscopic assignment and interpretation. Examples discussed include water, acetylene, methane, and silane. The author generally reviews the topic for pre-1985 literature, including 42 references.[32]

1.4.12 *Carbonyl Compounds (>C=O)*

a. Compounds containing carbonyl groups exhibit frequencies over a broad range of 5128 nm to 6452 nm (1950 to 1550 cm^{-1}). The position or frequency of the carbonyl C=O stretching vibration is affected by (1) the isotope effects and mass change of substituted groups, (2) bond angles of substituted groups, (3) electronic (resonance and inductive) effects, and (4) interactions of these effects. Substituents with higher mass decrease the C=O stretch frequency; increasing the mass or bond angles of substituents also decreases the frequency by up to 40 cm^{-1} at 1715 cm^{-1}; this is equivalent to an increase of 139 nm at 5831 nm. Similarly decreasing the substituent mass, or bond angles between the carbonyl carbon and its substituents increases the frequency by 25 cm^{-1} above the nominal 1715 cm^{-1} carbonyl C=O stretch frequency; equivalent to a decrease in wavelength of 84 nm at 5831 nm. More electronegative (electron withdrawing) substituents will increase the carbonyl carbon-oxygen stretch frequency by up to 100 cm^{-1} above the 1715 cm^{-1} nominal frequency; a decrease of 321 nm at 5831 nm. Conjugation of the carbonyl group to aromatic or olefinic groups tends to lower the frequency for both C=O and C=C by 30 to 40 cm^{-1} dependent upon the ring size of the substituent. For example, carbonyl conjugation with aromatic ring sizes of 6, 5, 4, and 3 carbons produces C=O stretch frequencies at 5935 nm (1685 cm^{-1}), 5928 nm (1687 cm^{-1}), 5931 nm (1686 cm^{-1}), and 5963 nm (1677 cm^{-1}), respectively. Electronic effects such as resonance and inductive type produce a double-bond character and increases the frequency of the carbonyl stretch.

1.4.12.1 *Special Cases of the Carbonyl Group*

The group called anhydrides (R-C=O -O- C=O -R), consists of two carbonyls connected by an oxygen. The frequency for the carbonyl stretch occurs from 5495 nm to 5714 nm (1820 to 1750 cm^{-1}); the in-phase (←C=O→ and ←C=O→) stretch for the two carbonyl groups in the anhydride occurs at 5495 nm (1820 cm^{-1}). In the out-of-phase carbonyl pair (←C=O→ and →C=O←) the frequency for the carbonyl is 5714 nm (1750 cm^{-1}). In cyclic anhydrides the in-phase frequency is 5348 nm to 5405 nm (1870 to 1850 cm^{-1}), and the out-of-phase occurs at 5556 nm to 5618 nm (1800 to 1780 cm^{-1}). Solvent effects like hydrogen bonding to the carbonyl oxygen lowers the carbonyl C=O frequency. As in other carbonyl containing molecules smaller mass substituents or decreasing bond angle of carbonyl carbon substituent bonds increase the frequency for the carbonyl. Large inductive effects from the addition of largely electronegative substituents also increases the frequency for the carbonyl. When a halogen is attached to the alpha carbon, which is bonded to the carbonyl carbon, the frequency is increased by 60 nm to 90 nm (20 to 30 cm^{-1}).

Acid dimers interact by strong molecular hydrogen bonding and lower the carbonyl frequency by 25 to 60 cm^{-1}. The interaction is between the carbonyl oxygen and the hydrogen associated with the acid molecules (R-C=O -OH.....O=C-R-OH).

1.4.13 *Amides (R-C=ONH-X)*

a. Amides are a special case of carbonyl compounds which are subject to frequency shifts due to hydrogen bonding. The frequencies for dilute solutions of amides can vary by as much as 250–275 nm (75 cm^{-1}) at ~5880 nm as compared to the same molecule with hydrogen bonding (such as a Nujol mull preparation).

Dilute solutions of amides exhibit carbonyl frequencies at 5764 nm (1735 cm^{-1}), primary, R-C=O -NH_2; 5865 nm (1705 cm^{-1}), secondary, R-C=O -NHR; and 6042 nm (1655 cm^{-1}), tertiary, R-C = O -NR_2. The respective hydrogen bonded vibrations occur at 5882 nm to 6061 nm (1700 to 1650 cm^{-1}) for primary; 5952 nm to 6116 nm (1680 to 1635 cm^{-1}) for secondary; and 6024 nm (1660 cm^{-1}) for tertiary amides.[33–34]

b. The N-H stretch for amides occurs as two bands between 2976 nm and 3150 nm (3360 and 3175 cm^{-1}). Geometrically, two forms of amides exist as either the trans or cis forms; the trans form demonstrating a frequency of 3008 to 3053 nm (3325 to 3275 cm^{-1}); and the cis form 3140 nm to 3190 nm (3185 to 3135 cm^{-1}). The geometries are as R-C=O -N-R′ -H where the R and R′ are either in the trans or cis configuration.[35]

The term "amide I" band refers to the C=O carbonyl stretch and from the section on carbonyls it was mentioned that the nominal C=O stretch frequency is 5831 nm (1715 cm^{-1}). The amide form of the carbonyl has the electronegative N atom, lowered by resonance with the nitrogen electrons and lowered by hydrogen bonding. The C=O nominal frequency for the amide I stretch is approximately 1650 cm^{-1} (6061 nm).

The term "amide II" refers to the NH_2 scissoring associated with primary amides (R-C=O -NH_2). The amide II frequency results from the combined interactions of the C-N stretch and the N-H in-plane bend; these vibrations interact to form bands at the 6494 nm (1540 cm^{-1}) region, and the 7634 nm to 8197 nm (1310 to 1220 cm^{-1}) region. A band at 6410 nm to 6515 nm (1560 to 1535 cm^{-1}) indicates a trans substituted amide; this band is lost in the cis-amide configuration.

The Carboxylate ion (O=C-O^-) has two identical C┄┄O, C┄┄O bonds; consisting of the in-phase ←C┄┄O→, ←C┄┄O→; and out-of-phase ←C┄┄O→, →C┄┄O← bands. The broad out-of-phase band occurs at 6192 nm to 6452 nm (1615 to 1550 cm^{-1}), whereas the broad in-phase band has a frequency of 7143 nm to 7692 nm (1400 to 1300 cm^{-1}). The in-plane band is not particularly useful in spectroscopy due to its variable intensity and position.

1.4.14 *Ethers (C-O-C)*

The ether group can be distinguished by the degree of saturation for the ether carbons. A saturated ether has the maximum number of hydrogens (6) associated with the ether carbons. Unsaturated ethers can either be single unsaturated or double unsaturated, depending upon the number of unsaturated methoxy carbons (=C-O-) within the ether group. Saturated ethers exhibit a very strong C-O stretch at a frequency of 1160 to 1050 cm^{-1} (8621 nm to 9524 nm); doubly unsaturated ethers exhibit a strong, slightly broadened C-O stretch frequency at 1275 to 1230 cm^{-1} (7843 nm to 8130 nm). Epoxides (C-O-C) demonstrate a symmetric stretch near 1255 cm^{-1} and an asymmetric stretch of 860 cm^{-1} (11,628 nm).

1.4.15 *Esters (R-CO-OR′)*

The ester group exhibits the frequencies of a mixed ether and it does show a very strong band at 7937 nm to 8696 nm (1260 to 1150 cm^{-1}); and a strong band at 8333 nm to 10,000 nm (1200 to 1000 cm^{-1}). These two bands correspond to the C-O for RCO-OR′ and the O-R′ of RCOO-R′.

1.4.16 *Anhydrides (R-C=O -C=O -R)*

Cyclic anhydrides show bands for doubly unsaturated ethers at 7937 nm (approximately 1260 cm^{-1}); open chain anhydrides show the saturated ether band at 9434 nm (1060 cm^{-1}).

1.4.17 *Peroxides (-C-O-O-C-)*

Infrared active bands are extremely difficult to identify for peroxides; there is an ether C-O stretch vibration near 9091 nm (1100 cm^{-1}) that can be useful for identification.

1.4.18 *X-H Functional Groups (O-H and N-H)*

X-H frequencies occur below 5000 nm (above 2000 cm^{-1}), except for Boron compounds which demonstrate bands from 4762 nm to 6173 nm (2100 to 1620 cm^{-1}). Interaction of X-H with other groups does not usually occur due to the high frequency of these bands and the low mass of hydrogen, therefore the frequency of X-H is specific and reliable for infrared spectroscopy. The X-H stretch follows the two-body harmonic oscillator approximation (discussed in another section of this book). The relationship of X-H to the harmonic oscillator allows the relationship to frequency of any specific X-H group to be related to the Periodic Table; the frequency of an X-H group increases as one goes upward in any column or to the right within any row of the Periodic Table. The first row of the Periodic Table gives H-H with a nominal fundamental stretch at 2404 nm (4160 cm^{-1}). The second row yields ν(Li-H) = 7353 nm (1360 cm^{-1}); ν(Be-H) = 5089 nm (1965 cm^{-1}); ν(B-H) = 4000 nm (2500 cm^{-1}) and 4762 nm to 6173 nm (2100 to 1620 cm^{-1}); ν(C-H) = 3384 nm (2955 cm^{-1}); ν(N-H) =

2941 nm (3400 cm^{-1}); ν(O-H) = 2770 nm (3610 cm^{-1}); ν(F-H) = 2532 nm (3950 cm^{-1}). In the third row, the more important ν(X-H) include ν(Si-H) = 4577 nm (2185 cm^{-1}); ν(P-H) = 4255 nm (2350 cm^{-1}); and ν(S-H) = 3876 nm (2580 cm^{-1}). The most important of these for the discussion of Near Infrared spectroscopy (see Volume 3) include O-H and C-H.[35–44]

1.4.19 *Hydrogen Bonding (X-H.......Y)*

A hydrogen bonded molecular system has four possible modes: (1) O-H stretch found at 2941 nm to 4000 nm (3400 to 2500 cm^{-1}), (2) R-O-H in-plane bend at 7143 nm to 7692 nm (1400 to 1300 cm^{-1}), (3) the R-O-H out-of-plane bend at 15,152 nm (660 cm^{-1}), and (4) the -O-H........O stretch at 57,143 nm (175 cm^{-1}).

Specific frequencies for molecules containing O-H with hydrogen bonding include: alcohols (R-O-H) with ν(O-H) = 2981 nm (3355 cm^{-1}), carboxylic acids (-COOH) with ν(O-H) = 3333 nm (3000 cm^{-1}), and amines (R-NH-X) with ν(N-H) = 3000 nm (3300 cm^{-1}). In general, hydrogen bonding has several main effects on frequency of X-H bands: (1) it lowers the frequency of the X-H stretch as compared to the gas phase and dilute liquid phase of a molecule; (2) it raises the X-H bend frequency; (3) it broadens all O-H stretch bands by up to 300 to 475 nm (300 to 500 cm^{-1}); and (4) it intensifies all bending and stretching X-H bands.[38–44]

1.4.20 *The O-H Stretch*

a. It is common knowledge among infrared spectroscopists that any fundamental band occurring above 2857 nm (3500 cm^{-1}) can be confidently assigned to the O-H stretch group. A sharp band exists for the stretching frequency of O-H for the infrared spectrum of O-H containing systems. Primary alcohols (-CH_2-OH) exhibit ν_s(O-H) = 2740 nm (3650 cm^{-1}); secondary alcohols (R-CHOH-R′) exhibit ν_s(O-H) = 2762 nm (3620 cm^{-1}); tertiary alcohols (R-CR′OH-R″) demonstrate ν_s(O-H) = 2770 nm (3610 cm^{-1}); phenols ν_s(O-H) = 2762 to 2793 nm (3620 to 3580 cm^{-1}); and carboxylic acids have ν_s(O-H) = 2817 to 2857 nm (3550 to 3500 cm^{-1}). To distinguish between primary, secondary, and tertiary alcohols using infrared spectra requires the use of these bands as well as the ν_s(C-O) = 8547 to 9615 nm (1170 to 1040 cm^{-1}).

b. The O-H in-plane bend occurs as ν_B(O-H) = 7143 to 7692 nm (1400 to 1300 cm^{-1}). The C-O-H out-of-plane bend ν_B(O-H) = 13,158 to 15,385 nm (760 to 650 cm^{-1}).[38–44]

c. Carboxylic acids (-COOH) form dimers (hydrogen bonding between molecules as (-C=O......H-O-C-) in condensed states that exhibit four distinct frequencies: (1) a very broad, very strong, O-H stretch at approximately 3333 nm (3000 cm^{-1} at a lower frequency than alcohols); (2) summation tones as one to three weak bands near 3846 nm (2600 cm^{-1}); (3) carbonyl (C=O) stretch vibration near 5797 nm (1725 cm^{-1}); and (4) a weak, broad band indicating a dimer condition near 10,471 nm (955 cm^{-1}).

1.4.21 *N-H Group Frequencies and Nitrogen Containing Compounds*[34–36]

a. The average frequencies for free N-H groups include primary amides (-C=O -NH_2, ν_s(N-H) = 2857 nm (3500 cm^{-1}) and 2933 nm (3410 cm^{-1}); and secondary amides (-C=O -NHC, ν_s(N-H) = 2899 nm (3450 cm^{-1}). Alkyl (aliphatic-NH-X) groups include alkyl-NH_2, ν_s(N-H) = 2946 nm (3395 cm^{-1}) and 3003 nm (3330 cm^{-1}); for alkyl-NH-X, ν_s(N-H) = 2981 nm (3355 cm^{-1}). In the case of aryl (aromatic-NH-X), the N-H stretch fundamentals occur as ν_s(N-H) = 2869 nm (3485 cm^{-1}) and 2946 nm (3395 cm^{-1}) for aryl-NH_2, and ν_s(N-H) = 2911 nm (3435 cm^{-1}) for aryl-NH-X. Note: the N-H frequency is lower and less intense than the O-H frequency; the hydrogen bonding effects with N-H are weaker and of less effect than in O-H.

b. Primary amines (R- NH_2) demonstrate two N-H stretching bands due to coupling; the in-phase (or symmetric), and the out-of-phase (or asymmetric). Aliphatic amines, (C_nH_{2n+1})-NHX, demonstrate in-phase frequencies from 2999 to 3053 nm (3335 to 3275 cm^{-1}) with medium strength; out-of-phase frequency occurs from 2933 to 2985 nm (3410 to 3350 cm^{-1}), also with medium strength. Aromatic amines exhibit in-phase and out-of-phase frequencies of 2920 to 3003 nm (3425 to 3330 cm^{-1}), and 2849 to 2950 nm (3510 to 3390 cm^{-1}) both with medium strength. The scissoring for primary amines occurs from 6042 to 6289 nm (1655 to 1590 cm^{-1}), as a medium to strong, broad band; the NH_2 wagging is a strong, broad band near 12,500 nm (800 cm^{-1}).

c. Secondary amines (R′R″-NH) exhibit a single, broad N-H stretch vibrational frequency near 3021 nm (3310 cm^{-1}); an N-H bend occurs as a weak-to-medium strength band near 6667 nm (1500 cm^{-1}).

d. To separate spectral features for the various amines, i.e., primary, secondary, and tertiary; it is helpful to point to obvious differences between the groups. For example, the number of N-H stretches for the three types are: primary (2), secondary (1), and tertiary (0). For N-H bending vibrations, primary amines have a band near 6042 to 6289 nm (1655 to 1590 cm^{-1}); secondary amines, near 6667 nm (1500 cm^{-1}); and this feature is absent in tertiary amines. Lastly, the C-H stretch vibration for CH_2 is to be found for all three types of amines from 3503 to 3683 nm (2855 to 2715 cm^{-1}); for primary amines, the C-H stretch will be found at higher frequencies nearer to 3448 nm (2900 cm^{-1}).

e. Amine salts as R-$NH_3^+$and R′R″NH_2^+ exhibit several weak bands near 3571 to 4545 nm (2800 to 2200 cm^{-1}). These salts exhibit a weak-to-medium, very broad band, centered near 4000 nm (2500 cm^{-1}), lower in frequency than the C-H stretch and up to 314 nm (500 cm^{-1}) in width at the baseline.

1.4.22 *Amines (R-NH-X)*

Amines are polar compounds, and both primary and secondary amines can form intermolecular hydrogen bonding. All classes of amines are water soluble (up to 6 carbon atoms) and will form hydrogen bonding in aqueous solutions.

References

1. H. Martens and T. Naes, *Multivariate Calibration*, 1st edn. (John Wiley and Sons, Chichester, 1992).
2. H. Mark and J. Workman, *Statistics in Spectroscopy*, 2nd edn. (Elsevier, Amsterdam, 2003).
3. H. Mark and J. Workman, Units of Measure in Spectroscopy, Part 1: Scaled Volume Fraction, *Spectroscopy*, **29**(2) (2014) 1–10.
4. ASTM E1421-99(2009), Standard Practice for Describing and Measuring Performance of Fourier Transform Mid Infrared (FT-MIR) Spectrometers: Level Zero and Level One Tests", ASTM International, 100 Barr Harbor Drive, PO Box C700, West Conshohocken, PA, 19428–2959 USA (2009).
5. Standard Reference Material (SRM)® 1921b Certificate, “Infrared Transmission Wavelength Standard”, National Institute of Standards and Technology, Gaithersburg, MD 20899, 7 pages, Certificate Issue Date: 19 February 2014.
6. L. M. Hanssen, C. Zhu, Wavenumber Standards for Mid Infrared Spectrometry, In *Handbook of Vibrational Spectroscopy*, J. M. Chalmers, P. R. Griffiths (Eds) (John Wiley & Sons Ltd.: Vol. 1, pp. 881–890, 2002).
7. D. Gupta, L. Wang, L. M. Hanssen, J. J. Hsia, and R. U. Datla, Polysytrene Films for Calibrating the Wavelength Scale of Infrared Spectrophotomers — SRM 1921, NIST Special Publication 260–122, National Institute of Standards and Technology, Gaithersburg, MD (April 1995), 25 pages.
8. S. R. Lowry, J. Hyatte, and W. J. McCarthy, Determination of Wavelength Accuracy in the Near Infrared Spectra Region Based on NIST’s Infrared Transmission Wavelength Standard SRM 1921, *Appl. Spectrosc.* **54**(3) (2000) 450–455.

9. W. J. (William James) Hershel, Finger-Prints, *Nature* **23** (Nov. 25, 1880) 76.
10. W. de W. Abney and Lt. Col. Festing, On the Influence of the Atomic Grouping in the molecules of Organic Bodies on their Absorption in the Infra-red Region of the Spectrum, *Phil. Transact.* **172** (1882) 887–918.
11. L. J. Bellamy and D. W. Mayo, Infrared Frequency Effects of Lone Pair Interactions with Antibonding Orbitals on Adjacent Atoms, *J. Phys. Chem.* **80** (1976) 1217–1220.
12. L. J. Bellamy, Some Wider Uses of Group Frequencies, *Appl. Spectrosc.* **33** (1979) 439–443.
13. J. Grasselli, R. Hannah, D. W. Mayo, and F. A. Miller, *Infrared Spectroscopy* (Bowdoin College, Brunswick, Maine; Volume 1, p. 33, 1994).
14. B. R. Henry, Use of Local Modes in the Description of Highly Vibrationally Excited Molecules, *Acc. Chem. Res.* **10** (1977) 207–213.
15. Von E. Fermi, Uber den Ramaneffekt des Kohlendioxyds, *Z. Physik* **71** (1931) 250–259.
16. J. Grasselli, R. Hannah, D. W. Mayo, and F. A. Miller, *op. cit., Infrared Spectroscopy* (Bowdoin College, Brunswick, Maine; Volume 1, p. 34, 1994).
17. J. F. Bertrán, L. Ballester. *Spectrochim. Acta,* **39A** (1983) 123.
18. N. B. Colthup, L. H. Daly, and S. E. Wiberley, *Introduction to Infrared and Raman Spectroscopy*, 3rd edn. (Academic Press, Boston, pp. 31–33, 1990).
19. D. Lin-Vien, N. B. Colthup, W. G. Fateley, and J. G. Grasselli, *The Handbook of Infrared and Raman Characteristic Frequencies of Organic Molecules*, 1st edn. (Academic Press, Boston, 1991).
20. J. Grasselli, R. Hannah, D. W. Mayo, and F. A. Miller, *op. cit., Infrared Spectroscopy* (Bowdoin College, Brunswick, Maine; Volume 1, p. 44, 1994).
21. H. L. Fang and D. A. C. Compton, Overtone Spectroscopy of Nonequivalent C-H Oscillators in 1-Alkenes and Dienes, *J. Chem. Phys.* **92** (1988) 7185–7192, 1988.
22. G. Socrates, *Infrared Characteristic Group Frequencies*, 2nd edn. (John Wiley & Sons, Chichester, 1997).
23. N. B. Colthup, L. H. Daly, and S. E. Wiberley, *op. cit., Introduction to Infrared and Raman Spectroscopy*, 3rd edn. (Academic Press, Boston, p. 262, 1990).
24. R. G. Bray and M. J. Berry, Intramolecular rate processes in highly vibrationally excited benzene, *J. Chem. Phys.* **71**(12) (1979) 4909–4922.
25. K. G. Spears and S. A. Rice, Study of the Individual Vibronic States of the Isolated Benzene Molecule, *J. Chem. Phys.* **55**(12) (1971) 5561–5581.
26. R. J. Hayward and B. R. Henry, Anharmonicity in Polyatomic Molecules, *Journ. of Mol. Spectrosc.* **46** (1973) 207–213.
27. J. Grasselli, R. Hannah, D. W. Mayo, and F. A. Miller, *op. cit., Infrared Spectroscopy* (Bowdoin College, Brunswick, Maine; Volume 1, p. 82, 1994).
28. N. B. Colthup, L. H. Daly, and S. E. Wiberley, *op. cit., Introduction to Infrared and Raman Spectroscopy*, 3rd edn. (Academic Press, Boston, p. 274, 1990).
29. R. J. Hayward, and B. R. Henry, Anharmonicity in Polyatomic Molecules, *J. Mol. Spectrosc.* **46** (1973) 207–213.
30. D. F. Heller, and S. Mukamel, Theory of Vibrational Overtone Line Shapes of Polyatomic Molecules, *J. Chem. Phys.* **70** (1979) 463–472.
31. M. S. Burberry and A. C. Albrecht, Local Mode Combination Bands and Local Mode Mixing, *J. Chem. Phys.* **71** (1979) 4631–4640.
32. M. S. Child, Local Mode Overtone Spectra, *Acc. Chem. Res.* **18** (1985) 45–50.

33. J. Grasselli, R. Hannah, D. W. Mayo, and F. A. Miller, *op. cit., Infrared Spectroscopy*, (Bowdoin College, Brunswick, Maine; Volume 1, p. 117, 1994).
34. A. Leifer and E. R. Lippincott, The Infrared Spectra of Some Amino Acids, *J.A.C.S.* **79** (1957) 5098–5101.
35. S. E. Krikorian and M. Mahpour, The identification and origin of N-H overtone and combination bands in the near infrared spectra of simple primary and secondary amides, *Spectrochimica Acta* **29A** (1973) 1233–1246.
36. R. J. Hayward and B. R. Henry, Anaharmonicity in Poyatomic Molecules: A Local-Mode Analysis of the XH-Stretching Overtone Spectra of Ammonia and Methane, *J. Mol. Spectrosc.* **50** (1974) 58–67.
37. B. T. Darling and D. M. Dennison, The Water Vapor Molecule, *Phys. Rev.* **57** (1940) 128–139.
38. E. R. Lippincott and R. Schroeder, One-Dimensional Model of the Hydrogen Bond, *J. Chem. Phys.* **23** (1955) 1099–1106.
39. J. N. Finch and E. R. Lippincott, Hydrogen bond systems–temperature dependence of OH frequency shifts and OH band intensities, *J. Chem. Phys.* **61** (1957) 894–902.
40. R. Schroeder and E. R. Lippincott, Potential Function Model of Hydrogen Bonds. II, *J. Chem. Phys.* **61** (1957) 921–928.
41. J. M. Stutman and A. S. Posner, Hydrogen Bonding in the Calcium Phosphates, *Nature* 193 (1962) 368–369.
42. J. N. Finch and E. R. Lippincott, Hydrogen Bond Systems: Temperature Dependence of OH Frequency Shifts and OH Band Intensities, *J. Chem. Phys.* **24** (1956) 908–909.
43. E. R. Lippincott *et al.*, Potential-function model of hydrogen bond systems, Hydrogen Bonding, Papers Symposium Ljubljana 1957 (1959) 361–374.
44. A. S. N. Murthy and C. N. R. Rao, Spectroscopic Studies of the Hydrogen Bond, *Appl. Spectrosc. Rev.* **2** (1968) 69–191.

Chapter 2

Theory of Infrared Spectroscopy, Light Interaction with Matter

Contents

2.1 Theory of Spectroscopy: Light Interaction with Materials

Light interaction using spectrophotometry for liquids, gases, semi-solids, slurries, solid composites, and ground particles is defined as the qualitative or quantitative measurement of the interaction of electromagnetic radiation (emr) with the particles themselves and with the atoms or molecules within a sample. The emr interacts mainly as scattering, absorption, reflectance, and emission with solid matter. A variety of spectrometer configurations are used to optimize the measurements of electromagnetic radiation as it interacts with solid matter.

2.1.1 *The Physics of Light Interaction with Solid Materials*

The main topics for this discussion are covered in significant detail in Chapter 2 for both Volumes 2 and 3. These topics listed in the index as well include: absorption, scattering, specular reflection, diffuse reflection, transmission, diffuse transmission, interactance, dispersion, diffraction, refraction, opacity, color, and absorbance. Many other concepts are described in these volumes; the reader is referred to the index for each specific topic. For infrared measurements the unique topics included here are film thickness measurements, the use of internal reflectance as ATR (attenuated total reflectance), and grazing angle infrared reflectance.

2.1.1.1 *Film Thickness Measurement*

A simplified formula for the spectrometric determination of film thickness under the measurement conditions where light from the spectrometer passes from ambient air onto a single film surface and where the underlying substrate reflects some of that light back to the spectrometer detection system is given below. In this case the measurement spectrum demonstrates an interference pattern consisting of a continuous wave function whose period is dependent upon the frequency of the energy striking the film surface. For these conditions the thickness of the film (τ) in centimeters is given by:

$$\tau = \frac{N}{2(\nu_2 - \nu_1)} \tag{2.1}$$

where N is the total number of complete waves (on the interference pattern). Note: One complete wave is measured from minima to minima or from maxima to maxima. The symbol ν_1 represents the lowest frequency over which the number of complete waves is measured; and ν_2 represents the highest frequency over which the number of complete waves is measured. For example, if 3 complete waves (Maxima to Maxima) are measured from 2000 to 1800 cm^{-1} on a single thin film producing an interference pattern superimposed upon a spectrum, the film thickness is 3/2(2000-1800) = cm = 0.0075 cm (or 75 micrometers).

The example shown in Fig. 2.1 is typical of normal film thickness problems. Three samples of the same film material are measured using near infrared spectroscopy. The three examples are shown below following baseline correction and after labeling the peak maxima for an ideal section of the spectrum where the most useful interference pattern is demonstrated. The thickness of Sample #1 is 6/2(1103) = 0.00272 cm = 27.2 microns; Sample #2 is 6/2(1097) = 0.00273 cm = 27.3 microns; and Sample #3 is 6/2(1090) = 0.00275 cm = 27.5 microns.

2.1.1.2 *Internal Reflectance as ATR (attenuated total reflectance)*

ATR (attenuated total reflectance) is used for surface analysis using physical contact with a sample surface. The technique utilizes a crystal of material for the actual physical contact with the sample. Typical crystal materials include diamond (Di), germanium (Ge), silicon (Si), and zinc selenide (ZnSe). ATR measurements can yield excellent quality spectra provided that the contact pressures of the ATR crystal and the sample are held constant; reproducible data requires reproducible contact

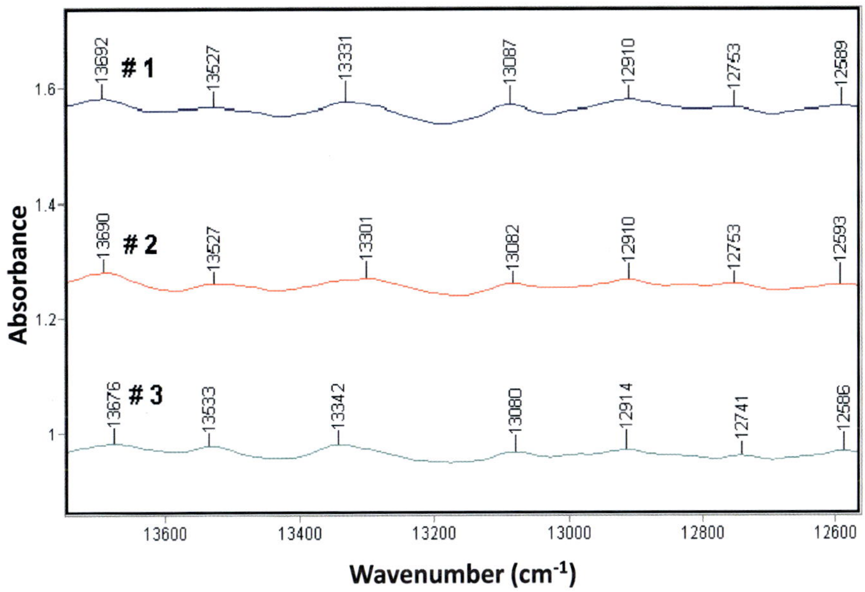

Fig. 2.1 Thin film thickness measurement spectra for example given in Equation 2.1.[1]

pressure. Commercial contact gauges are available. One percent reproducibility is typically achieved when keeping the contact pressure constant. ATR crystals are generally ZnSe with a refractive index of 2.42 and an angle of incidence of 45 degrees. Sample penetration is to a maximum of 40 microns using this ATR crystal. An air background is taken for ATR measurements. The depth of penetration (D_p) is calculated for ATR crystals using the relationship:

$$D_p = \frac{\lambda}{2\pi \cdot n_1 \left(\left| \sin^2 \alpha - \left(\frac{n_2}{n_1} \right)^2 \right| \right)^{\frac{1}{2}}} \tag{2.2}$$

Note: n_1 is the refractive index of the ATR crystal; refractive indices for typical ATR crystals are: ZnSe (2.42), Di (2.2), Si (3.6), Ge (4.0); n_2 is the refractive index of the sample (e.g., water at 20° C is 1.333); and α is the angle of incidence (and reflection) of the ATR crystal, typically 45 degrees. Note: $\sin^2 \alpha = [\sin(\alpha)]^2 = \sin(\alpha)^2 = 0.5$; for ATR the depth of penetration computations (in microns) for 45 degree incident angle ATR crystals and water as the sample (at 20° C) are: $\lambda_m/12.5$ for diamond; $\lambda_m/14.0$ for ZnSe; $\lambda_m/21.8$ for Si; and $\lambda_m/24.4$ for Ge. This relationship is illustrated in the following Figures 2.2 and 2.3. Note that λ_m is in microns for these calculations.

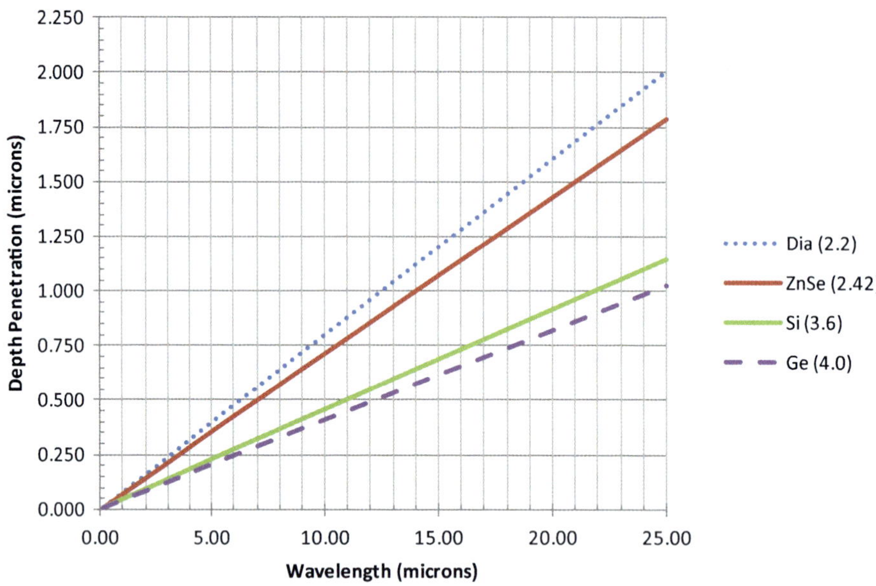

Fig. 2.2 Calculated depth of penetration (in microns) for ATR versus wavelength of incident energy (in microns) for 45° ATR crystals of different materials.

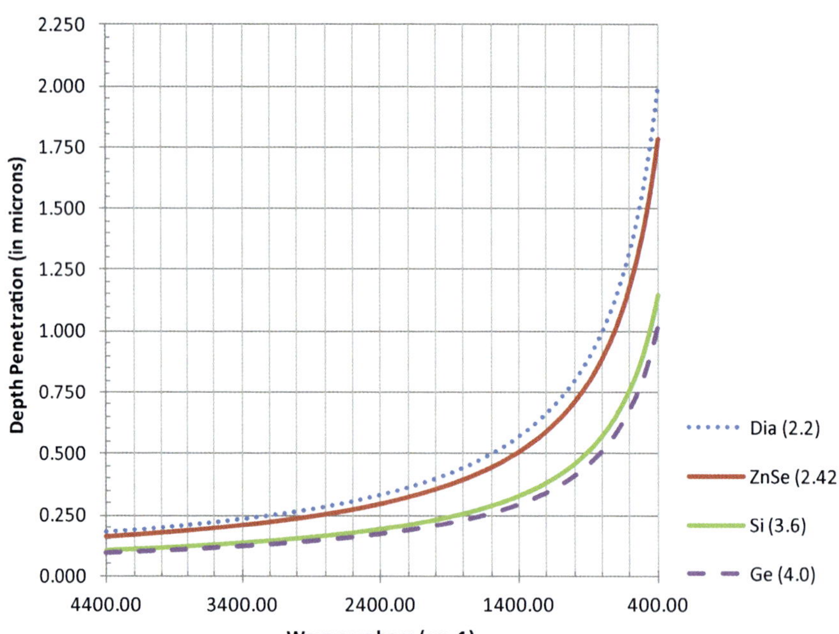

Fig. 2.3 Calculated depth of penetration (in microns) for ATR versus Wavenumber of incident energy (in $cm^{-1)}$ for 45° ATR crystals of different materials.

2.1.1.3 *Grazing Angle Infrared Reflectance*

A measurement configuration where light from the spectrometer strikes the sample at a high angle of incidence for measuring the coating or surface characteristics of a sample. This measurement technique has a sample penetration of typically less than 1 micron. This technique can be used to measure very thin surface characteristics of reflective samples. The technique is qualitative and is not particularly reproducible and thus not useful for quantitative analysis. The grazing angle objective typically is configured for a 65 to 80 degree angle of incidence. There is a five-fold pathlength increase when using grazing angle as compared to standard transmittance measurement geometry.

In closing, there are a number of excellent reference texts describing the early work and understanding of spectroscopic measurements. One such work noted here is reference (2).

References

1. Used with permission from: J Workman and A. Springsteen (Eds.), Applied Spectroscopy: A Compact Reference for Practitioners 1st edn. (Academic Press/Elsevier, 1998).
2. G. R. Harrison, R. C. Lord, and J. R. Loofbourow, *Practical Spectroscopy* (Prentice-Hall, Inc., New York, 1948).

Chapter 3

Infrared Instrumentation

Contents

3.1 Instrumentation for Measuring Infrared Spectra

Mid infrared (MIR) spectrophotometers used within the laboratory function using a variety of optical and mechanical design configurations. Spectrophotometers have a technical definition as a device designed to furnish the ratio, or a function of the ratio, of the radiant power of two beams as a function of spectral position (i.e., wavenumber or frequency). The two beams may be separated in time, space, or both. This definition, paraphrased from the ASTM E131 definition for spectrophotometer reveals enough information for us to determine the basic characteristics of measurements taken using this ratioing method.[1] The measurement for any spectrophotometer then is given by the ratio of the measurement of the sample at a wavenumber position or set of wavenumber positions divided by the measurement of a reference material (or dry air) at the same wavenumber positions. If the measurements are taken with a transmittance geometry, then this ratio is equal to T (i.e., the transmittance) with a value of from 0 to 1.0. If the measurements are taken with a reflectance geometry, then this ratio is equal to R (i.e., the reflectance) with a value of from 0 to 1.0. For transmittance reference measurements, a blank or dry ambient air is generally used. For reflectance reference measurements a highly reflective sample, with a defined R near 1.0 is used. This reflective reference is generally a gold sample, glass, or ceramic having a surface reflectance that has an R of 0.95 to 0.99 for the MIR wavelength region measured.

For quantitative spectrophotometric measurements Beer's law relationship (repeated as core information in different forms and contexts throughout this set of volumes) applies for a change in instrument response due to a change in concentration of analyte material present within a sample specimen. *Beer's law* assumes that the transmittance of a sample within an incident beam is equivalent to 10 exponent the negative product of the absorptivity, times the concentration of a molecule in solution times the pathlength of the sample in solution. However changes in sample and instrument temperature, atmospheric pressure within the instrument, and interferences caused by overlapping absorption bands will create deviations from the ideal case. For most analytical situations, Beer's law relationship holds for transforming infrared spectra useful for qualitative and quantitative measurements.

The Meaning of Beer's law for spectroscopic measurement is straightforward. Fundamental molecular vibrations of specific molecular groups or atoms absorb infrared energy at specific frequencies. The amplitude of the absorption at any particular frequency (or wavenumber) is determined by the absorptivity of the molecule, and the number of molecules encountered within the beam path of the measuring instrument. It is assumed that a change in spectral response is related to the concentration of specific molecule types as described by the Bouguer, Lambert, and Beer relationship most often termed Beer's law. The Beer's law relationship is described as the absorbance (A, Au, or signal strength) of an analyte being measured using a spectrophotometer, and is equivalent to the product of the absorptivity (ε) of a specific type of molecular vibration; the concentration (c) of the molecules in the measurement beam; and the path-length (l) of the sample holder within the measurement beam. This relationship between measured spectral signal and concentration of a molecule is most often expressed as:

$$\mathrm{A} = \varepsilon \mathrm{cl} \tag{3.1}$$

where ε is the molar absorptivity (referred to as molar extinction coefficient by earlier physicists) in units of Liter $\cdot$ Mole^{-1} $\cdot$ cm^{-1}; c is the concentration of molecules in the spectrometer beam in units of Mole $\cdot$ Liter^{-1} (as moles per volume, or weight percent per volume); and pathlength is the thickness in units of cm of the measured sample at a specific concentration. The absorptivity for any specific molecule type is calculated by careful measurements of the absorbance of a compound, generally diluted in a suitable organic solvent, and by applying the relationship:

$$\varepsilon = \frac{A}{cl} \tag{3.2}$$

Note that for transmittance (where T = 0.0 to 1.0) and percent transmittance (where T = 0 to 100.0) spectroscopy, a more complete delineation of the relationships between the various terms is contained in an expression such as:

$$T = \frac{I}{I_0} = 10^{-\varepsilon cl} \Rightarrow Abs. = A = -\log_{10}\left(\frac{I}{I_0}\right) = -\log_{10} T = \varepsilon cl \tag{3.3}$$

Here the symbols I and I_0 represent the attenuated energy detected after sample interaction and the initial energy incident to the sample, respectively. For reflectance (where R = 0.0 to 1.0) and percent reflectance (where R = 0.0 to 100.0) spectroscopy the various relationships are expressed as:

$$R = \frac{I}{I_0} = 10^{-\varepsilon cl} \Rightarrow Abs. = A = -\log_{10}\left(\frac{I}{I_0}\right) = -\log_{10} R = \varepsilon cl \quad (3.4)$$

Noting the relationship exists where the change in intensity (I) of the transmitted or reflected light from a sample is a function of the change in pathlength (l) of the sample as expressed by the absorptivity (ε) of a specific analyte (or molecular substance) and its concentration (c) by:

$$-\frac{\delta \mathrm{I}}{\delta \mathrm{l}} = \varepsilon \mathrm{cI} \cdot (\ln 10) \quad (3.5)$$

Modern spectrophotometers utilize these assumptions for making spectroscopic measurement and generally display spectroscopic data as transmission (T), reflection (R), and absorbance A (y-axis or ordinate axis) versus wavelength (nm, microns) or wavenumber (cm^{-1}) (as x-axis, or abscissa axis).

With a basic knowledge of Beer's law, one is able to calculate the absorptivity and thc other parameters of Absorbance, pathlength, and concentration. The relationship between Absorbance and Transmittance or Reflectance is given in Tables 3.1 and 3.2.

The use of the formula:

$$\mathrm{A} = \varepsilon \mathrm{cl} \quad (3.6)$$

allows the simple calculation of the absorptivity as:

$$\varepsilon = \frac{A}{lc} \text{ in units of L} \cdot \text{mol}^{-1} \cdot \text{cm}^{-1} \quad (3.7)$$

And one can readily calculate the reflectance of transmittance of any measurement given the absorbance of that measurement as follows.

Table 3.1 Showing the relationships between %T and %R, T and R, and A.

%Transmittance or %Reflectance	Transmittance or Reflectance	Absorbance
100.0	1.0	0.0
10.0	0.1	1.0
1.0	0.01	2.0
0.1	0.001	3.0
0.01	0.0001	4.0
0.001	0.00001	5.0
0.0001	0.000001	6.0

Table 3.2 Table showing the relationship of Absorbance unit (Au) values to Reflectance and Transmittance values and to %R and %T.

Absorbance	Reflectance/ Transmittance	% Reflectance/ % Transmittance
3.5	0.00032	0.032
3.0	0.00100	0.100
2.0	0.01000	1.000
1.5	0.03162	3.162
1.0	0.10000	10.00
0.5	0.31623	31.62
0.1	0.79433	79.43
0.01	0.97724	97.72

For conversion of Absorbance to Transmittance we use:

$$A = -\log_{10}\left(\frac{I}{I_0}\right) = \log_{10}\left(\frac{I_0}{I}\right) \tag{3.8}$$

$$-A = \log_{10}\left(\frac{I}{I_0}\right) \tag{3.9}$$

$$\frac{I}{I_0} = 10^{-A} \Rightarrow \frac{I_0}{I} = \frac{1}{10^{-A}} \tag{3.10}$$

$$\frac{I}{I_0} = 10^{-A} = T \tag{3.11}$$

$$\%\mathrm{T} = \mathrm{T} \times 100 \tag{3.12}$$

To convert Absorbance to Reflectance we use:

$$A = -\log_{10}\left(\frac{I}{I_0}\right) = \log_{10}\left(\frac{I_0}{I}\right) \tag{3.13}$$

$$-A = \log_{10}\left(\frac{I}{I_0}\right) \tag{3.14}$$

$$\frac{I}{I_0} = 10^{-A} \Rightarrow \frac{I_0}{I} = \frac{1}{10^{-A}} \tag{3.15}$$

$$\frac{I}{I_0} = 10^{-A} = R \tag{3.16}$$

$$\%\mathrm{R} = \mathrm{R} \times 100 \tag{3.17}$$

The goal in the design of a laboratory spectrophotometer is to maximize the energy (or radiant power) from a light source (usually a Globar emitting infrared energy or the energy of heat for MIR) through the optical components and onto the sample and then the detector (several detectors are useful for the MIR region). The optical throughput for a spectrometer is dependent upon multiple factors, such as: the light source area, the apertures present within the light path, optical element transmittance and mirror reflectance losses, the exit aperture, and the detector efficiency.

Several terms are useful in any discussion of spectrophotometry, these include *selectivity* or the specific sensor response to the *component of interest; sensitivity* or the *quantifiable level of response* from a sensor with respect to the concentration of a specified component of interest; and *detection limit* or the smallest concentration difference that can be detected above the background noise level of an MIR instrument. A quick estimate of the detection limit is three times the background noise as a signal and estimating the concentration for this signal level using a calibration curve.

3.1.1 *Types of Spectrophotometers*

3.1.1.1 *Discrete Photometers*

Discrete infrared filter photometers consist of an infrared irradiance source, discrete interference filters, a sample compartment, and detector; along with appropriate electronics for signal amplification and stabilization of detector signals. The optical configuration for a generic infrared filter photometer is shown in Figure 3.1.

3.1.1.2 *Interferometer Instruments*

Interferometry is often used to measure wavelengths; this type of measurement is conducive to spectroscopic measurements. In spectroscopy, the accuracy of wavelength measurements can be critical from 6 to 10 significant figures. Wavelength measurements from interferometer type devices are much more accurate than those generally made using dispersive type instruments, although the transmission or reflectance data may be less accurate. The most common type of interferometer is the Michelson type shown in Figure 3.2. A parallel beam of light from the spectrometer source (S) is directed into a beam splitter (BS) at an angle of 45 degrees from normal. The beam splitter consists of a 50% reflective surface. The transmitted portion of light from the beam splitter is directed to a highly reflective moveable mirror (MM), whereas the light reflected from the beam splitter strikes a fixed mirror (FM). From the perspective of the detector (D) the following phenomena are "observed". If the pathlengths to FM and MM are the same, an observer at the D position will see a bright fringe. If the pathlengths are not the same, the phase difference will determine whether the fringe observed at the detector is light or dark. The fixed

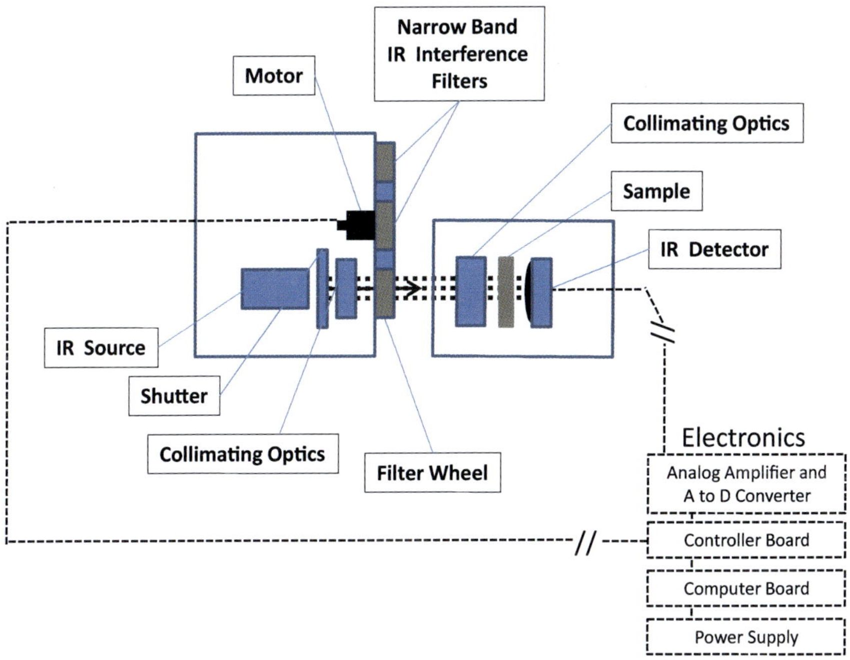

Fig. 3.1 Interference Filter Photometer Optical Configuration.

mirror is only adjusted for tilt to keep it normal to the incident beam. As the MM is moved, the path becomes different and the phase changes. A movement of MM (d) is related to the number of fringes in the interference pattern ($\#f$) observed at D by the relationship:

$$\frac{d}{\#f} = \frac{\lambda}{4} \tag{3.18}$$

when λ = the wavelength of parallel incident light entering the interferometer, and d is expressed in units identical to wavelength units. The optical design configuration for an interferometer based instrument is shown in Figure 3.2.[2]

3.1.1.3 *Open-Path and Emission Spectrometers*

Open path and emission spectrometers are designed to sense emitted or reflected energy from a remote sample. The light source is typically solar radiation, laser emission, or another hot (flame ionization, etc.) source that enters the instrument through collection or focusing optics and is processed much as laboratory data. The calibration of such instruments entails using standard sources and calibrated targets to assess the chemical nature of remotely sensed reflected or emitted energy. Such

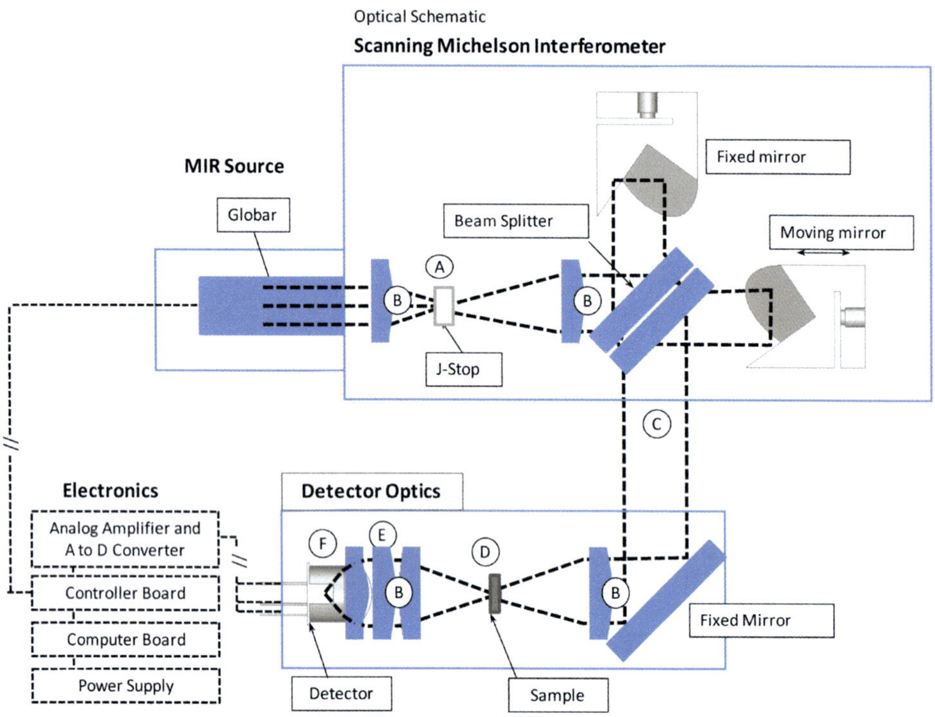

Fig. 3.2 Fourier Transform Spectrophotometer (Interferometer) Optical Configuration. Interferometer based system (Interferometer/FT-type). S = Source optics, A = J-stop (aperture), B = Collimating optics, C = Optical train, D = Sample cell, E = Focusing optics, F = IR Detector.

instruments may be based on Interferometry, or hybrid designs including telescopes for collection optics. Basically large space or land based telescopes are open path spectrometers with very remote energy sources. Figure 3.3 demonstrates an open path spectrometer design using a Michelson interferometer.

3.1.2 *Spectrometer Components*

3.1.2.1 *Light Sources*

All materials emit electromagnetic radiation, the nature of which depends upon the material and its temperature. Gases or vapors of atoms at less than atmospheric pressure emit radiation at discrete wavelengths when an electric current is passed through them. Excited gases containing molecules emit spectra consisting of multiple lines very close together which comprise emission bands. Solid materials emit or radiate continuous spectra across all frequencies. A more comprehensive list of energy sources and detectors encountered with MIR optical spectrophotometers are shown in Tables 3.3 and 3.4, respectively. Figure 3.2 illustrates a typical

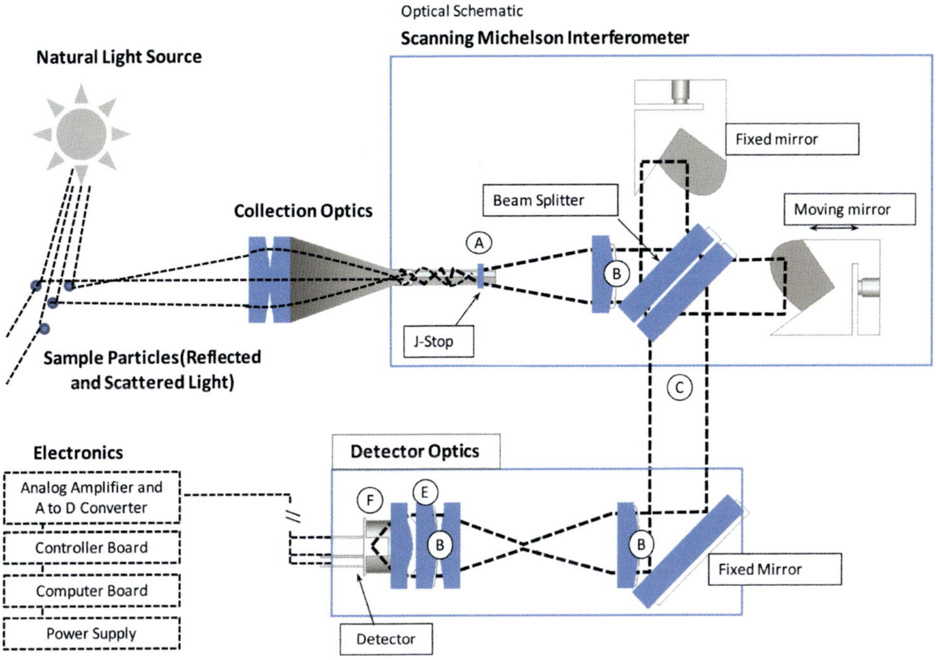

Fig. 3.3 Open Path Fourier Transform Spectrophotometer (Interferometer) Optical Configuration. Interferometer based system (Interferometer/FT-type). A = J-stop (aperture), B = Collimating optics, C = Optical train, E = Focusing optics, F = IR Detector.

Table 3.3 Infrared Energy Radiation Sources (microns).

Source emissivity ranges	Start	End	Difference
Globar	1	100	99
Nernst glower	0.3	35	35
Carbon arc	0.5	100	100
Mercury lamp	0.3	100	100
Near Infrared Region	0.78	2.5	2
Infrared Region	2.5	25	23
Raman Region	2.5	25	23

laboratory FTIR instrument configuration, Figure 3.3 is an open path FTIR system configuration, Figure 3.4 is an FTIR Microscope illustration, and Figure 3.1 shows a basic filter based MIR spectrometer.

The following define the physics of spectrometer energy sources used for laboratory and process instrumentation. Kirchhoff's law states that for a thermal radiator at constant temperature and in thermal equilibrium, the emissivity (e) at any given frequency is equal to the absorptance (a) for radiation from the same direction so

Table 3.4 Infrared Detectors (Most Commonly Used).

Detector sensitivity ranges (microns or μm)	Start (μm)	End (μm)	Difference (μm)
Germanium (Ge:X)	2	40	38.0
Germanium Cadmium (Ge:Cd)	2	24	22.0
Germanium Zinc (Ge:Zn)	2	40	38.0
Deuterated Triglycine Sulfate (DTGS/KBr)	0.83	25	24.2
Indium Antimonide (InSb)	1.8	6.8	5.0
Deuterated Triglycine Sulfate (DTGS/PE)	10	120	110.0
Mercury Cadmium Telluride (MCT)	1	17	16.0
Triglycine Sulfate (TGS)	10	120	110.0
Pyroelectric Lithium Tantalate ($LiTaO_3$) (PLT)	1.5	30	28.5
Near Infrared Range	0.78	2.5	1.7
Infrared Range	2.5	25	22.5

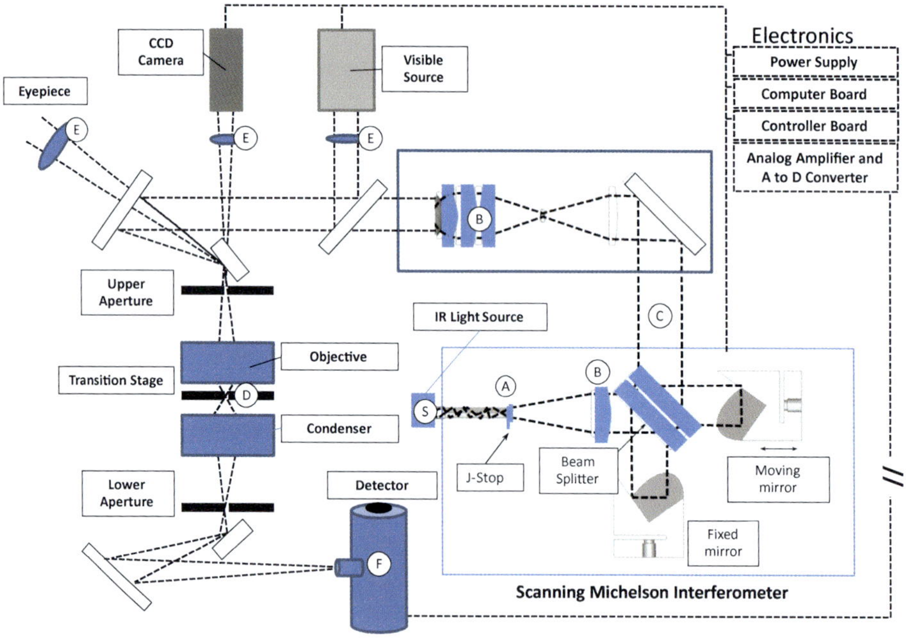

Fig. 3.4 Microscope Interferometer based optical system (Interferometer/FT-type). A = J-stop (aperture), B = Collimating optics, C = Optical train, D = Sample cell, E = Focusing optics, F = Detector or Photomultiplier Tube (PMT).

that $e = a$, where *absorptance* is defined as the ratio of energy absorbed by the surface to that of the energy striking the surface.

An ideal surface absorbing all the energy that strikes its surface could be termed an ideal *black body* (No energy is reflected or emitted, thus it appears perfectly black at all frequencies). It would also stand from Kirchhoff's law that the black body also would be an ideal emitter of radiation. The emissivity (ε) for any black body source is defined as the ratio of the emitted radiance (ρ) by the source of the radiance of a black body (ρ_β) at the same temperature and frequency (or wavelength). The following equation illustrates this principle.

$$\varepsilon = \frac{\rho}{\rho_\beta} \tag{3.19}$$

The spectral or emitted radiance (ρ) is defined as the radiant power (or photon flux density) emitted per unit source area per unit solid angle (typically denoted in W/sr/m^2 (i.e., watts per steradian per square meter).

Max Planck derived the principles of quantum mechanics and expressed this using the concept of light quanta as discrete energy packets that are transferred with an energy (E) proportional to the frequency of the electromagnetic radiation involved. Thus *Planck's law* has become one of the most widely known concepts in the physics of electromagnetic radiation. This expression of Planck's constant is given as follows:

$$\mathrm{E} = h\nu \tag{3.20}$$

where E is the energy content of each quantum of light; and Planck's constant (h) with uncertainty in parenthesis $6.62606957(29) \times 10^{-34}\,\mathrm{J} \cdot \mathrm{sec}$; and frequency (nu $= \nu$) is in sec^{-1}. From Planck's work, the Planck's radiation formula is used to calculate the radiance emitted from a black body (ρ_β) at a given wavelength (λ), emissivity (e), and temperature (T) as follows:

$$\rho_\beta = 3.745 \times 10^4 \times \lambda^{-5} \left(\frac{1.4388 \times 10^4}{e^{\lambda T} - 1} \right) \tag{3.21}$$

where ρ_β is in units of W/sr/m^2 (i.e., watts per steradian per square meter); or W/μm/cm^2 (i.e., watts per micron per square centimeter) denoting spectral radiance.

Further from Planck's radiation formula, other important relationships describing a black body spectrum can be derived. The Stefan Boltzmann law gives the total radiance emitted (ρ_β) from a black body as a well-defined function of the black body temperature. This relationship is summarized as follows:

$$\rho_\beta = S \times T^4 \tag{3.22}$$

where S is the Stefan's constant as 5.67×10^{-8} W/(m^2 °K^4).

The Wien's displacement law demonstrates the shift to shorter wavelengths of the maximum peak position for a black body spectrum as a function of temperature. This means that the higher the temperature of an energy source as a black body radiator the lower is the peak emissive energy. This is given by the following relationship:

$$\lambda_{\max} = \frac{2898\mu m \times {}^{\circ}K}{T} \tag{3.23}$$

where $\lambda_{\max}$ is the peak maximum position from the black body spectrum in micrometers (μm), and T is in °Kelvin. [Note: °K = °C + 273.]

To express Lamp output in terms of Amps, Volts and Watts the formula used is:

$$A \times V = W \tag{3.24}$$

$$\text{Amps} \times \text{Volts} = \text{Watts} \tag{3.25}$$

These are derived from the relationships,

Ohm's law:

$$V = IR \text{ or } W = A\Omega \tag{3.26}$$

Electrical Power Relationship:

$$P = IV \text{ or } W = AV \tag{3.27}$$

Noting that V = Voltage (volts or V); I = Current (amps or amperes or A); R = Resistance (ohm's or Ω); and P = Power (watts or W).

3.1.2.2 *Detectors (Performance and Sensitivity)*

There are two basic types of photon (photo) detectors, these being photoemissive and solid state. The photoemissive type is generally represented by the photomultiplier tube detectors; whereas the solid state type detectors are represented by photodiode detectors, pyroelectric detectors, and some photo detecting infrared detectors (e.g., photoconductive and photovoltaic), see Table 3.4. (See also Chapter 8 for greater details.)

In defining the physics of detector device performance or sensitivity, two basic terms deserve explanation. First the term specific detectivity or D-star (D*) is essential. This D* value is defined as the detectivity of a radiation detector as a function of the square root of the product of the active detector element area (A in cm^2) and the bandwidth (ω) in units of cycles per second (Hertz, Hz); divided by the noise equivalent power (NEP) in Watts of the detector element. This is summarized by the following formula:

$$D^* = \frac{\sqrt{A \cdot \omega}}{NEP} = \text{cm} \cdot \text{Hz}^{\frac{1}{2}}\text{W}^{-1} \tag{3.28}$$

The D* is reported in units of $\text{cm} \cdot \text{Hz}^{\frac{1}{2}}\text{W}^{-1}$ units and is used to compare photodetector responsivity as a function of wavelength. These are referred to as *Jones* units, for the namesake of R. Clark Jones.

For photo-detectors, detector responsivity (R) or input-output gain, is also used to express detector photo sensitivity (or efficiency) in units of Amperes per Watt (A/W) as a function of the frequency or wavelength of light incident to the detector. This is a measure of the electrical output (or detector efficiency or sensitivity) as a function of optical energy input. This relationship is expressed as follows:

$$R = Q\left(\frac{\lambda(\mu m) \cdot A}{1.23985(\mu m \cdot W)}\right) = \text{Amps/Watt} \tag{3.29}$$

This computational form is derived from the responsivity expression, equivalent to the quantum efficiency times the electron charge divided by Planck's constant times the frequency of the optical signal; the responsivity changes as a function of the wavelength of the incident energy.

Note that responsivity (R) in Amps/Watt may be converted to Quantum Efficiency (Q) in a dimensionless scale of 0 to 1. Note: The R term is expressed as (Amps $\times$ Watt^{-1}). This conversion of R to Q may be expressed as:

$$Q = \frac{1.23985R \cdot W}{A} \tag{3.30}$$

3.1.2.3 *Theoretical Limit of D* for a Detector System*

Noise from a detector often results from multiple factors, such as associated electronic and amplifier noise. However, there is a theoretical noise associated with the background fluctuation attributable only to the detector, irrespective of whether it is a photovoltaic or photoconductive detector. The limits of D* are somewhat different for photovoltaic or photoconductive detectors and may be expressed in terms of the wavelength of light (λ); Quantum efficiency (Q); and the flux of background radiation (F). The constants for Planck and the speed of light are also included in the expressions.

For the photoconductive detector:

$$D^* = \frac{\lambda\sqrt{Q}}{hc\sqrt{2F}} \text{ in units of cm} \cdot \text{ Hz}^{1/2}\text{ W}^{-1} \tag{3.31}$$

For the photovoltaic detector:

$$D^* = \frac{\lambda\sqrt{Q}}{2hc\sqrt{F}} \text{ in units of cm} \cdot \text{Hz}^{1/2}\text{W}^{-1} \tag{3.32}$$

This D* value is defined as the detectivity of a radiation detector as a function of the wavelength; the square root of the quantum efficiency (Q); and the background radiation flux (F). The background radiation (F) is computed in Watts per area-squared based on the detector temperature, typically in degrees Kelvin. In computing the signal to noise for a detector system, a higher shunt impedance indicates a lower overall dark noise for the detector such that a detector with the same peak response, and responsivity profile, with higher shunt resistance will yield a higher signal to noise for measurement.

MIR energy reflected or transmitted through a sample and reference material is captured by using a Mercury Cadmium Telluride (MCT), Deuterated Triglycine Sulfate (DTGS/KBr), or Indium Antimonide (InSb) detector (Table 3.4). The voltage signals generated by these detectors are amplified, converted from analog to digital signals, and used to quantify the amount of energy emerging from a sample. A more complete list of common detectors, common abbreviations, and their useful working wavelength ranges are included in Table 3.4. The reader is referred to the index for more details on the physics of infrared detectors.

3.1.3 *Michelson Interferometer Characteristics*

3.1.3.1 *History*

Albert Abraham Michelson first introduced the interferometric device in 1881,[2] and performed the famous Michelson–Morley experiments in collaboration with Edward Morley in 1887.[3] It is an oversimplification to state that the experiments measured the speed of light since the experiments were extremely important to determine the medium in which light traveled, referred to in that period as aether. The initial experiment was conducted to determine how this "aether", which supports the movement of light waves across the universe, actually behaves relative to earth. It makes for incredible scientific drama to study this subject, and the reader is encouraged to delve into this era of physics and learn about the fascinating history of light path analysis, its consequences, and the relationship to special relativity and the history of modern physics.[4] Michelson built his interferometric device to measure the speed of light in different directions from earth with extreme precision and accuracy. This was required to determine if the speed of light had directional dependence from a moving object, such as earth.

It was soon realized that the Fourier transform (FT) of the interferogram produced by the Michelson device is the spectrum of intensity as a function of frequency or wavelength.[5] This discovery combined with Fellgett's and Jacquinot's realization of the multiplex[6] and throughput advantages, respectively, led to the current use of Michelson and other types of interferometers for infrared spectrophotometers. For a review of this history and technical details see reference 7. The concept of FFT which formed the basis of FTIR spectrometers was introduced by Cooley and Tukey.[8] The basic Michelson design is used with FFT software in interferometric or FT-based spectrophotometers made in modern times. Various mechanical and optical design enhancements have been made to improve the robustness, speed, resolution, and overall performance of today's commercial FT spectrophotometer systems.

As briefly mentioned in the preceding paragraph, FT instruments have several theoretical or perceived advantages over other instrument designs. These include the Fellgett or Multiplex advantage. This principle enables all wavelengths of light to pass through the optical system simultaneously rather than one wavelength band at a time. In theory, more replicate measurements may be averaged in a shorter time. A second advantage is the Jacquinot or Throughput advantage. Theoretically, the interferometer provides more energy through the optics of an interferometer at equivalent resolution than for other design instruments; typically this gives high signal to noise at lower collection or scan times. Thus, for typical designs more energy from the source travels through the spectrometer to reach the detector. There is very small stray light in an FT instrument giving high linearity and dynamic range, particularly for highly absorbing samples. The Connes advantage provides accurate wavelength axis registration due to the use of a laser (HeNe) acting as an internal reference for the wavelength scale. Resolution for FT instruments is determined by the moving mirror velocity and the aperture size or J-stop (Jacquinot stop). The resolution is constant in wavenumber space, but varies in wavelength space. Conversely, dispersive instruments with fixed slit widths have a relatively constant resolution in wavelength space but different resolution in wavenumber space. Interferometric-based spectrophotometers have essentially no stray light so the requirement of using stray light filters does not exist. Thus, FT instruments have no discontinuities such as detector or filter cross-over points. Empirical performance measurements made using identical samples will demonstrate the effectiveness of different spectrophotometer designs features. This text reference set describes the methodology for testing instrument performance characteristics, such as: wavelength accuracy and precision; photometric accuracy and precision; wavelength and photometric linearity, noise characteristics, lineshape, and resolution. The reader is referred to the index for additional information.

3.1.3.2 *FTIR Spectrometer Basics*

Multiple interferometer types exist in modern spectrometers; the reader is referred to references 7 and 9 for additional details. The classical interferometer design is represented by the Michelson interferometer as shown previously in Figures 3.2 through 3.4, and 3.8. A moving mirror is displaced linearly at minute distances to yield an ever changing interference pattern (or interferogram) as a series of sine waves as the interference pattern is observed from a specific point of view. (Figure 3.5).

The interferogram is plotted as the intensity of light (y axis) versus the mirror position (x axis); thus, the signal from the interferometer is a function of time as the mirror is in motion at a constant velocity (Figure 3.6). The raw interferogram, as it is sometimes referred to, is converted to a spectrum using the Fourier transform

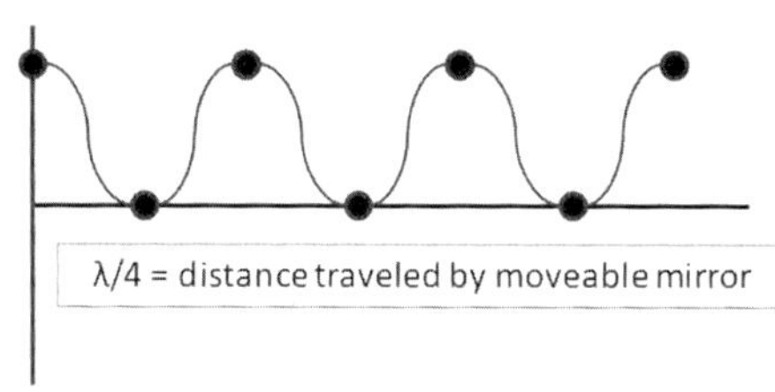

Fig. 3.5 Output of interferometer as a sine wave function denoting distance traversed by mirror as signal minima and maxima. A pattern is shown for a single wavelength.

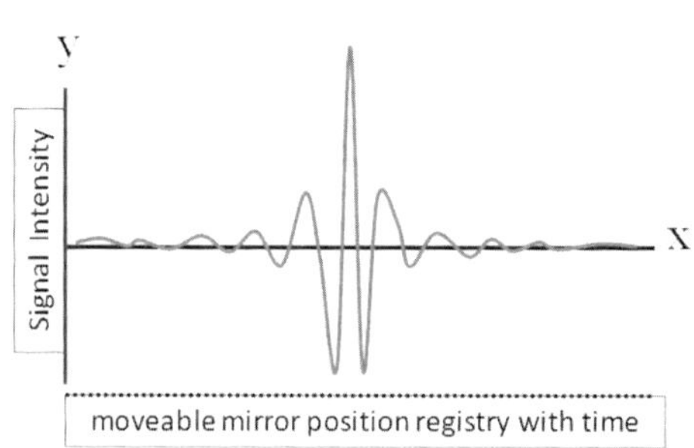

Fig. 3.6 An interferogram ZPD (zero path difference) center section is shown where light intensity amplitude is the ordinate (y-axis) and the moving mirror position is the abscissa (x-axis). The signal is a function of the time due to the fact that the mirror is moved at a constant rate. This raw interferogram is subjected to the multi-step Fourier transform resulting in a spectrum.

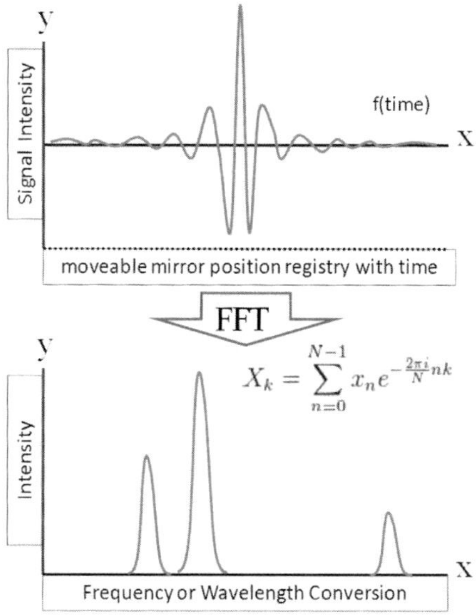

Fig. 3.7 The raw interferogram measured using the interferometer is subjected to the multi-step Fourier transform resulting in a spectrum. The sample spectrum minus the dark spectrum is used.

(Figure 3.7), and a spectrum is determined by ratioing a signal determined with a sample in the beam minus the signal from the dark current (or no light reaching the detector) to a signal determined with no sample in the beam minus the dark current (as the background spectrum). This relationship is defined as:

$$\text{Reflectance or Transmittance} = \left(\frac{I}{I_0}\right) = \left(\frac{X_s - X_D}{X_R - X_D}\right) \tag{3.33}$$

where X_S = Spectrum of inteferogram of Sample; X_D = Spectrum of interferogram of the Dark Signal; X_R = Spectrum of interferogram of the reference measurement; and X_D = Spectrum of the interferogram of the Dark Signal.

A laser beam of known frequency is used to signal a separate sensor to provide near perfect sampling of the mirror position. The resolution of the interferometer depends on the distance of motion per unit time (i.e., velocity) for the moveable mirror. The precise distance traversed by the moveable mirror can be determined using the number of interference fringes observed to pass a given field of view in time (t), given as the wavelength of light from the laser. The distance traversed by the mirror is given as:

$$d = \frac{f\lambda}{2} \tag{3.34}$$

where d is the distance traversed; f is the number of fringes in the interference pattern passing a specified field of view; and λ is the wavelength of the laser light used. The laser (usually a Helium:neon (HeNe) laser at ~633 nm) provides the precise positioning of the moveable mirror in the interferometer. The spectrum is generated from a sample measurement amplitude (Y-axis) plotted against the mirror position (X-axis). The sample (I) and empty instrument (Io) are measured and the interferograms are transformed and then ratioed into a spectrum (dark current is also removed from each). The available resolution is determined by the distance the mirror can travel. The wavelength accuracy is a function of the precision of the determination of the mirror position, which is usually quite excellent within most commercial FT instruments. The wavenumber precision is also usually excellent in an FTIR since the moveable mirror position is precisely known. High resolution involves moving the mirror at integral cycles over greater distances per unit time as compared to low-resolution measurements. A complete explanation of the physics of interferometry is available in references 7 and 9.

Resolution in an interferometer is mainly determined by the difference in optical distance that the two light beams travel in an interferometer (known as the optical path difference or OPD). The Jacquinot or J-Stop is an aperture placed in the beam to limit the divergence to be compatible with the selected resolution. Note that the Zero Path Difference, or Zero Optical Path Difference (ZPD, ZOPD) is the position of the moving mirror in an interferometer such that the optical path difference (OPD) for the two beams is zero; this produces the interferogram centerburst.

In FT spectrometers, as the resolution ($\Delta\nu$) increases (i.e., as the resolution goes to a smaller wavenumber value), the OPD increases, and the J-stop aperture (JS) decreases.

So the basic relationships are expressed as:

$$\Delta\nu \propto OPD, \quad \text{and} \tag{3.35}$$

$$\Delta\nu \propto \frac{1}{JS} \tag{3.36}$$

The J-stop aperture (JS) is related to focal distance (F); spectral resolution ($\Delta\nu$); and the maximum wavenumber measured (ν_{Max}):

$$JS \leq 2 \cdot F \left(\frac{\Delta\nu}{\nu_{\mathrm{Max}}}\right)^{\frac{1}{2}} \tag{3.37}$$

where F is the focal distance from source; $\Delta\nu$ is the spectral resolution; and ν_{Max} is the maximum wavenumber measured. The position of the J-Stop in a Michelson interferometer is shown in Figure 3.8.

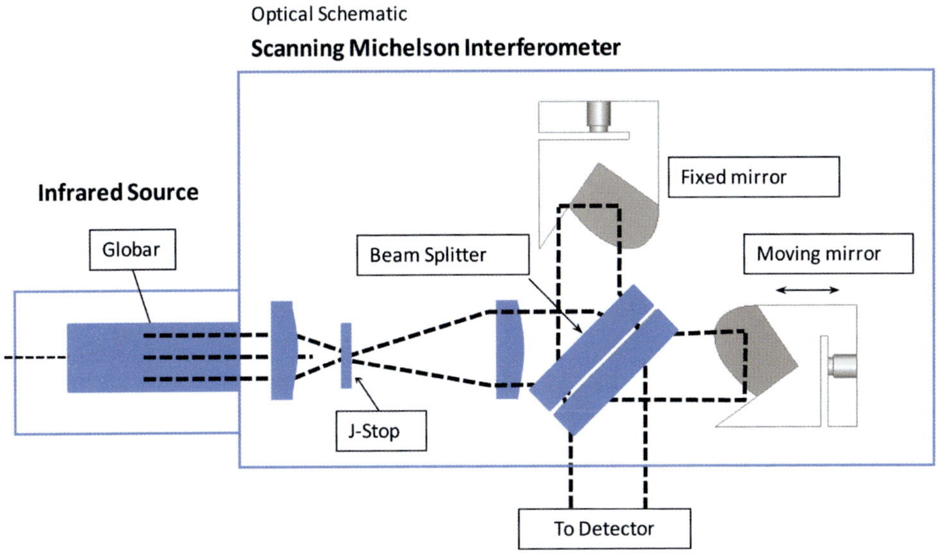

Fig. 3.8 Michelson Interferometer optics showing location of the Jacquinot or J-Stop and basic interferometer components.

3.1.4 *The Michelson Interferometer in Detail*

The Michelson interferometer, as commonly used for infrared measurements, includes: a source (Globar or hot infrared source), detector, beam splitter, and two mirrors, which form the two arms of the interferometer. This interferometer design has a maximum theoretical efficiency near 50%, as approximately half of the light emerging from the interferometer is actually directed back toward the source. The fixed mirror is a stationary mirror in one of the interferometer arms. Its purpose is to reflect the light back to the beam splitter along a fixed and constant path. The moving mirror, on the other hand, is designed to create a pathlength different from the fixed path arm of the interferometer.

The interferogram results because all the incident wavelengths are modulated differently due to velocity and mirror position. To avoid distortion in the interferogram during movement, a number of physical-mechanical ramifications must be addressed. These include: (1) The moving mirror must be maintained plane and parallel to the fixed mirror. This requirement represents a stringent limitation for the mechanisms involved with the motion. (2) Alignment errors will affect the intensity at particular points of the interferogram. (3) Constant velocity must also be maintained since the detector is sensitive to the modulation frequency and hence intensity can again be affected. (4) A constant sampling rate must be maintained since the intensity at particular points in the interferogram must be known. Errors in any of these will affect the intensity, shape and distribution of the interferogram and the transformed interferogram, (i.e. a single-beam spectrum). The light going to the beam splitter is usually collimated (i.e. the light is changed from divergent to more parallel rays with a collimating mirror, lenses or other optical components). Uncollimated light gives rise to poor resolution because it is out of focus. Currently several designs in Michelson interferometers are commercially available. These all share the same design and implementation goals for smooth, constant, vibration-free motion of the moving mirror.

3.1.4.1 *The Interferogram*

In operation, the moving mirror is driven at a constant velocity. If the signal at the detector is plotted, an output of intensity versus mirror displacement is obtained. This is commonly referred to as an interferogram. A typical interferogram from a continuous infrared source is shown in Figure 3.9. An interferogram from a single wavelength source, such as the calibration HeNe laser, is shown in Figure 3.10.

The interferogram contains all of the information found in the spectrum; this is the original data. The interferogram is the intensity in units of voltage as measured by the detector versus the position of the moving mirror, moving mirror wedge, or mirror disk (depending upon the specific interferometer design specifics). The most obvious feature of an interferogram for a broad or continuous source is the centerburst or zero path distance (ZPD) as shown in Figure 3.9.

The centerburst is the position where all frequencies are in-phase or nearly so, and corresponds to the positions where the two mirrors of the interferometer are

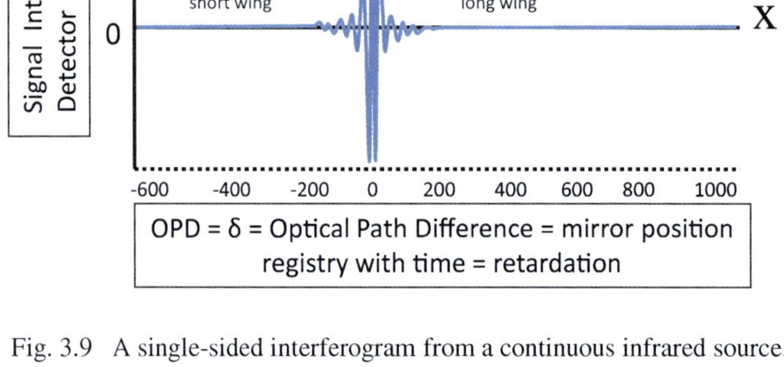

Fig. 3.9 A single-sided interferogram from a continuous infrared source.

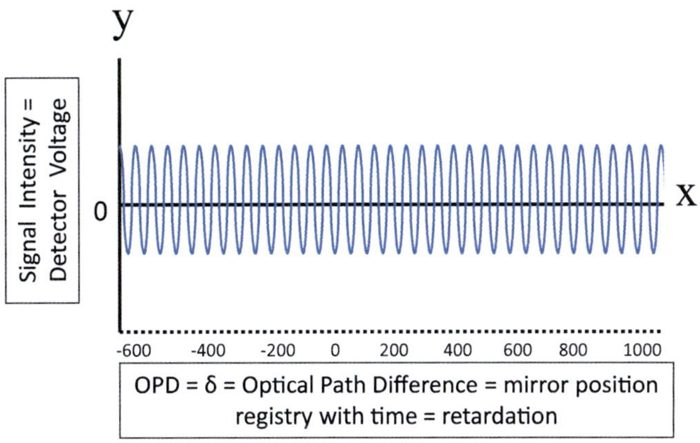

Fig. 3.10 A single-sided interferogram from a single wavelength HeNe laser source. The zero cross-over points represent the recorded data points.

equidistant from the beamsplitter (i.e., the retardation or difference in mirror distance from the beam splitter is zero). As noted in the Figure 3.9, the above interferogram is single-sided. The portion to the left is the short side that is needed for a part of the process known as phasing and may contain from a few hundred to over a thousand data points. The right side of the interferogram is the longer wing and can contain less than a thousand points for low resolution or many thousands of data points for higher resolution spectra.

At this point, let us consider how an interferogram is produced. If a line source, much as the one obtained from a HeNe laser is passed through an interferometer, the monochromatic light would generate a cosine wave of constant frequency and intensity for the distance the mirror is moved. The result is the simplest interferogram. This interferogram is generated by constructive and destructive interference of the light as the distance between the two arms of the interferometer is varied. This is depicted as wave λ_1 in Figure 3.11.

When the distances in the two arms are equal, the two optical paths followed by the beams are equal and reach the detector at the same time. At this point (ZPD or zero path difference), both beams combine constructively and are summed at the detector. When the moving mirror in the interferometer is moved $\frac{1}{4}$ λ distance of the source wavelength, the signal at the detector will be zero since the beam from one of the arms will be exactly $\frac{1}{2}$ λ (i.e., 2 times $\frac{1}{4}$) out of phase with that from the other arm. The beams here combine destructively. Moving the mirror another $\frac{1}{4}$ λ completes the cycle bringing the signal back to constructive addition at 2 times $\frac{1}{2}$ $\lambda = 1\lambda$ out of phase. Further movement away from the ZPD begins this cycle of

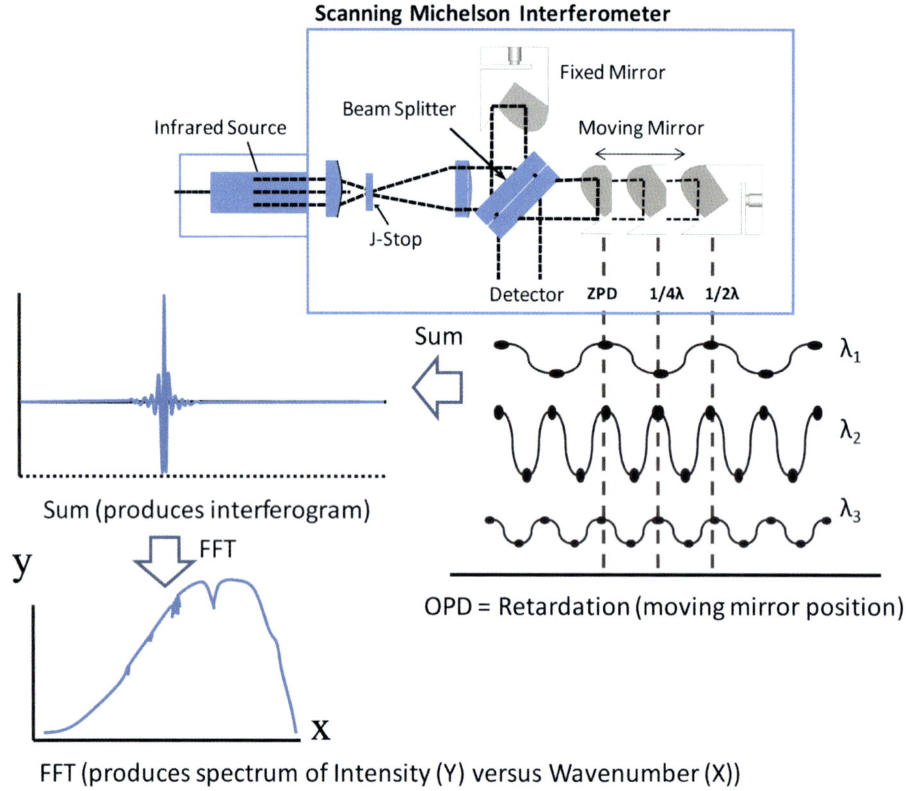

Fig. 3.11 A depiction of how an interferometer would process monochromatic light at different frequencies. The mirror position data is converted to an interferogram by summation of all frequency measurements, and the resulting inteferogram is further converted into a single beam spectrum using the FFT.

constructive and destructive summations again. The summation of λ_1, λ_2, λ_3 are shown in the representation of $\lambda_1 + \lambda_2 + \lambda_3$.

The line spectra on the right of Figure 3.11 vary both in wavelength and intensity. λ_2 is twice as intense and is positioned at one-half of the wavelength of λ_1. The result is that the intensity of the wave pattern and the frequency of cycles on the right is twice that of λ_1. Let us now consider another line source at twice the frequency or half the wavelength of the original line source. This wave is noted as λ_2 in Figure 3.11. When the intensity of the line source is the same as that shown in Figure 3.11, the ensuing result is that the wavelength of this line source is half that of λ_1, and the frequency is double that of λ_1. Halving both the wavelength and intensity and repeating the process in the above section gives the wave marked as λ_3. The wave pattern on the right shows one-half the intensity and twice the number of cycles as compared to λ_1. Halving both the wavelength and intensity and repeating the process in the above section gives the wave marked as λ_3. This line source demonstrates that the amplitude of the cosine wave on the right is equivalent to the intensity of the parent on the left, and the frequency of the cosine wave relates to the frequency or wavelength of the line source.

The process of passing multiple line sources through the Michelson interferometer to generate the interferogram is in essence equivalent to the mathematical function called the Fourier transform. The difference is that mathematically the Fourier transform represents an infinitely large number of data points while the interferometer limits the interferogram by the distance of the mirror travel. Thus, mathematically, a Fourier transform of a line source will generate a cosine wave. Conversely, performing a Fourier transform on a cosine wave will generate a line source.

To summarize the issues associated with the interferogram. The ZPD or zero path difference is the positioning of the two mirrors of the interferometer such that the distance (d) from the beam splitter to each of the mirrors is identical and thus the light path is identical for the two beams of light separated by the beam splitter. The term mirror displacement (MD) refers to the distance (d) of the moveable mirror from the ZPD position. Note that the extra distance traveled by the light beam is 2 MD (forward and backward path). This extra distance traveled is termed the OPD or optical path difference and is often designated as δ. So the relationship for OPD is as follows:

$$\delta = 2\text{MD} \tag{3.38}$$

when δ or the OPD is at integer multiples (n) of the wavelength passing through the interferometer there is both constructive and destructive interference of the light waves as the wave peaks (crests) and troughs add and cancel. When this occurs, maximum intensity is observed and so when n is an integer multiple of the wavelength constructive interference occurs as crests and troughs align and add as shown in Figure 3.12.

$$\delta = n\lambda \tag{3.39}$$

Where n = integer numbers as 0, 1, 2, 3, 4, etc.

When the OPD (or δ) is at half wavelength multiples of the wavelength, destructive interference occurs when crest and troughs are out of phase and cancel one another when added as in Figure 3.13.

$$\delta = \left(n + \frac{1}{2}\right)\lambda \tag{3.40}$$

Where n = integer numbers as 0, 1, 2, 3, 4, etc.

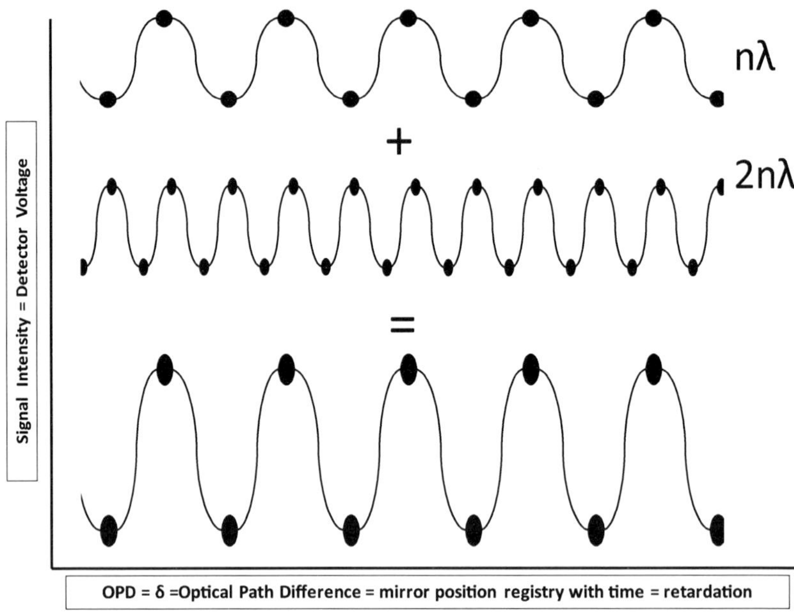

Fig. 3.12 Showing constructive interference as additive integer waves.

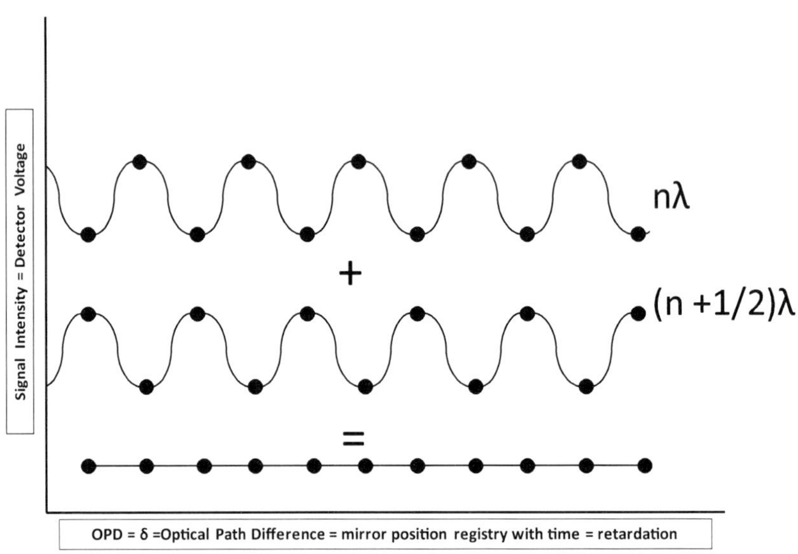

Fig. 3.13 Showing destructive interference as additive half wavelength multiples of waves.

The examples shown in the preceding two equations are the extremes of in phase and out of phase conditions. Most data points measured through the interferometer are in an intermediate phase of constructive and destructive interference and result (as a summation) in the more typical interference pattern shown in Figure 3.9.

With most OPD (δ) mirror positions being intermediate to either nλ or (n + 1/2)λ the interference pattern is mixed between constructive and destructive. The intensity in this normal scan sequence creates a cosine wave. The plot of this cosine wave is defined as the interferogram. A broad band source containing multiple or continuous wavelengths yields the typical interferogram with the maximum amplitude pulse at the ZPD. At other mirror distances the interferogram exhibits much diminished amplitude. With a single wavelength source the interferogram is a uniform cosine wave as illustrated in Figure 3.10.

3.1.5 *The Fourier Transform*

Performing a Fourier transform (FT) is the central process in changing an interferogram to a spectrum. However, to achieve a usable spectrum, a process called apodization must be done prior to the FT; and another operation called phase correction must be done after the FT. (This three-step process is sometimes called a complex Fourier transform.) The result is usually called a single beam spectrum. A second single beam spectrum with the sample in the beam is also needed. Dividing the individual data points of the second single beam spectrum (sample) by the corresponding data points of the first single beam spectrum (generally called a background spectrum) give the transmittance spectrum. The results are viewed as the usual plot of transmittance versus wavenumber. A computer system with the appropriate software packages is used to apply all the necessary mathematical functions to the data. The execution of these calculations by a computer is virtually invisible to the operator. However, to get the best data, a basic understanding of how the instrument works is helpful in collecting as well as processing the FTIR data.

A detailed explanation of this Fourier transform and its calculation is beyond the scope of this chapter. See reference 7 for more information. A Fourier series describes a periodic function as an infinite sum of sine and cosine terms whose frequencies are multiples of some fundamental vibration. The transform allows the analysis of non-periodic functions as an integral (also a summation) over a continuous range of frequencies as shown in Equation (3.41).

$$F(\omega) = \frac{1}{\sqrt{2\pi}} \int_{-\infty}^{+\infty} G(t)e^{-i\omega t}\,dt \tag{3.41}$$

In spectroscopy, the function $G(t)$ *is* the interferogram produced by the system, and $F(w)$ is the single-beam spectrum. The t in the term $G(t)$ relates time and frequency. The Fourier transform $F(\omega)$, a function of frequency, is related to $G(t)$, a function of time, as shown in Equation (3.41). In FTIR, t is really the position of the moving mirror from the ZPD (centerburst) position.[10]

An interferogram is Fourier transformed numerically by a computer using the Cooley-Tukey algorithm.[8] There are many programs in a variety of computer languages to accomplish this. To create an infrared spectrum, the interferogram measurements with no sample present (background), and when sample is present (sample) are measured. The interferograms are converted to single beam spectra using the FFT. The single beam spectra are ratioed as sample divided by background to construct the transmittance spectrum. This may be expressed as percent transmittance (from 0 to 100%) or Transmittance as (0.0 to 1.0). The Transmittance spectrum is converted to absorbance using $A = -\log_{10}T$ or $A = \log_{10}(1/T)$ as the conversion from T to A. The absorbance spectrum is linearized with respect to concentration. This process is illustrated in Figures 3.14 and 3.15.

3.1.5.1 *Selectable Resolution using FT Instruments*

Commercially available interferometer instruments have multiple resolution settings, generally as 0.05, 0.1, 0.25, 0.5, 1, 2, 4, 8, 16, 32, and 64 cm^{-1}. As the resolution setting is decreased from one setting to the next, the number of data points recorded doubles. For example, a 4 cm^{-1} setting may require 8192 points, and a 2 cm^{-1} setting 16384 points.

In terms of the measured spectrum, resolution can be defined as the minimum separation that two spectral features can have and still be distinguished from one another, as in Figure 3.16.

The Rayleigh criterion is used for this purpose (reference 11). This criterion describes limits of angular resolution based upon aperture, wavelength and the first-order Bessel function. So the basic equation for resolution is as:

$$\theta_A = \frac{1.22\lambda}{D} \tag{3.42}$$

where θ_A is the angular resolution (in radians); λ is the wavelength of light; D is the diameter of the aperture (J-stop in an interferometer or slit in diffraction grating system); 1.22 is derived from the first-order Bessel function.

For microscopes, spatial resolution θ_S is as:

$$\theta_S = \frac{1.22\lambda f}{D} \tag{3.43}$$

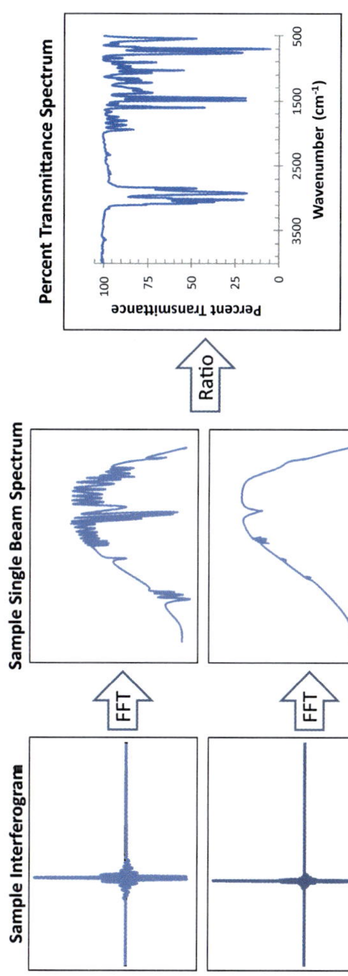

Fig. 3.14 Illustration of the collection of interferograms comprising an infrared spectrum.

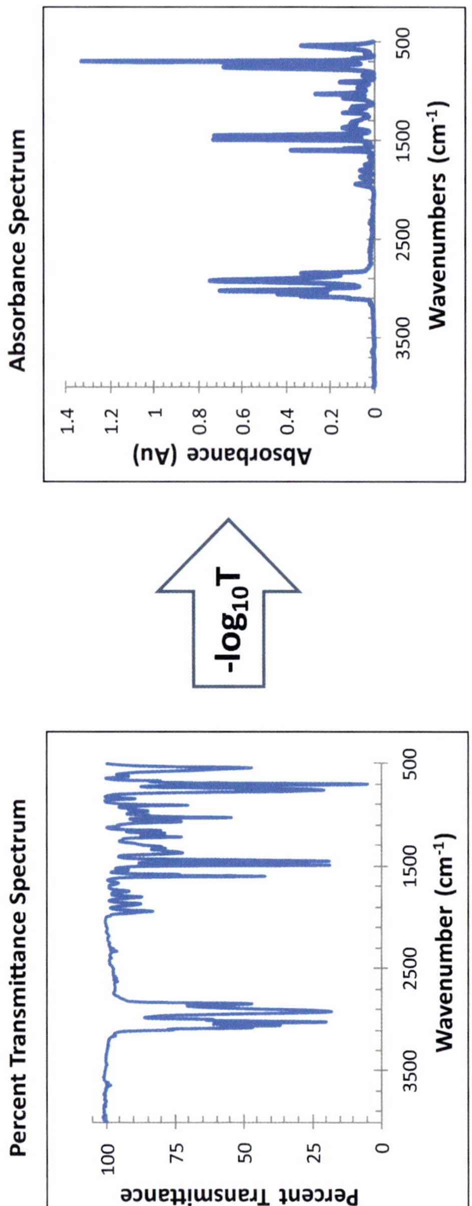

Fig. 3.15 Illustration of transmittance to absorbance spectrum transformation.

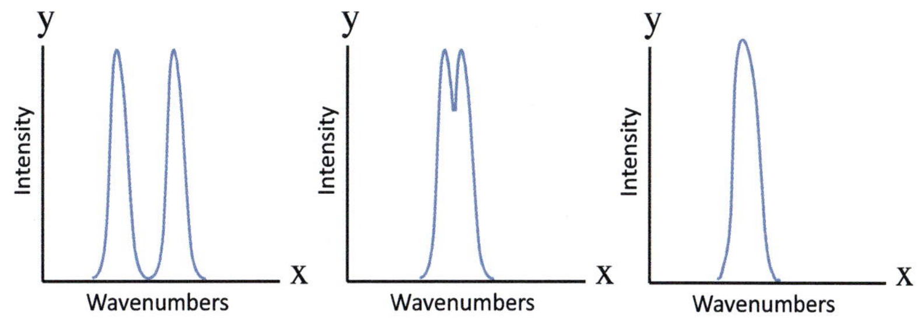

Fig. 3.16 Two bands fully resolved (left), partially resolved (middle), and unresolved (right). The Rayleigh criterion is used for computing the theoretical limits for resolution.

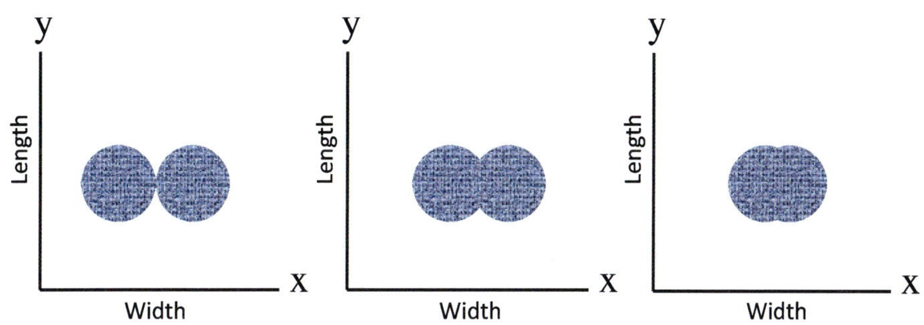

Fig. 3.17 Two objects fully resolved (left), partially resolved (middle), and unresolved (right). The Rayleigh criterion is also used for computing the theoretical limits for spatial resolution.

where f is the focal length of the microscope objective lens. The spatial resolution represents the diameter of the smallest object that is resolved (or distinguishable) on the image surface. Figure 3.17 illustrates the concept of spatial resolution using microscopic techniques with spectroscopy.

3.1.5.2 *Signal-to-noise ratio (SNR) in FTIR Spectrometers*[10]

In the following discussion, the SNR is defined as 100/N where N is the RMS noise. RMS noise[12] is usually measured between 2200 and 2100 cm^{-1}. The reason this region is used is that it is nearly free of interference from CO_2 and water vapor, and is close to the maximum energy of the spectrum as seen in a single beam spectrum. At the extremes of the spectral range, SNR is considerably higher.

The quantitative relationships between SNR, resolution, and measurement time in FTIR spectrometry are referred to as "trading rules". The factors that affect the SNR and dictate the trading rules are expressed in Equation (3.44), which gives the theoretical SNR of a spectrum measured with a rapid-scanning Michelson interferometer. The derivation of Equation (3.44) is given by

Griffiths and de Haseth.[7]

$$SNR = \frac{U_V(T)\Theta(\Delta\nu)t^{1/2}\zeta}{NEP} \tag{3.44}$$

Where:

$U\nu(\mathrm{T})$ = the spectral energy density at wavenumber ν from a blackbody source at a temperature T

θ = the optical throughput of the spectrometric system

$\Delta\nu$ = is the resolution of the interferometer

t = is the measurement time in seconds

ξ = the efficiency of the interferometer

NEP = Noise equivalent power

Note NEP represents the quantity of light incident onto the detector that is equal to the noise level of the detector; therefore the NEP represents the light incident when the signal-to-noise (S/N) is equal to 1.0. The equation to determine NEP for an MIR detector is as:

$$NEP = \frac{ID}{S/N \cdot \sqrt{nb}} \tag{3.45}$$

where:

I is the incident energy to the detector in Watts/cm2

D is the detector active area in cm^2

S is the signal in volts (V)

N is the RMS noise in volts (V)

Note that S/N is set to 1.0 for this calculation, nb is the noise bandwidth in Hz.

Although cumbersome, Equation (3.44) can and is used to evaluate SNR in the real world, for example see (reference 12). For this chapter, however, generalization will be inferred from this equation.

As shown in Equation (3.44), the SNR of a spectrum is proportional to the square root of the measurement time ($t^{\frac{1}{2}}$). For measurements made with a rapid scanning interferometer operating at a constant mirror velocity at a given resolution, the SNR increases with the square root of the number of scans being averaged.

Instrument software presents a range of resolution options. However, several factors must be considered when determining the optimum resolution for measuring an IR spectrum. These factors include: (1) the ability to distinguish between the spectral features of analytes and those of ambient interfering species in the atmosphere, such as water vapor and CO_2; (2) the trade-offs between resolution, IR peak absorbance, and SNR; and (3) practical considerations, such as measurement time, computational time to process the interferogram, and the size of the interferogram file for data storage. The proper selection of instrument resolution minimizes spectral overlap while maintaining a balance between the SNR, analysis time, and data storage requirements.

Water vapor and carbon dioxide do cause problems in many measurements as inferred in Item 1 in the preceding paragraph. A good instrument purge can make the difference between getting a good result or not. Measurements made in instruments without at least a sample compartment purge often must be repeated because the background or sample interferograms were not sufficiently similar. This is particularly true when measuring samples with low intensity features in the regions where water vapor and CO_2 bands are present. Again referring to the above paragraph, for normal laboratory use, data storage is not a problem; on the other hand for environmental work, such as open-path monitoring, special consideration for processing and saving many large files deserves attention.

The following table is a useful guide to noise-reduction versus number of scans.

Scans	S/N Ratio
1	1
4	2
16	4
100	10
400	20
1600	40
10000	100

Operators should be aware that every instrument has its upper-limit on the reduction of noise as function of the number of scans. This limit can be determined experimentally by following the above schedule as suggested in ASTM E 1421.[12]

3.1.5.3 *The Influence of Resolution on SNR*

For the following comments, optical throughput and the interferometer efficiency of the spectrometer are assumed to be constant for measurements made at both high and low resolution. Under these conditions, changing the nominal resolution from 2 to 4 cm^{-1} will improve the SNR and decrease the number of data points, and scan time by one-half. The SNR is proportional to the square root of the measurement time and most of the time proportional to the number of scans. As a consequence, the measurement time and number of scans required to maintain the original baseline noise must be increased by a factor of 4 each time the increment between data points is doubled, i.e., decreasing the resolution by one-half.

Resolution may also affect the peak absorbance of the bands being measured. For example, the special case, where a weak spectral feature has a 1/2-width (FWHH) much less than the instrumental resolution (that is, a narrow band), the peak absorbance will approximately double on increasing the resolution by a factor of 2. Therefore, the SNR ratio would appear to be the same for measurements taken at the higher and lower resolution settings, provided the measurement times are equal. For broad bands, the peak absorbance will not be affected very much by changes in resolution, and the lower resolution measurement would be preferred because SNR improves.

Instruments capable of medium and high-resolution measurements are equipped with adjustable or interchangeable apertures (in some software these are referred to as Jacquinot stops) in the source optics that reduce the solid angle of the beam passing through the interferometer. Spectra collected at settings less than 0.5 cm^{-1} resolution are generally measured with a variable throughput via apertures. In these cases, throughput is halved going from a setting of 0.5 to 0.25 cm^{-1} on most instruments. This results in an additional degradation in the SNR by one-half, which requires increasing the number of co-averaged scans by another factor of 4 to obtain the original SNR ratio. Thus, to maintain the original SNR, the total measurement time increases by a factor of 16.

For strongly absorbing bands, whether they are broad or narrow, calibration curves of absorbance values measured at different resolutions and plotted versus concentration must be developed to ascertain the optimum resolution needed.

Many chemometric analytical schemes tolerate lower resolution successfully. In general, the minimum limit of detection should be found for measurements made at the lowest possible resolution that adequately resolves the spectral features of the analyte from those of interfering species. Increasing the resolution beyond this point degrades the SNR. If possible, spectra should also be sampled at a resolution five times the FWHH of the absorption band of interest (see ASTM E168). In FTIR, the optimum resolution will be determined by the bandwidths of the absorbance bands in the spectra of the target compounds, the presence of interfering species, and the SNR ratio of the system. The optimum resolution will most likely vary with respect to specific analytes and measurement conditions.

3.1.6 *Electronic Components used in Spectrometry*

For references describing the electronic components of optical spectrophotometers, see modern textbooks on electronic instrumentation including references 4–6. Other volumes in this set discuss more details regarding electronic components for instrumentation, please see index for specific topics. Two special categories of electronic devices deserve mention when discussing spectrometers. The first category includes detectors and detector electronics that produce a current or voltage signal proportional to the amount of energy striking the detector. Detector stability is provided by appropriate electronic circuitry allowing the detector signal to be selectively amplified with the minimum amplification of noise, thus electronic circuitry enhances the signal-to-noise ratio of the detector signal.

Digital microcomputers comprise the second essential electronic element for modern spectrophotometers. With the addition of appropriate software, sophisticated instrument control and data processing, the usefulness and user friendliness of MIR instruments can be made faster, cheaper, and more accurate. A basic overview of electronic components is shown in Figure 3.18, and their associations with instrumentation as shown in Figures 3.1 through 3.4 within this chapter.

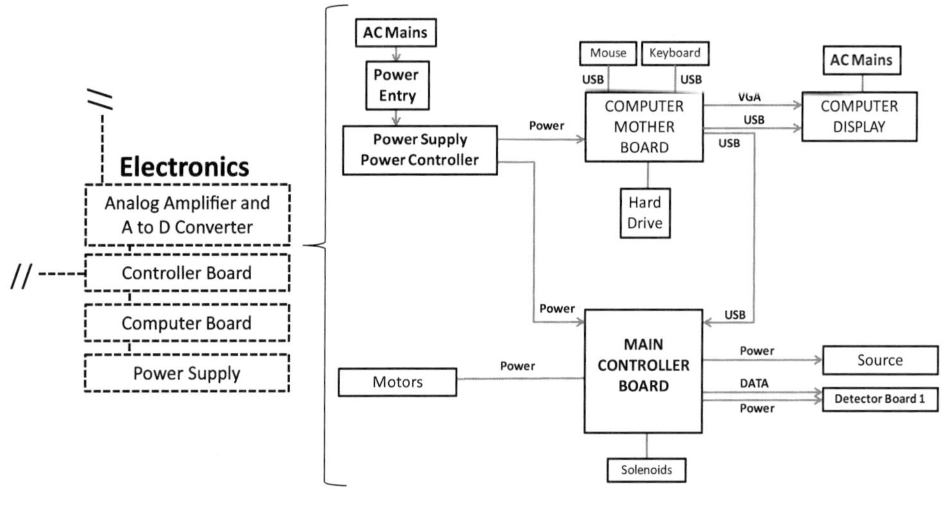

Fig. 3.18 Basic electronic systems for spectrometers.

3.1.7 *Properties of Spectrophotometers*

3.1.7.1 *Aperture Diameter*

To calculate the aperture (a) required by an optical system to resolve two objects with known linear separation, the Raleigh criterion for resolution is used as:

$$a = 1.22\lambda \cdot \alpha_R^{-1} \tag{3.46}$$

where α_R is the angle of separation from the measuring device exit aperture to the objects to be resolved (this is calculated as tan α = opposite/adjacent; where α_R is expressed in radians); λ is the wavelength of light observed from the objects; and a is the aperture of the optical system.

3.1.7.2 *Entrance and Exit Pupils*

The entrance pupil refers to the size and location of the entrance aperture between the light source and the remainder of an optical system. The exit pupil refers to the size and location of the exit aperture within an optical system or optical train just prior to the detector. Both the entrance and exit pupils affect the energy throughput and spectral resolution of a spectrophotometer.

3.1.7.3 *Band Pass (Bandpass) and Resolution*

The terms bandpass and resolution are used to express the capability of a spectrometer to separate spectral bands or lines which are separated by some finite distance on the spectral energy scale. For an instrument that measures energy over a prespecified spectral region of the electromagnetic spectrum, the bandpass of a spectrometer is used to describe which portion of the spectrum can actually be isolated by the spectrometer in a "pure" wavelength form. An illustration is often used to demonstrate the problem associated with measuring monochromatic light using conventional spectrophotometers. If the ideal spectrometer was used to measure a bright line emission spectrum at a single wavelength (λ_1), the spectrum would appear as a single line (Figure 3.19). What really occurs when such a spectrum is measured using a conventional spectrometer is a broad band spectrum as shown in Figure 3.20. The spectrum assumes a Gaussian-like (or bell-shaped) curve, actually closer to Lorentzian. This characteristic broadening of a line spectrum through the spectrometer is an illustration of the spectrometer bandpass. The actual bandpass for any instrument is assigned a value by determining the Full Width at Half Maximum (FWHM) height of the bell-shaped spectrum. Thus for the band in Figure 3.21 the FWHM could be empirically determined by finding the wavelength where

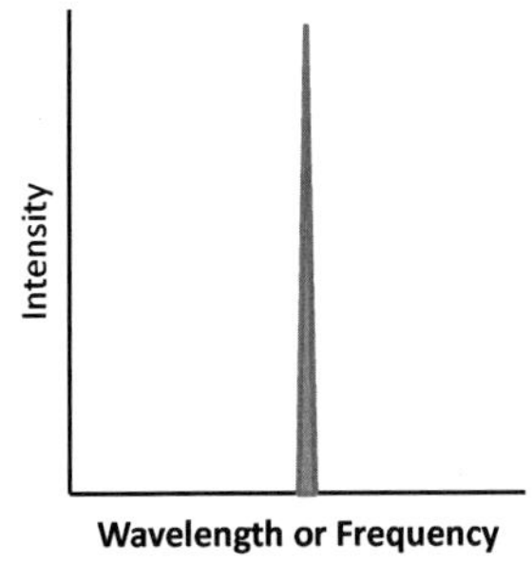

Fig. 3.19 The bright line emission spectrum at a single wavelength as it would appear in an ideal spectrophotometer.

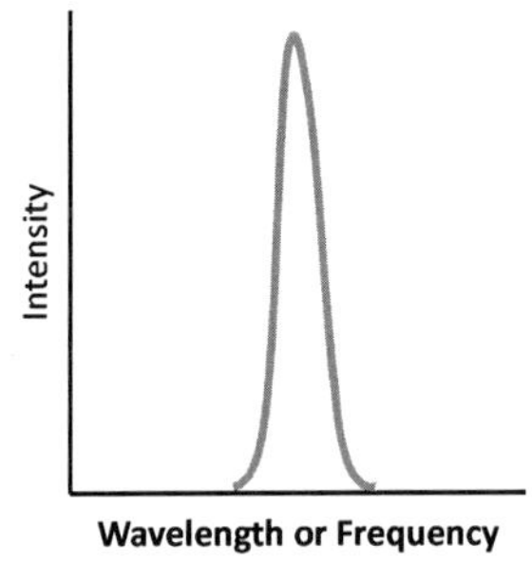

Fig. 3.20 Spectrum of a bright line emission source (e.g. deuterium lamp). The characteristic broadening is an illustration of the bandpass of a spectrophotometer.

maximum intensity occurs, and measuring the peak height at this position. This height measurement is divided in half and the bandwidth measured at this height on the band as illustrated in Figure 3.22.

The actual shape of a band is the result of several instrumental characteristics including the overall quality of the optics and detector systems, the mirror motions, and the aperture of the J-stop.

The resolution of a spectrometer can be defined as the minimum distance between two peaks that can be detected by the spectrometer under designated operational performance settings. Resolution is calculated by multiplying the slit width (generally expressed in mm) by the dispersion of the monochromator (in nm per mm) for a dispersive instrument. Due to practical issues and non ideal optics, the actual resolution of a spectrometer must be slightly less (poorer) than the theoretical value. For interferometric systems the mirror velocity and J-stop aperture determine spectral resolution.

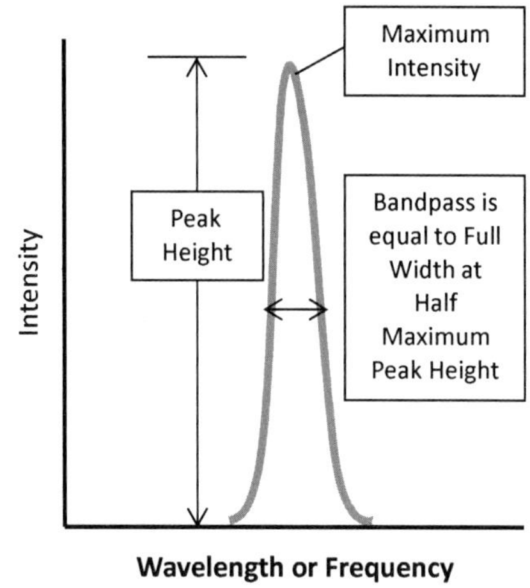

Fig. 3.21 Illustration of the determination of bandpass using the bell-shaped peak obtained by using a bright line source projected through a monochromator optical system.

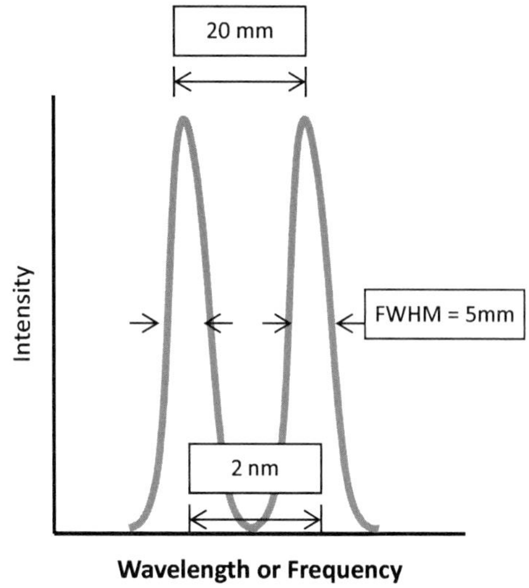

Fig. 3.22 The method to determine maximum resolution of a spectrometer under specific measurement conditions. In the example shown Bandpass = Resolution = FWHM × Dispersion = 2 nm/20 mm × 5 mm = 0.50 nm resolution. The manual method for estimating spectral resolution is shown in the figure.

To summarize, bandpass and resolution are identical in practice. Only the resolution specification of a spectrometer is the expression of bandpass under the specified measuring conditions of an instrument dependent upon the grating dispersion and slit width settings (for dispersive instruments), or minor velocity and J-stop settings (for interferometer based instruments).

The empirical (or measured) resolution of a spectrometer is determined by measuring the FWHM in mm for two narrow bands which are completely resolved (to the baseline) using the spectrometer. The spatial difference between the maximum absorbance (λ_{max}) is determined between the bands (in mm); simultaneously noting the difference between the λ_{max} points in nm. The various measurements required for this calculation are shown in Figure 3.22 and illustrated by the relationship:

$$\text{The bandpass} = \text{resolution} = \frac{\text{band.difference.in.nm}}{\text{band.difference.in.mm}} \times \text{FWHM in mm} \quad (3.47)$$

3.1.7.4 *Numerical Aperture*

The numerical aperture, designated as NA is a measure of how much light can be collected by any optical system. The NA is expressed as the product of the refractive index of the material where incident energy is directed into (n_i) times the sine of the ray angle maximum difference from normal incidence; this function is given as:

$$\text{NA} = n_i \cdot \sin \theta_{max} \quad (3.48)$$

3.1.7.5 *Attenuation*

Attentuation is a term often used to describe the light transmittance losses within (or through) an optical system. Losses in transmitted light through spectrometers are due to absorption, reflection, scattering, and optical misalignment; the losses can vary with temperature and wavelength. The quantity of optical loss can be expressed as an attenuation rate in decibels (dB) of optical power per unit distance (cm). The usual attenuation occurs through inefficiencies in source launch optics, lenses, mirrors, bandpass filters, slits and J-stop features, detector focus, fiber optic losses, and the like. Typical changes in the energy transmitted through a spectrometer over time result from source aging, launch optics, temperature variations resulting in mechanical alignment changes, inefficient optical couplings within an optical path, aging of mirrored surfaces, and typical soiled optical surfaces (e.g., lenses, mirrors, windows).

The losses in energy transmitted through a spectrometer can be calculated by using Beer's law, which states that the irradiance of the energy through an absorbing medium falls exponentially with the distance of transmission through a medium. The relationship for this is as:

$$I_d = I_0 \cdot 10^{-ad}, \quad \text{and it follows that} \quad (3.49)$$

$$T = \frac{I_d}{I_0} = 10^{-ad}, \quad \text{and} \quad (3.50)$$

$$-\log\left(\frac{I_d}{I_0}\right) = \alpha d, \quad \text{and so} \tag{3.51}$$

$$\alpha = -\log\left(\frac{I_d}{I_0}\right) \cdot d^{-1} \tag{3.52}$$

where I_d is the irradiance measured at distance (d) from the source through the optical system (train) in cm; I_0 is the source irradiance at $d = 0$ from the source; and α is the attenuation loss (absorption) coefficient in units of dB/cm through the optical system. Attenuation losses are wavelength dependent, thus the value for α is a function of the incident wavelength.

3.1.8 *Instrument Comparison and Evaluation Methods*

A number of standard measurements may be completed that allow detailed comparison of overall instrument performance. These are delineated in this section. The following series of tests is used to qualify instrument performance and to determine which issues are problematic due to deficiencies in instrument design features. These tests are related to alikeness in measurement performance between instruments and to overall accuracy, and precision (as repeatability and reproducibility). Note that the terms optical density (O.D.) and absorbance units (A, Au, or AU) are synonyms. (Note that "Optical density" is found in historical documents, and is still used in physics, and biomedical and optical engineering, but not often in analytical chemistry.) The techniques used for evaluation of spectrometer performance are described in detail within Chapter 6, entitled *Data Analysis and Presentation.*

Recommended tests for FTIR instruments include ASTM E1421, which specifies Level zero and Level one tests. Level zero tests are comprised of routine checks of instrument performance to determine significant changes and record a history of data from each instrument to be used for later comparison purposes. Level one tests are used to provide quantitative data on instrument function and performance with the purpose of diagnosing any significant instrument performance issues. (See Chapter 6 for details).

3.1.9 *Advanced FTIR Measurement Techniques*

Advanced methods are available using modern FTIR spectrophotometers. Included are traditional continuous single channel scan modes, continuous dual-channel scan modes, and step-scan modes. The following Table 3.5, with references, describes the current measurement modes available using full featured, research grade FTIR spectrophotometers.

Table 3.5 Infrared measurement modes, Basic and Advanced (expanded from Chapter 1).

Instrument measurement mode	Description of measurement	Experiment type	Comments
Extended spectra range	28,000 to 10 cm^{-1} or 350 to 1,000,000 nm	Study full range of electronic transitions to molecular vibrations and rotational functionalities of molecules	Multiple detectors, beam splitters and energy sources used
High Resolution	0.02 to 0.09 cm^{-1}	For gas phase measurements and studies over infrared range	Requires longer measurement times to achieve sufficient signal-to-noise
Low Resolution	16 to 32 cm^{-1} (as high as 132 cm^{-1})	Allows mimicking of low resolution spectrometers for database collection, calibration transfer, or instrument design simulations	Provides different spectra for applications studies relating to other low resolution instruments
Scan	Absorbance* vs. wavenumbers	Traditional scan	Optional scan range, number of replicate scans, and spectral resolution
	Reflectance or Transmittance vs. wavenumbers	Traditional scan	Optional scan range, number of replicate scans, and spectral resolution
Timedrive	Absorbance* vs. time at each wavenumber	Slow kinetics and reaction rate experiments	May be used to study fermentation or chemical processes that are on second time scales
Step-Scan Phase-Resolved	Amplitude Modulation, Phase Modulation, or Sample Modulation	Electroluminescence measurements	Polymer stretching, and electrochemistry measurements, or PAS
Step Scan Time-Resolved	Microsecond to nanosecond time resolution for timed events	Nanosecond chemical kinetics, liquid crystal dynamic studies, polymer stretching experiments	Also useful for PAS or (photoacoustic spectroscopy) for depth profile measurements
Step-Scan Space-Resolved	Imaging spectra for spatial data acquisition for focal-plane imaging	IR imaging	Combines chemical function and structural information
Polarization Modulation	Modulation of light polarization for measuring polarization effects	Dual-channel polarization modulation measurements	For IRRAS, VCD, and VLD with low absorbance values near 10^{-4}A
Individual wavenumber step-scan	Individual Absorbance(s)* at selected wavenumbers(s)	Polymer rheological studies are possible	Shows the structural changes that occur in materials when stress (stretching) is applied

(*Continued*)

Table 3.5 (*Continued*).

Instrument measurement mode	Description of measurement	Experiment type	Comments
Two-dimensional Correlation[13]	Correlates multiple spectra for conditions of applied stress and relaxation to determine rheo-optical properties. Also for correlation of spectra from different regions such as NIR-IR.	Allows correlation comparison of different spectral ranges to correlate the functional group positions for more detailed chemical (structural) information	Provides more information on molecular structure than single first-order spectra
Quantitative methods	Concentration of analyte vs. absorbance	Uses calibration database of spectra with reference values for univariate or multivariate calibration	Requires quantitative software suite
Qualitative methods	Correlation or spectral matching for Identification	Uses spectral libraries for material identification	Requires qualitative software suite
Kinetics	Kinetic rates of reaction	Single-channel rapid scan for kinetics measurements	Scan rates of 75 to 100 spectra per second at reasonable resolution (e.g., 4 or 8 cm^{-1})

*As $-\log_{10}$ (transmittance) or $-\log_{10}$ (reflectance).
**Note the following terms:

IRRAS (Infrared reflection absorption spectroscopy) for studying adsorbed or thin layers of materials on metal surfaces.[14,15]

VCD (Vibrational circular dichroism) is used to detect and record left handed and right handed circularly polarized light to detect and study enantiomers (i.e., stereoisomers or molecules that are mirror images of one another).[16]

VLD (Vibrational linear dichroism) is used to measure the difference in infrared spectra between two orthogonally polarized infrared energy beams. It is used to measure the vibrational orientation of molecules on surfaces or in biological systems.[17–19]

PAS (Photoacoustic spectroscopy) for depth profiling or measuring spectra at different depths within a sample to study structural layering.

References

1. ASTM (American Society for Testing and Materials) Method ASTM E131 - 10, Standard Terminology Relating to Molecular Spectroscopy, ASTM International, 100 Barr Harbor Drive, PO Box C700, West Conshohocken, PA, 19428-2959 USA (2010).
2. A. A. Michelson, The Relative Motion of the Earth and the Luminiferous Ether, *American Journal of Science* **22**. (1881) 120–129.
3. A. A. Michelson, E. Morley, On the Relative Motion of the Earth and the Luminiferous Ether, *American Journal of Science* **34** (203) (1887) 333–345.

4. D. M. Livingston, The *Master of Light: A Biography of Albert A. Michelson*, 1st edn. (The University of Chicago Press, 1979).
5. A. A. Michelson, *Light Waves and Their Uses* 1st edn. (The University of Chicago Press, 1902).
6. P. B. Fellgett, *Theory of Infra-Red Sensitivities and its Application to Investigations of Stellar Radiation in the Near Infra-Red,* PhD thesis (University of Cambridge, Cambridge, UK, 1951.
7. P. R. Griffiths, J.A. de Hasseth, *Fourier Transform Infrared Spectrometry,* 2nd edn. (Wiley-Blackwell, 2007).
8. J. W. Cooley and J. W. Tukey, An algorithm for the machine calculation of complex Fourier series, *Math. Comp.*, **19** (1965) 297–301.
9. J. M. Chalmers, P. R. Griffiths, *Handbook of Vibrational Spectroscopy* 1st edn. (John Wiley & Sons, 2002).
10. Roy Cain, personal communications, *A User's Guide to Fourier Transform Infrared Spectrometers*, (ASTM, March 2006). Roy was an active leader in ASTM activities and was a recipient of the ASTM Fellow, Award of Merit in (1999).
11. Lord Rayleigh, F. R. S. Investigations in optics, with special reference to the spectroscope, *Philosophical magazine*. Series 5, **8**(49) (1879) 261–274.
12. ASTM (American Society for Testing and Materials) Method ASTM E1421 - 99(2009), "Standard Practice for Describing and Measuring Performance of Fourier Transform Mid Infrared (FT-MIR) Spectrometers: Level Zero and Level One Tests", ASTM International, 100 Barr Harbor Drive, PO Box C700, West Conshohocken, PA, 19428-2959 USA (2009).
13. I. Noda, A. E. Dowrey, C. Marcott, Recent Developments in Two-Dimensional Infrared (2D IR) Correlation Spectroscopy, *Appl. Spectrosc.* **47**(9) (1993) 1317–1323.
14. K. Kunimatsu, W. G. Golden, H. Seki, M. R. Philpott, Carbon monoxide adsorption on a platinum electrode studied by polarization modulated FT-IRRAS. 1. Carbon monoxide adsorbed in the double-layer potential region, and its oxidation in acids, *Langmuir* **1**(2) (1985) 245–250.
15. T. Buffeteau, B. Desbat, D. Blaudez, J. M. Turlet, Calibration Procedure to Derive IRRAS Spectra from PM-IRRAS Spectra, *Appl. Spectrosc.* **54**(11) (2000) 1646–1650.
16. L. A. Nafie, T. B. Freedman, Vibrational circular dichroism: an incisive tool for stereochemical applications, *Enantiomer* **3**(4–5) (1998) 283–297.
17. L. A. Nafie, M. Diem , D. W. Vidrine, Fourier transform infrared vibrational circular dichroism, *J. Am. Chem. Soc.* **101**(2) (1979) 496–498.
18. C. Marcott, Linear Dichroism of Polymer Films Using a Polarization-Modulation Fourier Transform Infrared Technique, *Appl. Spectrosc.* **38**(3) (1984) 442–443.
19. P. L. Polavarapu, D. F. Michalska, D. M. Back, On the Measurement of Vibrational Circular Dichroism of a Single Enantiomer on Fourier Transform Infrared Spectrometers, *Appl. Spectrosc.* **38**(3) (1984), 438–442.

Chapter 4

Sampling Considerations for Infrared Spectroscopy

Contents

This chapter in part is used with permission and was first published as J. Coates, A Review of Sampling Methods for Infrared Spectroscopy, in J. Workman and A. Springsteen, Applied Spectroscopy: A Compact Reference for Practioners (Academic Press, Boston, 1998).

4.1 Introduction

The quality of an infrared spectrum is strongly dependent on the method of sample preparation and the optical interface between the sample and the infrared instrument — usually defined by a sampling accessory. It must be appreciated that the physical state and how the sample is treated will influence the appearance of the spectrum. Furthermore, different sampling accessories that use different principles of optical measurement produce variations in the appearance of the final spectrum for the same sample. Taking care and time at the moment of sample preparation, and understanding the impact of different measurement techniques, can save time and much anguish later.

This chapter is intended to serve as a basic guide to provide direction for the optimum method of sampling for a given type of sample. It is designed differently than traditional texts in that it is focused on the practical rather than the theoretical aspects of the subject. To that end, the text is divided into four functional parts. The first part deals with the sample itself, where the emphasis is placed on the physical state of the sample, the potential impact of the chemical properties of the sample, and the sampling approaches that are amenable to that type of sample. Having established the available options, that is, the methods and procedures for sample handling, the individual techniques are reviewed briefly in this chapter. Most times, it is preferred to use the sample in its natural, unmodified state. However, sometimes this is not practical because the sample is not in a convenient form to measure. In order to provide options for such situations, a section dedicated to procedures for sample modification is also provided. The final section deals with the actual acquisition of the spectrum and the options available for handling the spectral data from its raw form to the final presentation.

There are no guarantees that a specific recommended procedure will work because individual samples can and often do behave differently from the norm. In addition to the specified procedure, it is important to document all known facts about the sample, as well as any abnormalities observed during sample preparation. The text sections dealing with sample type versus available procedures should be used to provide a definition of the sample characteristics and the method to be used. Often there is more than one method available for a given sample type. The best advice is keep it simple; often, a straightforward transmission measurement will provide the best quality data. The actual choice may well depend on the final application — that is, the purpose for recording the spectrum in the first place — such as a qualitative identification of an "unknown", a quality comparison for raw material screening, or a quantitative measurement for chemical composition.

For many applications, the basic, unprocessed spectral data should be adequate. If any form of spectral processing is used, such as baseline correction, digital smoothing, or spectral subtraction, it is important to appreciate the impact and the consequences of the specific data manipulation. Sometime, irreversible distortions are introduced into the spectrum that can lead to erroneous quantitative calculations or incorrect conclusions in spectral interpretation. Also, when sharing spectral data, or storing or archiving spectra, it is essential to note the form of processing used and any numerical terms or factors that are applied. If spectral subtraction is used, it is strongly recommended that the original sample spectrum, and the spectrum used for subtraction are also made available. Refer to Section 4.6 for recommendations for spectral acquisition, data and presentation formats, and comments regarding spectral data manipulations.

4.2 General Considerations

Infrared is a form of electromagnetic radiation nominally extending from the visible (approximately 700 nm) to just before the microwave region — practically defined as approximately 200 microns. In this text, it will be described as radiation or light, and although "light" is normally used to describe visible radiation, there is sufficient overlap in optical characteristics of materials to justify the use of this as a generic term. Most infrared interactions with matter are recorded as absorptions of energy, either as a direct or indirect measurement. Emission measurements may be performed, but these are less frequent for analytical chemistry applications. Measurements may be made via light transmission through the sample or by reflection directly from the sample surface or from an interface. The absorption cross section of molecules in the mid infrared can be extremely high, and this often dictates which method should be used for recording spectra.

It is assumed that the reader has a basic knowledge of optical spectroscopy and the molecular theory of chemistry. Infrared (IR) spectroscopy is one of the analytical techniques available for the measurement of absorption or emission of radiation as a consequence of vibrations within a molecule between the component atoms. For an understanding of underlying theory, the reader is directed to a standard reference text on organic chemistry or spectroscopy[1,2] or to texts focused on the structure/spectra relationships of molecular compounds.[3–5]

The term infrared spectroscopy can be generally applied to measurements in at least three spectral regions: the near infrared, the mid infrared, and the far infrared. Commonly, when the term infrared is applied, this implies the mid infrared spectral region, today defined by the frequency range 4000-400 cm^{-1} (2.5–25 microns,

2500–25,000 nm in units of wavelength). This range tends to be arbitrarily defined, but the upper spectral limit does define the extent of all the fundamental molecular vibrations, with the exception to that of hydrogen fluoride.

The lower limit is normally defined by the sampling method or matrix or by the optics of the instrument. In the latter case, the standard potassium bromide (KBr) beam splitter substrate of a Fourier transform IR (FTIR) instrument provides an optical cutoff at approximately $400\,cm^{-1}$. Fundamental vibrations do extend below this cutoff, and these are normally associated with either low-energy lattice vibrations or the vibrations of chemical bonds involving the heavier elements. The region below $400\ cm^{-1}$ (this limit is subjective) is referred to as the far infrared.

The spectral region above $4000\,cm^{-1}$ (2500 nm) and extending out to the visible region of the electromagnetic spectrum (approximately 700 nm) is known as the near infrared (NIR). Commercially, the instrumentation for this region has been differentiated from instrumentation covering the fundamental regions (mid- and FIR), and so NIR as a technique has been treated separately from traditional IR spectroscopy.[6] Today, these artificial barriers are disappearing, and it is more appropriate to discuss all spectral regions under the generic term IR spectroscopy. From a sampling perspective, there are differences between NIR and the mid infrared spectral region. These are due to various factors, such as large differences in absorptivities and the efficiency of coupling of the radiation with the material as a function of wavelength.

The NIR spectral region is composed of absorptions related to vibrational overtones and combination absorptions. The absorptivity associated with these is less, by one or more orders of magnitude, than the fundamental absorptions. As noted earlier, the magnitude of the absorption may be used to define the best procedure for sampling. It is commonly accepted that NIR is easier for sample handling than traditional IR spectroscopy. However, this is a subjective statement, and with modern sampling technologies it can be challenged for a broad range of sample types, dependent on the final application. In this text, the focus will be on sampling technology as developed for the mid IR spectral region; in some cases, the concepts can be extrapolated to the NIR region. For the most part, sampling for NIR analysis involves transmission or reflectance (normally diffuse reflectance) or a combination of both. For a general overview of the technique and its related sampling procedures, the reader is referred to Volume 3, Chapter 4, and to a standard text on the subject.[6]

In many cases, factors such as price, convenience of operation, optical performance/efficiency, and final application may be the final arbiter in the selection of a sampling method. The actual application can be an important deciding factor. A host of techniques are available for simple qualitative analysis for material identification, whereas for a high-precision/high-accuracy quantitative method the field may be narrowed to only one or two techniques. Likewise, if the application is very specific, such as a microscale analysis or the characterization of a chromatographic fraction directly from a chromatograph, then a specialized approach is dictated. Infrared spectroscopy, both near- and mid IR, are both used for factory-based measurements during manufacturing. These may be manual or automated, off-line batch-oriented or on-line for continuous monitoring. In all scenarios, the need for a rugged, simple to implement (and use), and reproducible technique will limit the available technologies. For background reading, the reader is referred to references (7–13).

4.3 Sample Type versus Sampling Method

One of the most important practical aspects of infrared spectroscopy is that it may be applied to almost any type of sample in any physical state, form, or modification. In this section, sampling will be discussed in terms of physical state and/or physical nature of the sample. One point that will be emphasized throughout is the importance of documentation of all experimental procedures, sampling methods and conditions, and any abnormalities observed during sample preparation procedures.

If the sample, (solution or mixture) is known to contain water, then not only should this information be recorded but also extreme care must be taken to ensure that the sampling method is compatible with water being present. Comments are provided in this chapter regarding the removal of water and other volatile materials that can interfere with an analysis. Also, if a sample is seen to be made up of more than one phase, it is helpful to obtain spectra from each separated phase. Techniques such as filtration, decanting, and centrifuging should be used if practical to ensure representative sampling of a particular phase. Record the nature of the phase being sampled (liquid or solid), and note whether it is an upper or lower layer (this obviously provides some idea of relative densities).

4.3.1 *Gases and Vapors*

The sampling for gases and vapors is relatively straightforward, and the main issues in selecting the best sampling conditions are pressure, temperature, and the relative concentrations of the analytes. For concentrated samples it is normal to use a transmission cell with a short pathlength, typically in the range of 1–10 cm, the final choice being dependent on the concentration of the most important analytes. For gases, it is important to record the pressure and partial pressures that are used, especially when performing a quantitative measurement. This also applies if a nonabsorbing makeup gas, such as nitrogen or air, is used; the type of makeup gas, if used, should always be noted.

Another important application of infrared gas analysis is for trace analysis, such as for the analysis of a dilute mixture (in the ppm range) or an environmental specimen. In such cases, individual analytes are measured from the high ppb levels to the 10's or 100's of ppm. For such analyses, extended pathlengths are required, and typically multipass gas cells from 1 to 20(μm) in pathlengths are used. One very specific application is an open-path measurement for ambient air monitoring in manufacturing plants or in toxic waste sites in which no cell is used. Instead, a source and interferometer combination are focused on a remote detection system with the aid of special telescope optics. In such cases, several hundreds of meters of effective pathlength are used. See Chapter 3 of this volume for information on open path instrumentation.

Two special forms of gas or vapor analysis involve the interfacing of an infrared instrument to a gas chromatograph (GC) or a thermogravimetric analysis (TGA) instrument. In both cases, relatively high concentrations of gas or vapor are produced within an extremely small volume. In such applications, the key is to remove dead volume between the two instruments (GC-FTIR or TGA-FTIR) and to have an optimum optical coupling with the sample at minimal sample volume. In the case of gas chromatography, the separated fractions can be measured either with a heated lightpipe or by a cooled trapping method. It is important to appreciate that these last two are very different methods of measurement; both are measured in the solid phase compared to the gas phase used with the lightpipe.

4.3.2 *Liquids*

IR is not compatible with quartz, glass, and silicate cell windows. IR absorbs water intensely across region.

4.3.2.1 *General*

Traditionally, liquids have been considered to be relatively easy except for low-molecular-weight compounds, which often suffer from volatility problems and extremely high absorptivity that typically require extremely small optical pathlengths (often 10 microns (μm) or less). The main methods for liquids are transmission in a sealed or semi permanent cell or by Attenuated Total Reflectance (ATR; internal reflectance). Today, ATR-based methods are becoming a preferred method because of the relative simplicity, ease of cleaning, and good match to the required sample pathlength. For volatile liquids, there is the option to handle the sample in vapor form. A heated gas cell is required for this approach.

For liquids in general, it is very helpful to note if the sample is known to be a single material, a mixture, or a solution. If the sample is a solution, the major solvents, if known, should also be recorded, preferably during the same time frame. In this way, it is possible to differentiate the solute and solvent, especially with the aid of computer subtraction techniques. Other issues relevant to knowledge of the sample include the following. It is important to know if the sample is "wet", especially if salt optics (KBr, NaCl, or CsI) are used, and it is beneficial to know if there are volatile components present.

4.3.2.2 *Mobile*

If the sample is a nonvolatile or semi volatile liquid, then it is possible to use a short pathlength transmission cell — with either a sealed or semi permanent design. Pathlength is an important issue, and it is important to select a pathlength that provides adequate intensity to minor absorptions, while ensuring that the information is not lost in the major bands due to a "bottoming out" at 0% T. A new class of ATR accessories for handling liquids has become available during the past decade. These include ATR liquid cells, based on traditional ATR designs; cylindrical ATR cells; horizontal ATR trough cells; and dip probes. For routine sample handling, the horizontal ATR format has become the most popular. This format is preferred for ease of use and ease of cleaning. In the case of an enclosed or sealed cell, make sure that the cleaning solvent is effective for removing the sample (test this fact outside of the cell) and is compatible with the window materials. Note that with cylindrical ATR cells, there is a tendency for sample and/or solvent to become trapped between the sealing O-ring and the surface of the Internal Reflectance Element (IRE).

One of the most important factors to consider is the compatibility of the cell optics with the sample, especially with an ATR sensing element (the IRE), because of their relative high cost. Accessories featuring diamond-based optics are now available, and these provide corrosion and chemical resistance for virtually any type of sample.

4.3.2.3 *Volatile Materials*

There are several important issues to consider with volatile liquids, such as the need to use a sealed cell, potential loss of sample during analysis, protection of the instrument (in particular sensitive optics), and general fire and safety precautions. The traditional approach for volatile liquids and solutions (based on volatile solvents) is to use a sealed, fixed pathlength transmission cell. If the sample is extremely volatile, then thermal control of the cell (cooling) may be advised to prevent leakage or damage to the seals of the cell. Various configurations of liquid ATR cells are available, as previously noted. One of the best options for samples of high-volatility is to use the high-pressure flowthrough style of cells, with thermal control (cooling, if available).

As commented earlier, it is possible to analyze the sample in the vapor phase by placing it within a heated gas cell. The latter suggestion is useful when the total material is volatile or when there is the need to obtain information about volatile components only. If a vapor-phase approach is used, the sample temperature in the gas cell must ideally be known because this will impact the appearance of the final spectrum. Note that the appearance of a vapor-phase spectrum may be significantly different than the corresponding spectrum in the condensed phase. Several commercial spectral collections of common organic compounds in the vapor phase are available in both hard copy and digital formats.[14,15]

4.3.2.4 *Viscous Materials*

A simple and convenient method for viscous, nonvolatile materials is to produce a capillary film between a pair of transmission windows or, if the material is very viscous, to form a thin smear film on a single window. It is recommended that barium or calcium fluoride, or comparable window materials, be used for aqueous-based materials. It is very difficult to fill traditional sealed cells with viscous liquids, attempts to force-fill the cell can result in permanent damage, particularly to the seals. Also, once filled, the cells can be extremely difficult to clean. If the sample is extremely viscous, such as tar or resinous material, it is possible to dissolve the sample in a suitable volatile solvent and evaporate a film of the material from solution on the surface of a single infrared window. The main caution here is to ensure that all traces of solvent are removed, noting that there will be a tendency for solvent residues to be retained. Heating under an infrared lamp or in a vacuum oven will help this process.

As previously discussed, liquid ATR cells may be used and, in general, the horizontal ATR format is the most appropriate and convenient-in use and for cleaning. Alternatively, either a boat cell cylindrical ATR accessory or an ATR dipping probe may be used. In all cases, pay careful attention to cleaning. Trace contamination of the surface of the ATR element will result in a major interference to the sample spectrum.

4.3.2.5 *Pastes, Emulsions, and Slurries*

Dependent on the constituents; pastes and slurries generally may be handled in a similar manner as proposed for viscous liquids. Always be aware of potential homogeneity problems. Also be cautious of the effect of any abrasive materials, retained in suspension, on optical window or crystal surfaces. This is particularly so for any system designed for continuous flow. Constant abrasion will result in light losses (due to light scattering) and eventually permanent damage to the optical surfaces. Emulsions containing water should be handled with either barium or calcium fluoride (or a comparable non hygroscopic material) windows or with an ATR cell. Make sure that separation does not occur at the window or the IRE interface. Sometimes the polar nature of the optical element will cause segregation to occur. Common examples are emulsions, such as mayonnaise, and protein-containing materials; in this later case, the protein may actually form a coating on the optical surface. Also, some emulsions are stabilized by amino or ammonium compounds and these will attack many of the window materials, especially those containing heavy metals (such as BaF_2 and ZnSe).

4.3.3 *Solids*

4.3.3.1 *General*

Of the three physical phases, solids pose the greatest challenge for infrared spectroscopy. The main issues are how to get effective interaction between the infrared radiation and the sample; how to minimize artifacts and distortions caused by refractive index, scattering, and polarization effects; and how to sample reproducibly and in a manner that is representative of the sample. A very wide range of sampling techniques are available for solids, and their suitability is dependent on the nature and physical form of the sample. These include transmission, reflectance (various forms), emission, and absorption (photoacoustic) measurements.

4.3.3.2 *Powders*

A standard approach for grindable powders is the preparation of a compressed potassium halide (KBr or KCl) pellet or a mineral oil mull. In both situations, the objective is to grind the sample, ideally, down to a sub-micrometer particle size and to disperse the finely ground material in a matrix with a closely matched refractive index. This approach is intended to minimize or eliminate the impact of light scattering caused by the particulate nature of the sample. If the particle size is similar or greater in magnitude to the wavelength of the infrared radiation, then a significant portion of the light is lost (not transmitted) due to light scattering. Reduction of the particle size below this level, by up to an order of magnitude, will significantly reduce the light losses, and this is further aided by dispersing the material in an index-matching matrix (the alkali halide or the mulling agent).

In the case of compressed halide pellets, the sample is first ground to the desired nominal particle size and is then mixed (with partial grinding) with the halide salt (usually KBr but sometimes KCl). The mixture is then compressed in a special high-pressure die, with a nominal pressure of approximately 10 tons (for 13-mm-diameter pellet). It is important to ensure that there is no ionic interaction between the sample and the potassium halide. Also, be aware of materials that undergo phase changes when subjected to the extreme pressures of the compressed pellet technique. Make sure that the potassium halide is kept in a clean, dry area, and if there is any doubt about its quality, run a blank pellet without sample. For materials that react with alkali halides, or for spectra to be recorded below $400\,cm^{-1}$, polyethylene powder may be used as an alternative compressed pellet medium.

In preparation of an oil mull, the sample is ground in the same way as described previously, and a small quantity of an oil, usually a white mineral oil (such as medicinal white oil), is added and mixed with the sample to form a stiff paste — to a consistency of petroleum jelly. The paste is distributed between a pair of infrared windows (usually KBr or NaCl) to form a thin translucent film. If the film appears to be totally opaque, then there is a probability that the particle size is still too large. The mull spectrum will obviously contain hydrocarbon (C-H) absorptions of the mulling agent. A second mulling agent, known as flourolube (a perfluorinated hydrocarbon oil), which does not create such an interference, may be used to provide transparency in these regions.

Coarse or hard powders are not well served by either the compressed pellet or mull techniques mainly because of difficulties associated with grinding these materials. In such situations, the best approaches require the use of an accessory, such as diffuse reflectance or photoacoustic. Be aware that diffuse reflectance is also particle-size dependent. Mixing the sample with a potassium halide may be beneficial to reduce anomalous optical behavior at the surface of the sample. This is especially helpful with inorganic compounds or other materials that have high absorptivities or have a high index of refraction. As a rule of thumb, organic compounds can be measured in about a 10% mixture and inorganics with approximately 1 or 2% mixture (in KBr or KCl).

Note that both diffuse reflectance and photoacoustic may be applied to most forms of powdered solids. As a rule, photoacoustic measurements which are the only form of true absorption measurement are not significantly influenced by sample morphology.

An alternative procedure for powders is ATR, either with a traditional accessory or with a flat plate-style horizontal accessory equipped with a pressure applicator. Note that the use of pressure is recommended to ensure intimate contact between the sample and the IRE surface. Normally, the sample must conform to the surface of the IRE, and because the strength of the IRE is typically limited the procedure is recommended only for soft powders. However, with the recent introduction of high-pressure horizontal ATR devices equipped with small-area diamond IREs, it is possible to handle most types of powdered material.

4.3.3.3 *Amorphous Materials (Organic), Continuous Sheets, and Polymeric Films*

Moldable materials may be prepared as self-supporting films for transmission measurements. One must be aware of the probability for the occurrence of interference fringes in the spectrum caused by internal reflections of the front and back surfaces of the film. Materials not pre-existing as films, such as polymer pellets, may be molded or hot-pressed; accessories are available for producing films with well-defined pathlengths.

As an alternative to the hot-press film method, materials that are soluble in a volatile solvent may be cast into a film on a transmission window (or on the surface of an IRE of an ATR accessory) from solution. If this approach is used, ensure that all residual solvent is removed by evaporation usually under an infrared lamp or a vacuum oven. If this is carried out too fast, small bubbles will form in the film. If this occurs, photometric distortions can occur in the recorded spectrum. Always record the solvent used in the preparation of a cast film in case residual solvent remains.

Certain materials already exist as a sheet or film, including plastics, metals (with coatings), paper, etc. If the material is transparent (to IR radiation) then it may be analyzed directly as a self-supported film by transmission. Alternatively, one of the reflectance methods may be used. For polymers and materials such as paper, the ATR approach may be used — in particular, the horizontal ATR equipped with a pressure applicator. This will work for coated or uncoated samples.

When dealing with a coating on a metal surface — a metal sheet or a foil — either the ATR or an external reflectance/specular reflectance device may be used. Under these conditions the measurement is defined as a reflection/absorption or a transflectance measurement. This method works best for thin to medium film thicknesses. If the coating is very thin, a grazing angle reflectance measuring device may be used. Very thick films may cause problems by transflectance due to anomalous optical effects at the surface, resulting in spectral distortions (a function of the refractive index of the material). If distorted spectra are obtained, one should consider using an alternative sampling method, such as ATR. If an alternative is not available, it is possible to apply a function known as the Kramers — Kronig transformation to separate the refractive index and the absorption components of the spectrum.[16] This will provide a more conventional form of the absorbance spectrum, relatively free from distortion.

Less common situations to consider are when studies of surface contamination are required and when the surface cannot be separated and brought to the instrument. In the first example, a practical solution is to grind a small amount of alkali halide powder (KBr or KCl) at the point of the contamination. This will result in the transfer of some of the contaminant to the alkali halide, and this in turn may be examined either by the compressed pellet technique or by diffuse reflectance. A way to assist this surface extraction is to add a few drops of a solvent, such as methylene chloride, in the area of the contamination. In the second situation, sampling may be performed by gently abrading the surface with silicon carbide paper. This action will transfer a small amount of the surface to the abrasive paper, and this in turn may be examined with the aid of a diffuse reflectance accessory.

Finally, one technique that can work for a limited number of applications, in which thin films are to be measured on a surface, is emission spectroscopy. This can be a difficult measurement to control, although some accessory companies do offer devices to retain the sample and to provide controlled heating of the sample (for emission to be effective, it is necessary for the sample temperature to exceed that of the instrument's detector).[17] Generally, this is not a simple measurement, and factors such as self-absorption from the film and absorption from the substrate can constitute significant interferences.

4.3.3.4 *Lumps, Granules, or Pellets*

This class of solids is an extension of the sample types already discussed, and many of the procedures already highlighted may be used here. If the material dissolves in a suitable solvent, then a cast film may be prepared on an IR transmitting window or on a surface of an ATR element. Moldable materials, such as polymer pellets, may be prepared as hot-pressed films, with care taken to ensure that material does not thermally degrade. Grind-able materials can be handled as previously discussed for powders using the compressed halide pellet, mineral oil mull, or diffuse reflectance methods to acquire the spectrum.

Elastomeric or rubber-like materials are typically difficult to use with transmission-based techniques; however, dependent on the ingredients, they are ideal for ATR-based measurements. Most elastomers conform well with the IRE surface, providing good intimate contact. Elastomers with a high filler content, and in particular with dispersed carbon black, will cause problems because of the absorption characteristics of the carbon. In such cases, either ATR with a high refractive index IRE (such as germanium) or a photoacoustic measurement may be employed. In the event that such an approach is not available, then a destructive method such as pyrolysis can be used.

Fibrous materials may be analyzed by a number of different methods, including the standard method for sample preparation — compressed halide pellets. In this latter case, normally the fiber structure is destroyed during sample preparation. Alternatively, one can consider using diffuse reflectance, photoacoustic, or infrared microscopy. With due consideration to the preparation procedures, the fibers may be retained in the original form with these methods. In some cases, such as with IR microscopy, it is possible to study the orientation characteristics of the fiber material and to correlate this information to certain mechanical properties of the fiber. The use of IR polarizers is implied with this type of measurement. Diamond compression cells work well for single filament fibers by both transmission (diamond anvil cell) and ATR.

4.3.3.5 *Intractable Materials*

As noted previously, elastomeric materials filled with inorganics or carbon black can be problematic, and they often are considered to be intractable. If all accessory-based methods fail, then pyrolysis for characterization of the base polymer may suffice. Note that if the pyrolysis procedure is taken to completion (such as in a TGA apparatus) then an analysis on the residue or ash can lead to information on any inorganics present (excluding carbon). If only a small fragment of material exists then a diamond compression cell may provide adequate information, as long as the carbon content is not too high. A good alternative for filled polymers is to use photoacoustic detection, even for high-carbon-containing compounds. With this procedure there is a size constraint — it must be possible to place a representative sample within the accessory and this is typically constrained to a few milligrams.

There is a broad range of materials classified as insoluable and intractable; for example, cured resins, thermosetting polymers, wood, and other cellulosic materials, certain inorganics, etc. These materials may be handled by one of the techniques already mentioned, such as photoacoustic, external reflectance, IR microscopy, or even pyrolysis (or TGA). Potential problems may be experienced with reflectance methods (as discussed earlier) in terms of spectral distortions. In such cases, it may be necessary to apply the Kramers — Kronig transformation procedure.

One other technique, previously described, is the abrasion of the surface sample with a silicon carbide paper. The abraded material may be collected as a dust and analyzed as a compressed alkali halide pellet, or by diffuse reflectance. Alternatively, the spectrum of the sample from the surface of the abrasive paper may be obtained with a diffuse reflectance accessory.

4.4 Sampling Method versus Sample Type

This section provides similar information and cross-referenced information to that of the previous section, but in this case it is more procedure oriented. Hints and tips are provided for the actual method rather than focusing on the sample and its properties.

4.4.1 *Transmission/Transmittance Windows*

The most basic sampling device is the IR transmission window; see Table 4.1 for a summary of common IR transmitting materials. This is primarily used for cast films, capillary sandwich films, or smear films. These are summarized as follows:

a. Smear films: best suited for viscous and grease-like materials (including solid samples prepared as a mull)
b. Cast films: formed by evaporation of a material from solution; can be applied to any nonvolatile liquids, and soluble amorphous materials
c. Capillary films: used for viscous or semi-viscous, nonvolatile liquids

The following are words of caution about the preparation of evaporated films from solvent-based solutions. If KBr or NaCl windows are used, ensure that any solvents or solutions used are water free. When the film is cast, allow the solvent to evaporate slowly (to prevent moisture condensation) and finish evaporation under a heat lamp to remove any residual solvent. If a solid is cast from solution, make sure that crystallization has not occurred. Note that for some crystalline solids a preferred orientation can occur (such as along the crystal lattice lines of the salt window), causing intensity variations as a function of position of the evaporated film. Generally, in the preparation of any film — cast, smear, or capillary — ensure

Table 4.1 Common infrared window materials (Transmittance ranges).

Material	Useful range (cm^{-1}, transmission)	Refractive Index (at 1000 cm^{-1})	Water Solubility (g/100 ml, H_2O)
Sodium chloride (NaCl)	40,000–590	1.49	35.7
Potassium bromide (KBr)	40,000–340	1.52	65.2
Cesium iodide (CsI)	40,000–200	1.74	88.4
Calcium fluoride (CaF)	50,000–1140	1.39	Insoluble
Barium fluoride (BaF_2)	50,000–840	1.42	Insoluble
Silver bromide (AgBr)	20,000–300	2.2	Insoluble
Zinc sulfide (ZnS)	17,000–833	2.2	Insoluble
Zinc selenide (ZnSe)	20,000–460	2.4	Insoluble

that the film is regular and free from voids, air bubbles, or pinholes. Failure to do this will result in photometric errors in the recording of the spectrum.

4.4.1.1 *Transmission Cells (Liquids)*

Several forms of transmission cells exist, and these are a combination of some form of framework or assembly (cell body) and a pair of transmission windows (Figures 4.1 and 4.2). These may be defined in terms of demountable cells, semi permanent cells, fixed pathlength cells, and variable (continuously) pathlength cells. Demountable cells can be used for neat (undiluted) liquids or solutions and the benefit of this style of cell is that it is easily disassembled for cleaning. Fixed pathlength or semi permanent cells are typically used for solutions or nonviscous liquids. The pathlength, defined by a spacer (Figures 4.1 and 4.2), may be calibrated by the interference fringe method. When filling an assembled cell via the filling ports, it is important to ensure that the sample is drawn through with gentle suction. Attempts to push the sample through can result in cracking of the windows due to excess pressure buildup, especially with viscous liquids.

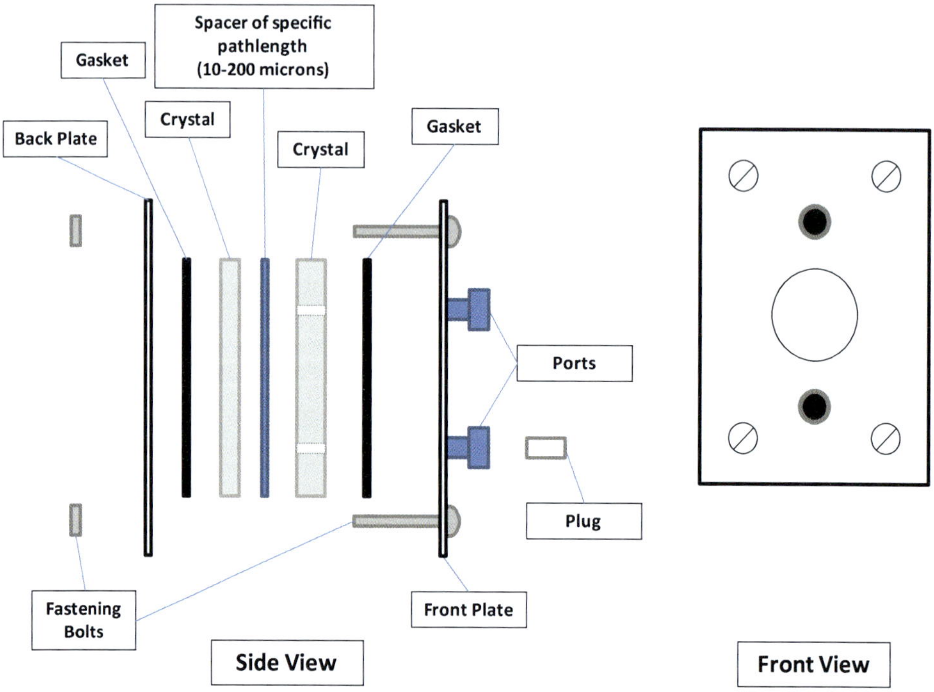

Fig. 4.1 A cross-sectional exploded view, and front view, of an infrared liquid transmission cell.

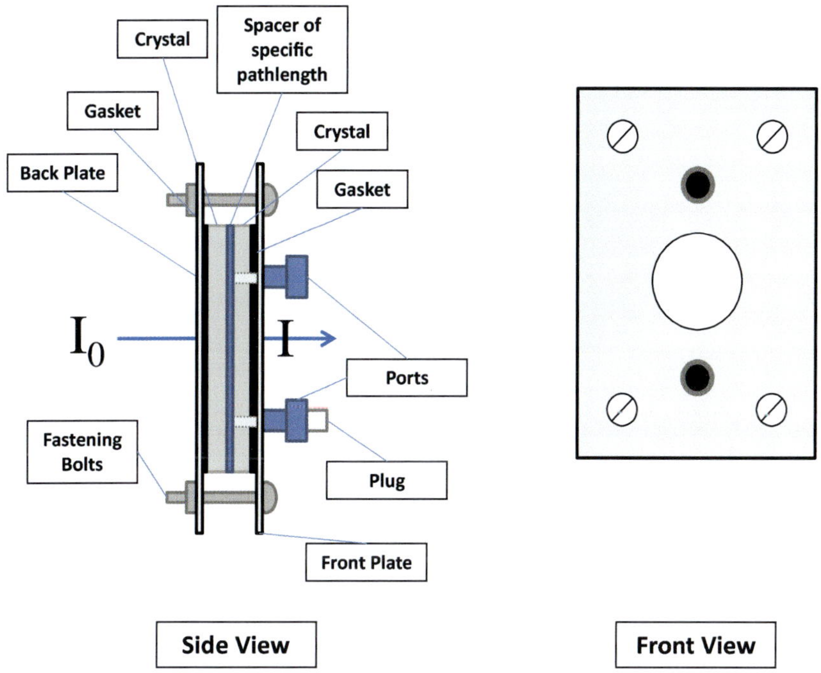

Fig. 4.2 A cross-sectional assembled view, and front view, of an infrared liquid transmission cell.

Be aware that cells made from nontraditional window materials, such as zinc selenide, zinc sulfide, germanium, or other medium to high refractive index materials, will usually result in the production of interference fringes, even in the presence of the sample.

A product produced by 3M, known as the IR card, uses a thin polymer film (polyethylene or PTFE) as a substitute for a traditional window. This is a relatively inexpensive method of sampling. However, if used, it is normally necessary to record a reference spectrum of the card substrate material in the absence of the sample; the sample is then obtained following spectral subtraction of the substrate spectrum. A novel alternative, formerly produced by Janos Technology, is the ECRAN screen card. This card uses a fine mesh, with essentially no infrared signature, to support the sample.

4.4.1.2 *Gas Cells*

a. Short pathlength Gas cells may be considered a special form of transmission cell. The difference is primarily in the construction of the body of the cell and its extended pathlength. In essence, a traditional gas cell (short pathlength) is a tube (with filling ports) with windows mounted and sealed at both ends. Like liquid cells, gas cells

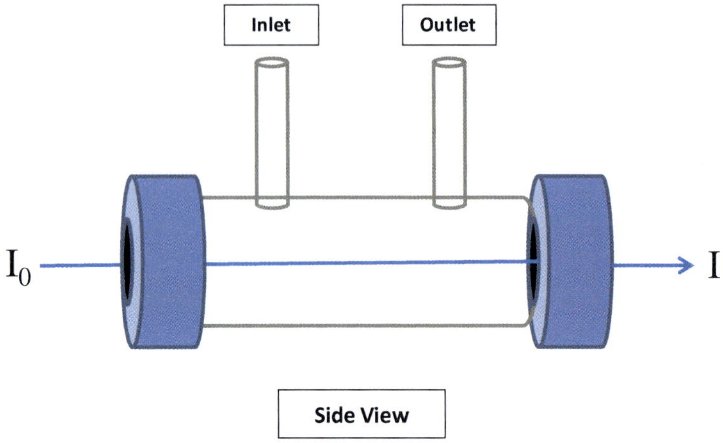

Fig. 4.3 Typical short pathlength gas cell (side view).

are available in demountable (removable windows) and permanent formats (bonded windows). Materials of construction can be important, especially if corrosive gases or vapors are studied. The most inexpensive cells feature glass tubular bodies, and these are ideal for many sample types. For more rigid construction, cell bodies made from stainless steel, Monel, and Hastelloy C are available.

Short pathlength gas cells, with pathlengths in the range of 1–20 cm, are used for the analysis of gases at high concentrations (Figure 4.3). The cells are usually filled by evacuating with a vacuum pump and backfilling with the sample to the desired pressure. Remember that pressure and concentration for a gas are equivalent, so pressure regulation is important. Note that if the sample pressure is less than atmospheric pressure, and ambient pressure is required, the cell may be "topped-up" with a non-absorbing gas such as dry nitrogen. Air may be used as a makeup gas, but it must be remembered that untreated air contains water vapor and carbon dioxide, both of which can constitute an interference. Also, note that some pressure broadening and intensification of the sample absorbance bands may occur with the introduction of makeup gas. If analyzing gas from a dynamic system, the cell may be placed within a sampling loop and the gas sampled by continuous flow through the cell. Once the cell is flushed by at least four cell volumes, the cell may be isolated and placed into the instrument.

Temperature is also an important parameter, and the heating of cells may be required for samples that condense at ambient temperatures. Various forms of heated gas cells are available for such applications. If the cell is heated, it is necessary to record the temperature because this will impact the quantitative aspects of any gas-phase measurement (because of the P–V–T relationship of gases). Measurements at high temperatures will also result in a broadening of the spectral lines.

b. Long pathlength Long pathlength gas cells, with pathlengths in the range of 1–20 meters, are used for the analysis of gases at very low concentrations — typically in the range of 1–1000 ppm. Such cells differ in design to the conventional short pathlength versions. The extended pathlength is generated inside the cell by internal optics that provide a folded optical path with multiple reflections of the infrared beam. Some cell designs use adjustable optics that may be tuned to different pathlengths. Because the optics are enclosed within the cell and are in contact with the sample, it is necessary to provide some degree of protection to the optical surfaces. Normally, these surfaces are gold coated, which provides a high degree of both reflectivity and protection.

Filling procedures for long path cells are similar to those described for short pathlength cells. However, it should be realized that most long pathlength cells have an internal cell volume of 1 liter or more. Control of pressure is very important. At ambient pressures or below, many gaseous species provide a linear response to changes in concentration, up to approximately 1000 ppm, assuming that no gas-phase interactions occur between component species. Above this concentration, pressure effects can result in nonlinear response, which can be accommodated by suitable calibration procedures (accounting for temperature and pressure effects).

It is essential to ensure that no condensed material is formed on the internal optical surfaces prior to use. For some samples, it may be necessary to use the cell at elevated temperature to prevent condensation from occurring. If this is the case, it is important to be certain that the cell is constructed for operation at higher than ambient temperatures. This includes the materials used for construction (such as bonding materials for the internal optics) as well as the ability of the cell to retain optical throughput (due to dimensional changes).

4.4.1.3 *Compressed Pellets*

Compressed pellets or "KBr disks" are a standard procedure for the preparation of powdered or friable solids (Figure 4.4). The correct procedure requires careful attention to the experimental technique and, if possible, some understanding of the sample characteristics. The following are the guidelines for this technique:

- The sample is ground to a sub-micrometer particle size(less than 0.1 microns) in either a pestle and mortar (preferably made from agate) or a mechanized grinder.
- The ground sample is mixed uniformly with dry ground KBr or KCl powder.

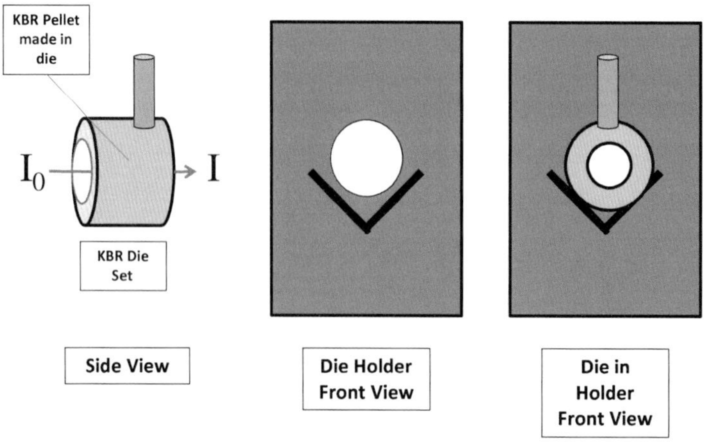

Fig. 4.4 KBr Pellet Cell.

- The mixture is placed in a clean, dry pressure die and compressed at a nominal 10-ton pressure for 5–10 minutes. A vacuum may be applied to the die to remove traces of moisture.
- Once pressed, the resultant pellet is placed carefully into a holder, which in turn may be placed in the instrument. Care must be taken not to touch the surface of the freshly pressed pellet. Be aware that the fresh pellet surfaces are very sensitive to moisture.

Note the following:

1. The need to grind the sample down to particle sizes of below 1 micron (μm) is important to reduce or eliminate the effects of light scattering (Figure 4.5). Time taken on the preparation of a pellet is usually time well spent. A poorly prepared pellet will usually provide low transmission and even a distorted spectrum (if the sample is not adequately ground).
2. Make sure that the KBr and KCl powder used is pure and is stored in a clean, dry container. Make sure that it does not come into contact with any plasticized polymer surfaces.
3. Be aware that some samples will react with KBr via a mechanism of halogen ion exchange. Also, note that some oxidizing agents may react with KBr, liberating free bromine — indicated by an orange appearance to the pellet.
4. A normal ratio of KBr to sample is approximately 300:1 — a good pellet is usually obtained from a mixture of 1 mg of sample with 250–350 mg of KBr.

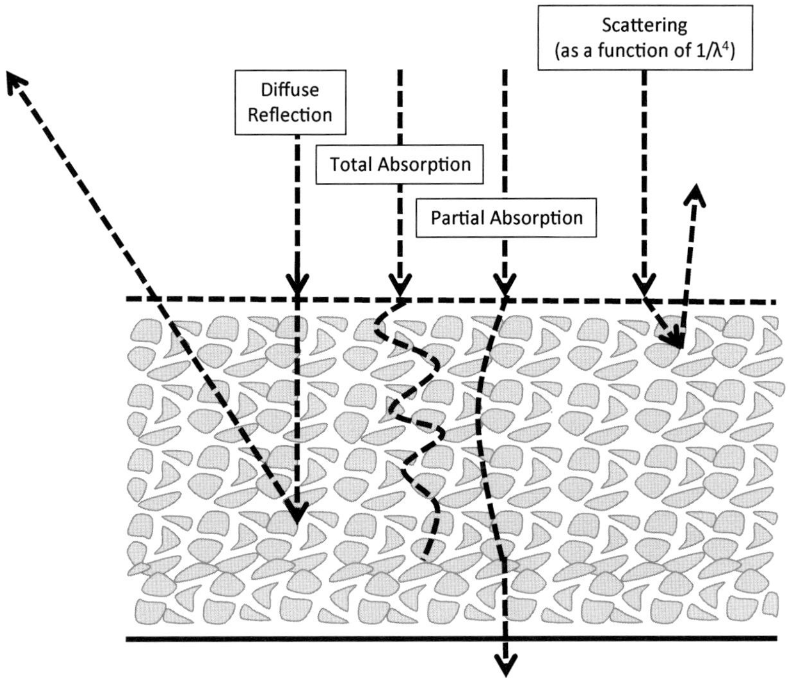

Fig. 4.5 The effect of scattering of light transmitted in solid sampling applications.

5. Trace quantities of sample can be handled from solution, where the solution is dropped on the KBr powder and the solvent is allowed to dry. Take care to prevent condensation from forming and ensure all traces of residual solvent are removed by heating (possibly under a heat lamp or in a vacuum oven).

4.4.1.4 *Oil-based Mulls*

The oil mull technique is a companion method to the alkali halide compressed pellet (Figure 4.6). In this procedure, the sample is ground, as previously described, to a submicrometer particle size (the previous issues regarding light scattering apply here). A few drops of mulling agent, and a medium viscosity oil, are added and the sample is reground/mixed to produce a paste with a consistency of petroleum jelly (Vaseline). This resultant paste is applied to a KBr window as a smear on a single window or as a capillary film between two windows. The normal mulling agent is a hydrocarbon white oil (light paraffin or medicinal white oil), also known by the trade name of Nujol. Another common mulling agent is Fluorolube, a perfluorinated hydrocarbon oil.

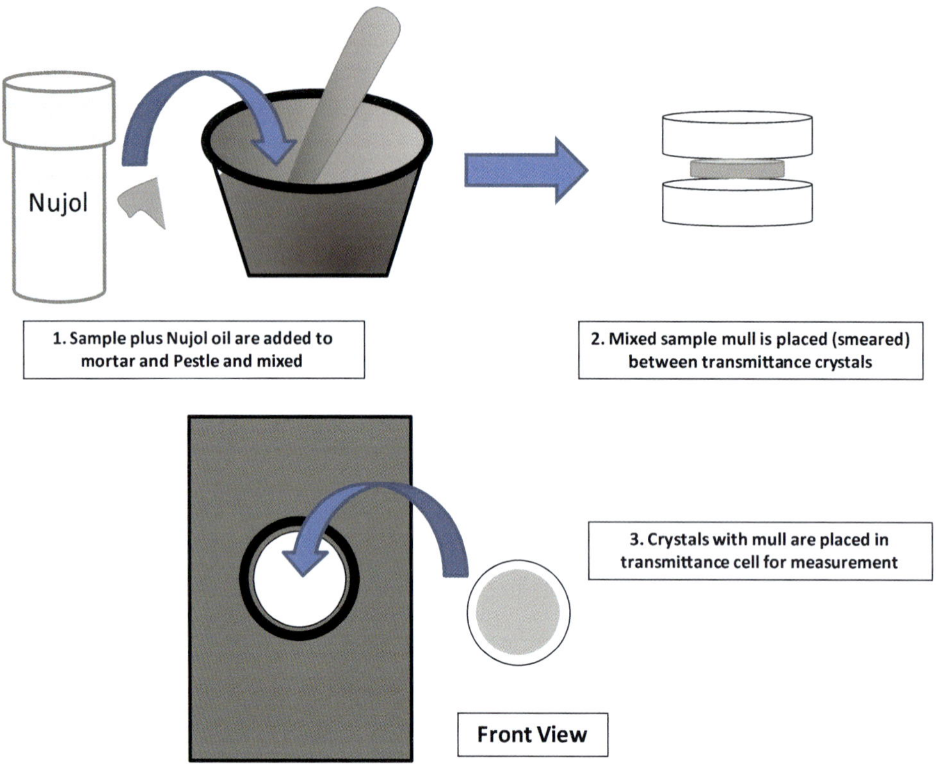

Fig. 4.6 Nujol Mull sample preparation overview.

4.4.2 *Self-Supporting Films*

Some samples exist in the form of film. Such samples can be analyzed directly by transmission, assuming that the film is not too thick (10–100 microns, dependent on the material) and that the material naturally transmits IR radiation (Figures 4.7 and 4.8). Regular, well-formed films can produce interference fringes, which are formed by internal reflections between the surfaces of the film. This effect may be reduced by roughening the surface of the film or by placing the film between a pair of infrared windows. Note the roughening will cause light scattering to occur with a subsequent loss in sample transmission.

Other samples, such as polymers in powder or pellet form, can be converted into a film. The normal methods of film production are either casting from solution or by hot-pressing between heated metal platens. Commercial accessories are available for producing hot-pressed films with a well-defined thickness. The thickness may be checked after preparation by use of a micrometer gauge. Special card mounts can

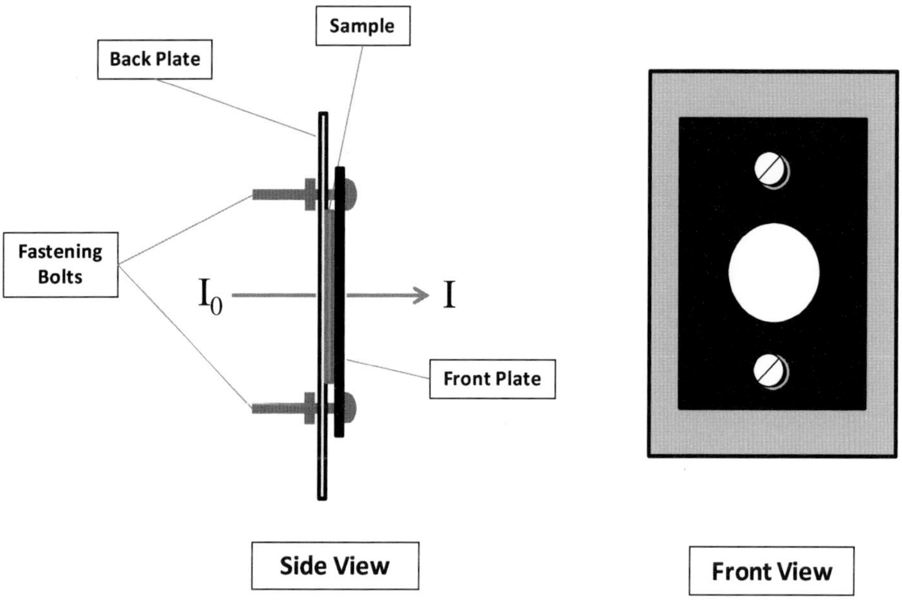

Fig. 4.7 Thin Film holder or thin pellet cell.

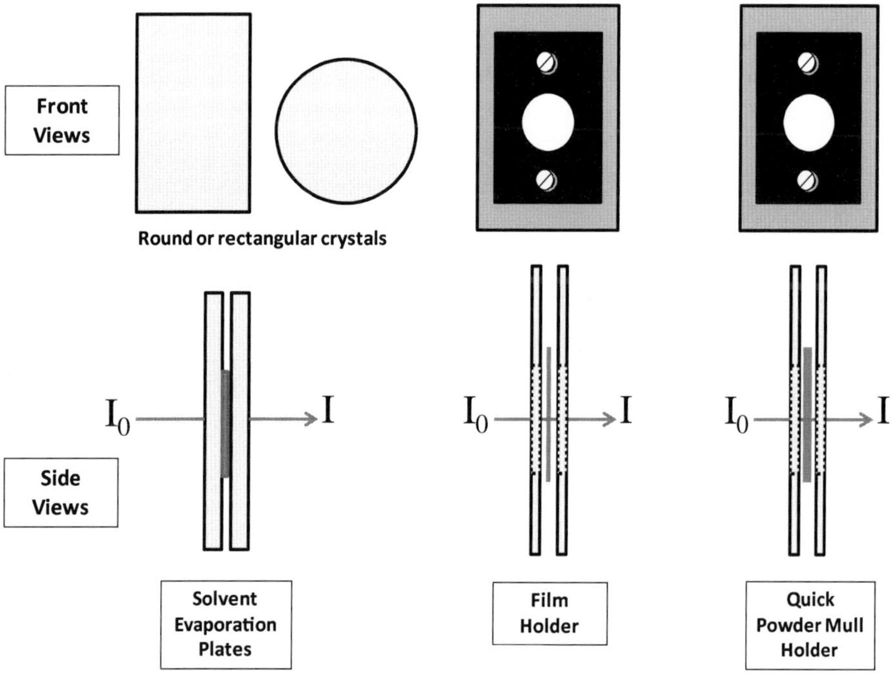

Fig. 4.8 A variety of thin film holders for transmittance measurements.

be purchased from accessory manufacturers for permanent mounting of prepared films. This is an ideal approach for sample documentation.

4.4.3 *Internal Reflectance or Attenuated Total Reflectance (ATR)*

There are many types of ATR accessories. The old-style versions are the fully mountable types, with the ATR element (IRE) mounted vertically (traditional). The more modern variants include the horizontal ATR and the cylindrical ATR (such as the Circle and Tunnel cells). The "traditional" accessory is suitable for studying continuous surfaces, such as sheets or small blocks of compliant materials, and today it has relatively limited application. The horizontal accessories are ideal for liquids and pastes (trough versions) as well as soft powders and sheet films (flat plate versions). If a powder or film is to be analyzed with a horizontal ATR, pressure must be applied. The accessories are normally supplied with an optical pressure applicator for this purpose — often equipped with a micrometer gauge for reproducible setting of the applied pressure. The cylindrical ATR accessories are normally limited to mobile liquids.

The internal reflectance technique has been available in several forms for many years.[18,19] It is also known as ATR and multiple internal reflectance or frustrated MIR (MIR/FMIR).

The measurement principle is based on the transmission of light through an optical element made from a high refractive index material. The angle of incidence used in the accessory is set up to generate internal reflection. The geometry of the element (known as an IRE) is normally designed to provide multiple bounces (reflections) on the internal faces (Figure 4.9). If an infrared-absorbing material is placed on the air — IRE interface at the point(s) of internal reflectance, an interaction occurs between the infrared light and the sample. This results in an attenuation of the internally reflected light as a function of the absorption of the sample, hence the name ATR. A spectrum similar in appearance to a transmission spectrum, except for intensity variations as a function of wavelength (λ), is produced.

ATR (attenuated total reflectance) is used for surface analysis using physical contact with a sample surface. The technique utilizes a crystal of material for the actual physical contact with the sample. Typical crystal materials include diamond (Di), germanium (Ge), silicon (Si), and zinc selenide (ZnSe). ATR measurements can yield excellent quality spectra provided that the contact pressures of the ATR crystal and the sample are held constant; reproducible data requires reproducible contact pressure. Commercial contact gauges are available. One percent reproducibility is typically achieved when keeping the contact pressure constant. ATR crystals are generally ZnSe with a refractive index of 2.42 and an angle of incidence of

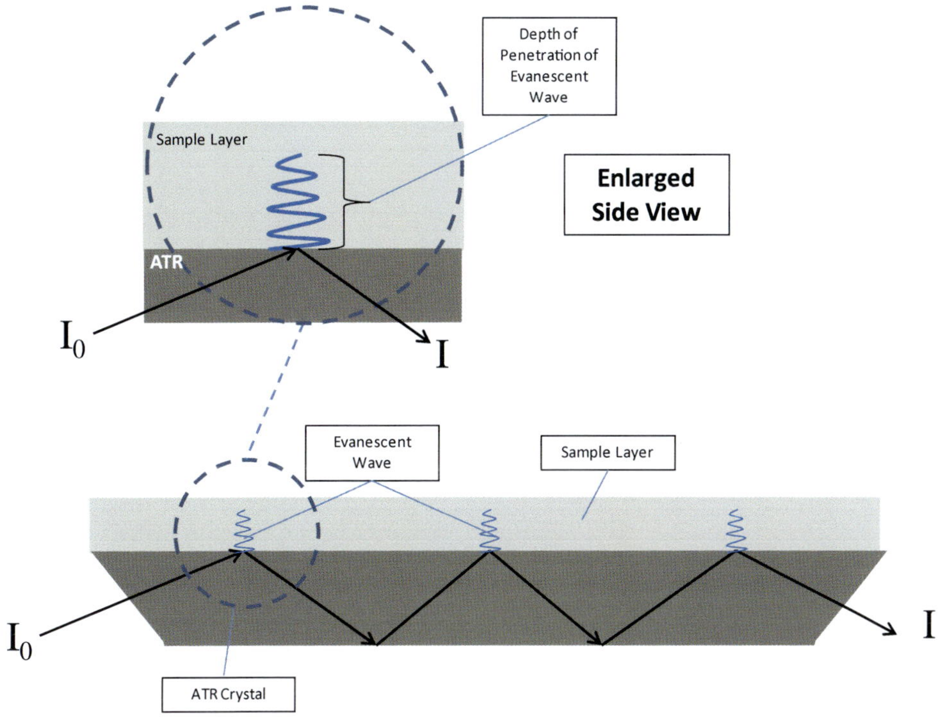

Fig. 4.9 A conceptual cross-sectional drawing of a multireflection IRE.

45 degrees. Sample penetration is to a maximum of 40 microns (μm) using this ATR crystal. An air background is taken for ATR measurements. The depth of penetration (D_p) is calculated for ATR crystals using the relationship:

$$D_p = \frac{\lambda}{2\pi \cdot n_1 \left(\left| \sin^2 \alpha - \left(\frac{n_2}{n_1} \right)^2 \right| \right)^{\frac{1}{2}}} \tag{4.1}$$

This equation is most often seen in the form of:

$$D_p = \frac{\lambda}{2\pi \left(\left| n_1^2 \sin^2 \alpha_1 - n_{21}^2 \right| \right)^{\frac{1}{2}}} \tag{4.2}$$

The critical angle for an ATR cell is computed as:

$$\alpha_c = \sin^{-1} \left(\frac{n_2}{n_1} \right) \tag{4.3}$$

Note: n_1 is the refractive index of the ATR crystal (i.e., IRE element); refractive indices for typical ATR crystals are: ZnSe (2.42), Di (2.2), Si (3.6), Ge (4.0); n_2 is

Table 4.2 Common Infrared Window Materials used for an ATR Internal Reflectance Element (IRE).

Material	Useful range (cm^{-1}, transmission)[a]	Refractive Index (at 1000 cm^{-1})	Water solubility (g/100 ml, H_2O)
Zinc sulfide (ZnS)	17,000–833	2.2	Insoluble
Zinc selenide (ZnSe)	20,000–460	2.4	Insoluble
Cadmium telluride (CdTe)	20,000–320	2.67	Insoluble
AMTIR ($Ge_{33}As_{12}Se_{55}$ Glass)[b]	11,000–625	2.5	Insoluble
KRS-5 (TlBr-T11)[c]	20,000–250	2.37	0.05
Germanium (Ge)	5,500–600	4.0	Insoluble
Silicon (Si)	8,300–660	3.4	Insoluble
Cubic zirconia (ZrO)	25,000–∼1,600	2.15	Insoluble
Diamond (C)	45,000–2,500 1,600–<200	2.4	Insoluble
Sapphire (Al_2O_3)	55,000–∼1,800	1.74	Insoluble

Note: Materials such as cubic zirconia, diamond, and sapphire may be used for transmission windows for special applications.
[a]The useable range is normally less than the transmission range because of the higher optical attenuation that results from the extended optical pathlength of an IRE.
[b]AMTIR: Infrared glass made from germanium, arsenic, and selenium.
[c]Eutectic mixture of thallium iodide/bromide.

the refractive index of the sample (e.g., water at 20°C is 1.333); n_{21} is the ratio of n_2/n_1; and α is the angle of incidence (and reflection) of the ATR crystal, typically 45 degrees. For ATR the depth of penetration computations (in microns) for 45 degree incident angle ATR crystals and water as the sample (at 20°C) are: $\lambda_m/12.5$ for diamond; $\lambda_m/14.0$ for ZnSe; $\lambda_m/21.8$ for Si; and $\lambda_m/24.4$ for Ge. Note that λ_m is in microns (μm) for these calculations. In a multiple internal reflection configuration, the effective pathlength of the sampling surface is the product of the number of reflections at the measurement surface and the depth of penetration. Also, refer to Table 4.2 for common IRE materials. Note for computations in the above equations: $\sin^2\alpha = [\sin(\alpha)]^2 = \sin(\alpha)^2$; and $\sin^{-1}(x) = \arcsin(x)$. Therefore, for a 45° angle: $\sin^{-1}(0.71) = \arcsin(0.71) = 45°$; and $\sin^2(45) = 0.50$.

It is essential that the sample is placed in intimate contact with the IRE. For compliant samples, pressure may be applied to ensure adequate contact. Air voids between the sample and IRE will reduce the intensity of the spectrum. Make sure that even pressure is applied across the sample; excess or uneven pressure can permanently damage the IRE.

Samples with hard, granular, or textured surfaces may not provide adequate contact with the IRE. Examples are certain fibrous materials, some fabrics, or noncompliant foams such as polyurethane foams. Such materials will often provide

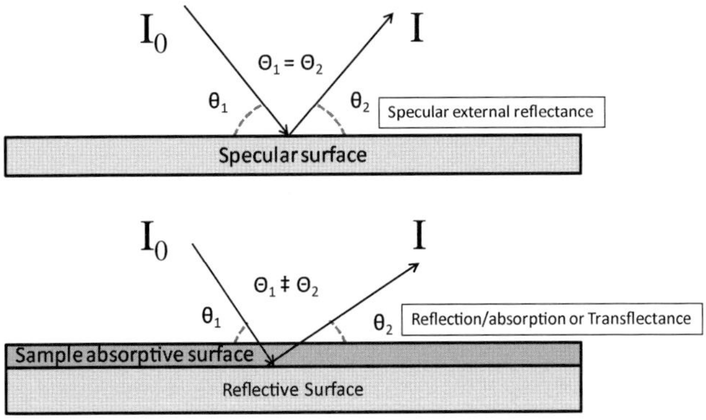

Fig. 4.10 External reflectance regimes.

a poor spectrum by the ATR method. One solution here is to utilize one of the diamond-based ATR systems. These accessories permit the use of significantly higher pressures, and under these conditions even relatively rigid samples will deform to provide surface contact with the diamond IRE.

Note: *It must be appreciated that the ATR technique is a surface phenomenon. This requires that good cleaning practices are used, and it is essential that no surface contaminants are present prior to use.*

4.4.3.1 *External or Specular Reflectance*

The external reflectance method provides a mirror-style reflection from the surface of the sample. It is used only for samples that have flat, reflective surfaces. The most common applications of external reflectance are for direct measurement for films, coatings or surface contaminants on metal surfaces. These measurements are sometimes called transflectance or reflection/absorption (Figure 4.10).

When studying a surface film by transflectance, the light passes through the material twice — to and from the reflective surface; the resultant spectrum is therefore at least twice the normal intensity of the normal transmission through the film (coating). Both the angle of incidence and the reflected angle from the surface are important primarily because they impact the intensity of the final spectrum. For thicker films, one typically uses an angle close to that of normal incidence. Conversely, for very thin films or surface contaminants, a grazing angle of incidence (as high as 80–85 from normal incidence) should be used.

External reflection can be applied to samples that have a reflective surface but are not coated on a metal surface. Such samples usually provide a distorted (mixed-mode) spectrum — a convolution of the refractive index and absorption spectra.

It is very difficult to provide a meaningful interpretation from such spectral data. A numerical operation, known as the Kramers — Kronig transformation,[16] can be performed on such data to provide a separation of the refractive index and absorbance components to yield a spectral representation equivalent to the absorbance form of the data. Note, however, that this is not always a simple operation, and complete separation may not be achievable.

4.4.4 *Diffuse Reflectance (DRIFTS)*

Diffuse reflectance (also known as DRIFTS or Diffuse Reflectance Infrared Fourier Transform Spectroscopy) in some ways can be considered to be a parallel technique to external reflectance. However, in general it is more versatile.[20] The method is used when a light-diffusing surface, such as a powder or a roughened surface, is to be studied. For the study of many types of powdered materials, it is possible for the sample to be analyzed directly. Although the technique has a wide range of applications in the mid IR, it is more widely used in NIR. Optical effects which primarily occur to the surface of the sample (Figure 4.11), can impact the quality of the final spectrum. These tend to limit some of the mid IR applications, especially for measurements involving the sample as received. In general, such limitations do not exist with NIR measurements. A conventional diffuse reflectance geometry is shown in Figure 4.12.

Note that particle size is important. Reducing the particle size by grinding will often improve the quality of the final spectrum. Also, a uniform particle size or a narrow particle size distribution is beneficial. For some organic solids, and most inorganic materials, it is necessary to dilute the sample with a non-absorbing matrix, such as KBr or KCl powder. This helps to reduce spectral distortions that originate from anomalous surface reflections (Figure 4.11). For organics, the dilution should be approximately 10:1 KBr to sample and for inorganic materials, dilution ratios as high as 100:1 may be necessary. Mixing with a non-absorbing matrix is recommended if a distorted or a poor-contrast spectrum is obtained from the sample. It is often beneficial to experiment with the measurement, with or without dilution to ascertain the best method of sample preparation.

Nonpowdered samples can be studied directly from the sample surface as long as the surface is irregular (granular) and not polished. In some cases, a thin film of KBr powder placed on the surface of the sample may help to improve the final spectrum quality; this works well with fibers, fabrics, and paper samples. In this case, the KBr film reduces the incidence of unwanted surface reflections and helps ensure better coupling between the sample and the infrared beam by promoting penetration beyond the surface.

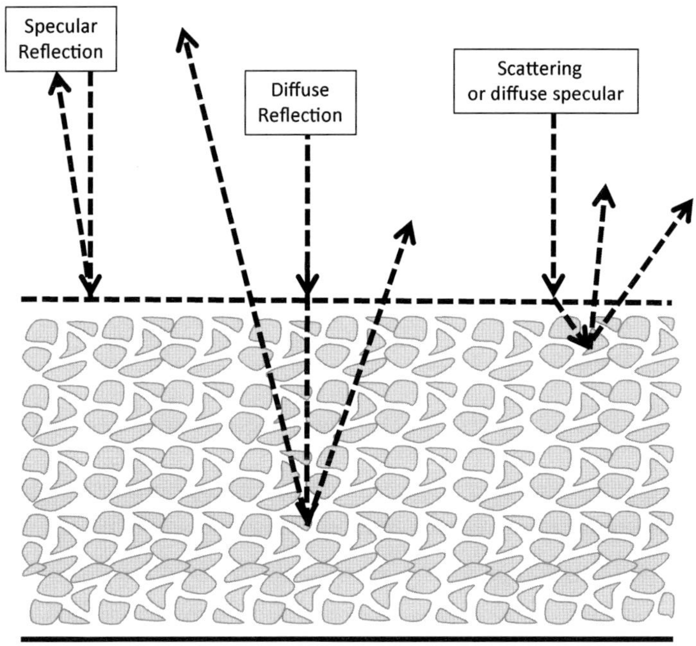

Fig. 4.11 Reflection modes encountered with diffuse reflectance.

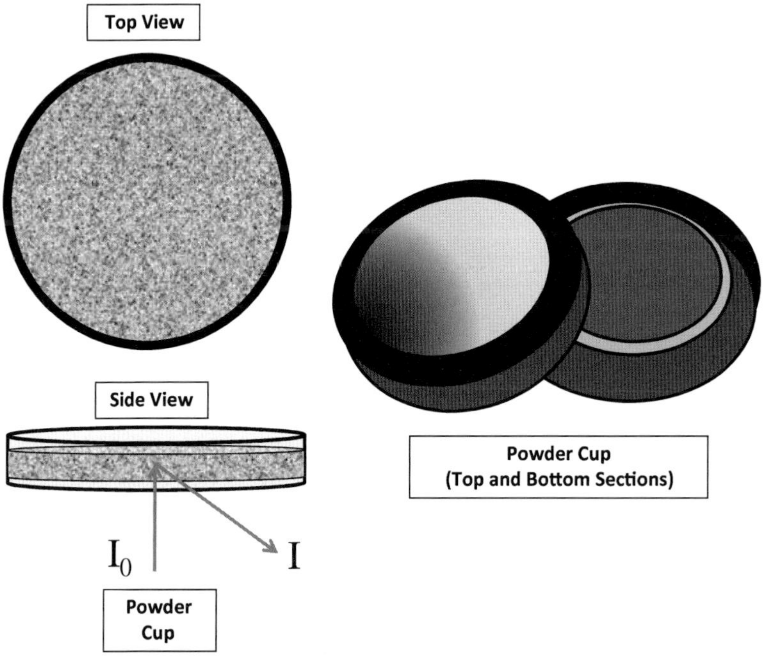

Fig. 4.12 Diffuse reflectance cell for powder samples.

The spectrum produced from a diffuse reflectance measurement does not bear a direct numerical relationship between peak intensity and composition, in contrast to traditional transmission measurements in which the absorbance form of the data is directly proportional to concentration. With diffuse reflectance, spectral distortions are generated by a constantly varying effective pathlength, which is defined by the depth of penetration of the beam into the sample. This is dependent on the absorptivity of the sample at a particular wavelength (frequency) and the measured intensity throughout the spectrum varies accordingly. A correction that can be applied to the diffuse reflectance spectrum, which can help linearized the data, is the Kubelka-Munk function:

$$f(R_\infty) = \frac{(1 - R_\infty)^2}{2R_\infty} \tag{4.4}$$

where $f\ (R_\infty)$ is the corrected spectrum and R_∞ is the ratio of the sample diffuse reflectance spectrum and a nonabsorbent reference (normally KBr or KCl), both measured at infinite depth. In the mid IR region, infinite depth corresponds to approximately 3 mm.

The consequence of the $1 - R_\infty$ is to invert the spectrum and provide an absorbance-like format. It will be noted from the previous expressions that this corrected function can be correlated to concentration. Note that this function is normally only applied to undiluted (or concentrated) powdered samples.

Diffuse reflectance can be used as a trace analysis method. In this case, the material to be studied is applied as a small quantity of a solution in a volatile solvent to a powdered KBr matrix. The solvent is evaporated and the sample is allowed to coat the KBr crystals. The normal sampling cup dimensions are approximately 13 mm in diameter $\times$3 mm deep; however, smaller diameter cups are available with diameters of 1.5–2 mm. These are closely matched to the typical instrument beam diameter after imaging with the diffuse reflectance accessory primary focusing mirror (typically a 6:1 off-axis ellipsoid). With this microsampling extension, it is possible to obtain good spectra from sample loadings as low as a few 100 μg.

One novel approach to the measurement of difficult solid samples is to use silicon carbide abrasive paper. The silicon carbide surface provides an irregular reflective surface that scatters all incident radiation. If the sample to be analyzed is rubbed against the abrasive surface, a small amount of sample is transferred to the surface. In reality, the rubbing action distributes the sample as a thin film across the irregular, reflective surface. The sample, in turn, is measured directly on the surface by the use of a diffuse reflectance accessory; note that this is really a transflectance measurement (measurement of a coating on a roughened surface) and not a classical diffuse reflectance measurement. Any spectral contributions of the silicon carbide can be removed by a ratio of the spectrum of a clean portion of the abrasive paper.

4.4.5 *Photoacoustic Spectroscopy (PAS)*

Analysis with photoacoustic detection is the only method that involves the direct measurement of the absorption of the infrared radiation. The technique involves the conversion of the infrared energy into thermal energy (within the sample), followed by thermal energy transfer from the sample phase into a gas phase (immediately above the sample) using a solid PA cell (Figure 4.13). In the case of gaseous samples the transfer of infrared to thermal energy occurs directly in a PA gas cell (Figure 4.14). The efficiency of the process is linked to the absorptivity of the sample and to the thermal conductivities of the sample and the gas phases. High sample absorption combined with good thermal conductivities (both phases) provide a situation conducive to effective photoacoustic measurements. Note that because absorption is the primary process, carbon black, a universal absorber, is used as a reference material.

The technique is sometimes referred to as the method of last resort. In practice, it is a versatile technique and is ideal for difficult samples, such as heavily filled polymers, coatings on metal wires or metal powders, or samples generally described as "intractable".[21,22] The sample is placed in a small chamber, which is sealed at the time of measurement. Prior to sealing, the sample and chamber are purged with a

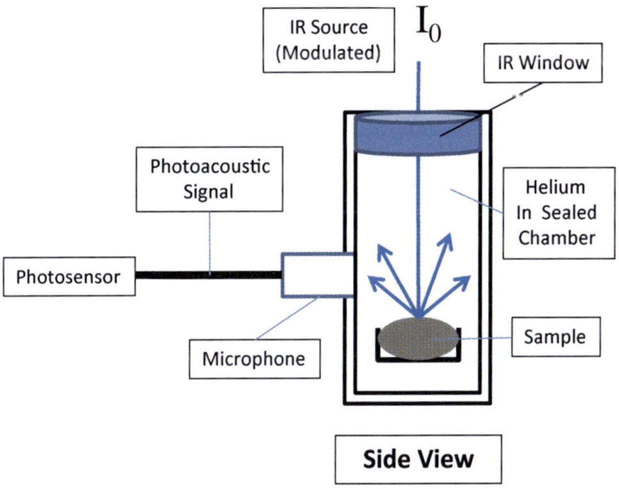

Fig. 4.13 Photoacoustic solid cell configuration.

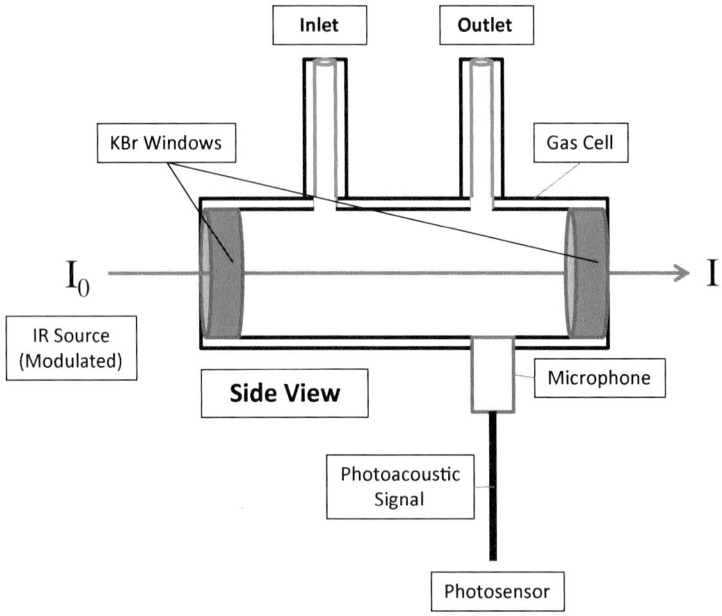

Fig. 4.14 Photoacoustic Gas Cell Configuration.

clean dry gas — dry nitrogen or, preferably, helium or argon (both gases have better thermal conductivities than nitrogen).

It is always advisable to adhere closely to the manufacturer's instructions, especially in regard to the use of a dry purge gas. Failure to do this will result in data dominated by the spectrum of water vapor. Photoacoustic measurements are based on thermal processes, which are intrinsically slow, and in order to maximize sensitivity, slow interferometer scan rates are recommended for best performance. Note that the sample size is normally limited to a few milligrams. Always document the purge gas that is used and the scan speed of the instrument because these impact the intensity and quality of the final spectrum.

4.4.6 *Microtransmission or Microtransmittance Cell/Pellets*

In cases in which insufficient sample is available for the traditional sampling procedures, a micro approach may be adopted. Microversions of compressed halide pellets (down to 0.5 mm), transmission cells, and reflectance accessories can be used. Dependent on the method and the sample type, the sampling size can be reduced to a few micrograms or even a few hundred nanograms, for solids and approximately 1 mg (10^{-3} mL) for liquids. Reducing the physical sample size reduces the sample weight required for a given spectrum intensity; the ground rules are defined by the constraints of the Beer — Lambert law, which establishes the basic relationship in terms of units of weight/area.

In concept, the smaller the sample size, the lower the amount of material required (a logical conclusion); however, the trade is optical throughput. Traditional instruments are limited by the image size of the infrared beam at the sampling point, which is typically between 6 and 8 mm in diameter. An improved matching between sample size and beam size can be achieved by the use of a beam condenser accessory, providing beam size reductions on the order of 4:1 or 6:1 times. With such accessories it is possible to handle physical sample sizes in the range of 0.5–0.25 mm in diameter. For sample sizes smaller than this, it is often necessary to use an infrared microscope.

When working with any of the microsampling methods. Cleanliness is of the utmost importance. Traces of impurities picked up during sample preparation can have higher concentrations and therefore greater impact on the final spectrum, than the sample itself. This attention to cleanliness must extend to the purity of any solvent used during sample preparation and also preparation materials, such as KBr, if micropellets are to be prepared.

4.4.7 *Diamond-based Accessories*

Diamond is considered to be an ideal infrared sampling optic for several reasons: its high mechanical strength (resistance to abrasion), its chemical inertness, and its reasonable optical transmission over most of the mid IR spectral region (with the exception of a small area in the region of 2000 cm^{-1}). Several commercial accessories are available featuring diamond optics; these include diamond compression cells (for high-pressure transmission measurements) and diamond ATR sampling devices in both horizontal and cylindrical ATR formats.

4.4.7.1 *Compression Cell: Diamond Anvil Cell*

Large-scale diamond (high pressure) cells have been available for some time however, during recent years, a microversion of the cell, also known as the diamond anvil cell, has been produced (Figure 4.15). This small diamond anvil cell is ideal for the study of physically hard (but compliant) samples, certain intractable samples, and samples that are optically too thick for normal transmission measurements. Note that this small form of the accessory is intended for samples 1 mm or less in size. The aperture through the diamonds is small, and therefore the use of a beam condenser accessory is recommended for optimal throughput. Alternatively, the cell may be used in combination with an IR microscope.

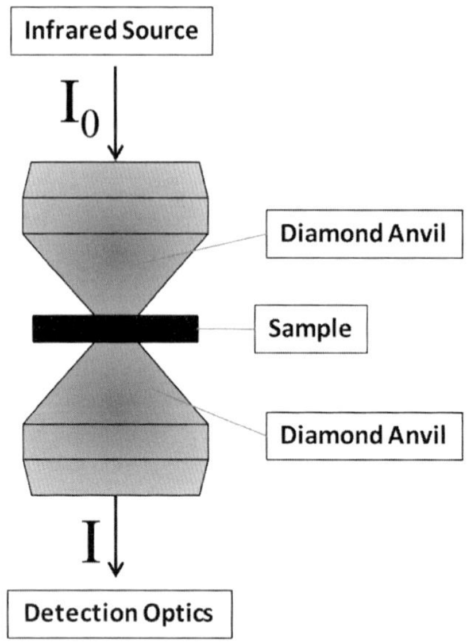

Fig. 4.15 High Pressure Diamond Anvil Cell.

4.4.7.2 *Diamond ATR*

Several diamond ATR-based accessories are available. A small cylindrical ATR, featuring a Diamond IRE, used in flow cell or immersion probe geometries, provides an ideal tool for the continuous monitoring of chemically reactive media. An alternative configuration, based on a small diamond sensing surface combined with a zinc selenide support optic, is available in both insertion probe and horizontal ATR formats (Figure 4.16). The probe formats are ideal for industrial process monitoring applications. The horizontal ATR format provides a more or less "universal" sampling platform for both liquids and solids. Its benefits over traditional horizontal ATR accessories include the strength of the diamond, for handling hard noncompliant surfaces; and the chemical inertness, for handling corrosive media such as strong acid and strong alkaline solutions.

4.4.8 *Infrared Microscopes*

Infrared microscopes (Figure 4.17) have become popular accessories during the past decade. These accessories are extremely versatile and can be used for a wide range of samples. Their applications extend far beyond the discussion in this chapter, and the reader is directed to reference texts dedicated to the subject.[23,24] In terms of optimum use and operation, users are advised to follow the manufacturer's recommended procedures. It is important to always use matching apertures for

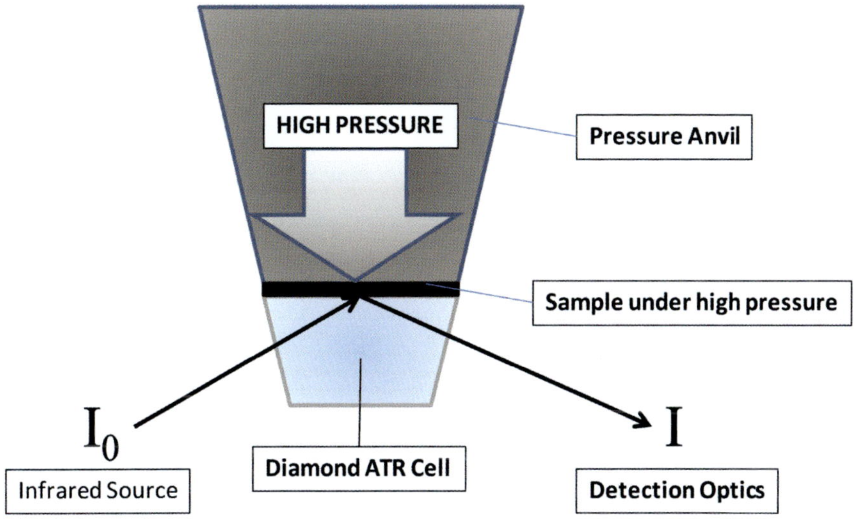

Fig. 4.16 Diamond ATR cell.

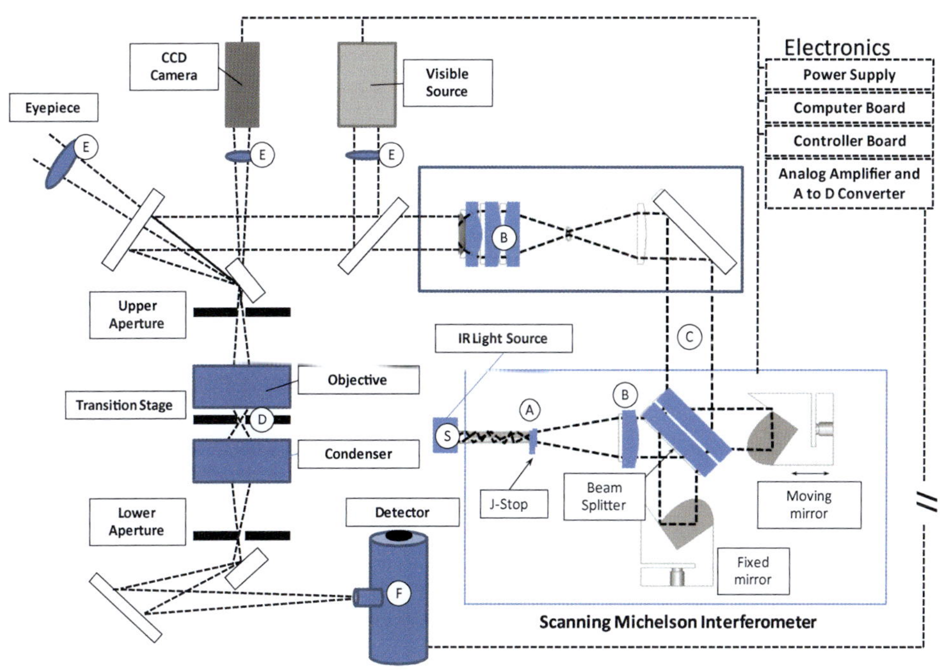

Fig. 4.17 Schematic of infrared microscope, Fourier transform (FT) spectrophotometer (interferometer) based microscope optical configuration. S = Source optics, A = J-stop (aperture), B = Collimating optics, C = Optical train, D = Sample, E = Focusing optics, F = IR Detector.

sample and background scans and always document the aperture size used. Also, most commercial microscopes may be used in transmission and reflection modes and with or without infrared (and visible) polarizers. A broad range of new applications have been developed around the imaging ability of the modern microscopes in combination with computer video graphics and computer controlled sampling stages.

4.4.9 *Gas Chromatography-Infrared (GC-IR)*

The combination of gas chromatography and infrared is ideal for the analysis of complex mixtures.[25,26] In one standard configuration, the eluted components are transferred, down heated transfer lines, to a long pathlength heated lightpipe. The lightpipe, which may be approximately between 10 and 20 cm long, has a small internal diameter (1 or 2 mm) and is internally coated with gold. Issues to consider are the need to minimize the opportunity for thermal degradation of the eluted components while maintaining a continuously heated path through the interface between the GC and the IR instrument and ensuring that there is minimal dead volume in the transfer lines and in the lightpipe (and its connections). Things to remember and document are the type of column used for the separation, the temperatures used throughout (injector block, column, and lightpipe cell), the number of spectra co-added around a separated GC peak, and the solvent used for injection.

Note the following:

1. Make sure that there is no column bleed or that any bleed is compensated.
2. Remember that injectors and columns are usually very hot. For some thermally unstable compounds, these can appear as highly reactive surfaces. As a result, what goes on the column may be different from what elutes; be prepared for surprises — pyrolysis and thermal decomposition can occur.
3. Co-elution can occur, and therefore the final coadded spectrum may still be a mixture. Evaluate possible differences between the spectra obtained from the leading and trailing edges of the eluded GC peak.

4.4.10 *Liquid Chromatography-Infrared (LC-IR)*

In recent years, a new focus has been applied to the combination of liquid chromatography (LC) and FTIR. This is of particular interest for the analysis of polymeric materials (for molecular weight studies), pharmaceutical products, and natural and biological products. The main problem has always been how to record spectrum of the eluded component, which is at relatively low concentration, in the presence of a high concentration of solvent. This is particularly a problem with reversed-phase separations involving binary (e.g., water–methanol) and ternary (e.g., water–methanol–acetonitrile) solvent systems. New accessories have been developed for off-line measurement of eluded fractions in which the components are deposited on the surface of a rotating reflective plate. The deposited components are measured in an automated specular reflectance unit to produce a series of spectra, which are processed to generate a reconstructed chromatogram.

One variant, close to LC, that has been successfully interfaced to infrared spectroscopy is supercritical fluid chromatography (SFC). In this case, one of the key mobile phases, super critical carbon dioxide, is virtually transparent for most of the mid IR region. This may be operated directly on-line with a high-pressure liquid flow cell.

4.4.11 *Thermogravimetric Analysis-Infrared (TGA-IR)*

Thermogravimetric analysis is a standard laboratory procedure for studying the thermal degradation characteristics of materials in oxidative, inert, and reducing environments. The technique may be coupled on-line to an infrared instrument to permit monitoring of the gaseous degradation products. This is an important combination for evaluating thermal degradation, oxidation mechanisms, and pyrolysis mechanisms for polymers. After a thermal cycle, there is always the opportunity to analyze the residue (ash) to determine the presence of inorganic compounds in composite materials.

4.4.12 *Miscellaneous Infrared Sampling Techniques*

There are a plethora of sampling methods, and this chapter has presented by no means an exhaustive listing. If a unique sampling procedure is used, make sure that it has been thoroughly documented and commented on. Note, in particular, any unique sample treatments and always record the instrument operating parameters.

4.5 Sample Modification and/or Purification Methods

Where possible, use a sampling method that does not involve any sample pretreatment or modification; that is, analyze the sample "as is." It is recognized that this is not always possible. For routine sampling, some of the newer diamond-based accessories allow many different sample types to be analyzed in an original, native form. However, it is recognized that not all laboratories have this facility, therefore, sample pretreatment may be the only answer. A few of the common sample pretreatments are noted in the following Sections. If any of these procedures are used, it is important to ensure that a comment is provided relative to the use of the method within the sample documents. Failure to do this can lead to erroneous results or misleading answers.

4.5.1 *Solution-based Measurements*

In the past, solution-based measurements were popular mainly because sampling was limited to the simple procedures involving transmission measurements. One possible benefit is that both solids and liquids may be studied in solution. In this case, the optical effects caused by differences between these two types of sample phase are removed. The problem has always been the selection of an appropriate solvent. All liquids have an infrared spectrum, and almost all liquids have relatively intense and complex spectra. There are few materials that have simple spectra that have clear "windows" for broad-range transmission measurements. As a general comment, the materials that are the best solvents, particularly for polar compounds, by nature are strong infrared absorbers. Therefore, it is very difficult to find convenient solvents that have good solvent characteristics, that are relatively involatile, and that are convenient to handle (e.g., are nontoxic).

Favorite materials in the past were carbon tetrachloride, chloroform, and carbon disulfide. Carbon tetrachloride and carbon disulfide are only modest solvents, in terms of solvent power, but have very simple spectra. Also, conveniently, the regions of major absorption of these two compounds are very different, and by use of a two-solvent strategy it is possible to construct a single-solution spectrum from two fragment spectra, recorded individually from each solvent. Unfortunately, these materials are banned from laboratory use in many countries, and so their use as IR solvents has almost ceased.

Chloroform is a good solvent, and its use has been constrained but not necessarily banned in all countries. It is important to be aware that commercial grades of chloroform contain stabilizers (to inhibit reactive carbine formation); Low-molecular-weight olefins or ethanol are typically used for this purpose. Other halogenated solvents, such as methylene chloride (dichloromethane), perchloroethylene, and certain halothanes (freons or perchlorofluorocarbons), can be used as infrared solvents.[27] However, although the freons are relatively innocuous and do not pose a safety hazard, their use is regulated in North America and some other countries on environmental grounds. Even so, standard methods, particularly environmental procedures for extraction and enrichment, are still in use based on these solvents. When using volatile halogen-based solvents, it is important to ensure that the cells do not leak and that the instrument is protected either by an active purge or by environmental seals. Halogenated solvents decompose in the region of the hot infrared source, and failure to protect the instrument can result in excessive corrosion of exposed metal components within the instrument, including aluminum-coated optical surfaces.

Hydrocarbons, such as hexane, heptane, and isooctane, may be used, but they have reduced solvent powers and are typically limited in use for the dissolution of nonpolar materials (the old rule of like-dissolves-like applies here). More polar solvents, such as alcohols (methanol, ethanol, isopropylalcohol, etc.), dioxane, tetrahydrofuran, sulfolane, and dimethyl formamide, may be used for special applications, but it must be realized that the spectra of these solvents are very intense and often quite complex. As a result, there are limited windows of transparency at which measurements may be made. A number of excellent spectral collections have been published that feature solvent spectra,[27–31] and it is recommended that the reader consult one of these texts before proceeding with a solvent-based method.

One important solvent that is often overlooked is water. In the past, this has been considered an inappropriate sampling medium. This was primarily because of the high intensity of the water absorptions and lack of suitable sampling devices. However, with modern ATR cells, designed for liquid applications, it is practical to perform analyses on aqueous solutions. With careful control of temperature, it is possible to determine materials in water solutions at fraction percentage concentrations.

Two important issues for the measurement of materials in solution that must be taken into consideration are temperature and solvent — solute interactions. The impact of temperature was implied previously. One of the main reasons for considering thermal control when measuring solutions is that it is common practice to subtract (in absorbance) or ratio out the spectral components of the solvent from the solution spectrum. This exercise will be successful only if the temperatures of both the solution and the solvent are identical, or very similar, when the spectra are recorded. Wide differences in the temperatures will result in significant artifacts in the solvent-subtracted spectra due to spectral mismatch of the solvent components.

Solvent — solute interactions are also important, and it is important to realize that the spectrum of a solute in solution may appear to be very different from the spectrum of the solute in its free state. This especially applies to the spectra of solids recorded from solutions. Not only can the spectra change but also the bandwidths and band positions can change as a result of mutual interaction between solute and solvent. This will apply to certain absorptions of the solvent spectra, especially with polar compounds, and in particular at high solute concentrations. Be aware that solutes such as certain inorganic salts may not have a significant infrared spectrum (such as sodium chloride in brine or saline solutions) but the presence of the salt will impact the spectrum. This is the result of ion solvation, which perturbs the spectrum of the solvent (water in the case of salt), and this is quite significant at high concentrations.

4.5.2 *Sample Drying or Solvent Removal*

For some applications in which samples are in solution, or are "wet", it is desirable to remove the volatile solvent components prior to analysis. In the case of organic solvents, it is common to evaporate the solvent by heating either on a steam bath or in a rotary film evaporator. In such situations, the final traces of volatile material are typically removed in a vacuum oven. Caution: Make sure that oil traps are used on all vacuum lines; pump oil vapors from a vacuum pump will contaminate samples in a vacuum oven. In cases in which a thermally sensitive material must be removed from an aqueous solution, freeze-drying is a practical alternative for isolating the sample if suitable equipment is available.

A special case of drying is the removal of water from a wet sample. This may be important for two reasons: (i) to remove a major spectral interference caused by the presence of the water and (ii) to protect any moisture sensitive components within the instrument that may be used in the sampling accessory (the cell) or in extreme conditions. For organic liquids, it is recommended that a solid inorganic drying agent such as anhydrous sodium sulfate be used. A practical alternative is to use a 3A molecular sieve, but this normally requires that the liquid be stored over the drying agent for a long period.

Powdered solids may be dried in a desiccator or a vacuum oven. There are a wide variety of desiccants in common use. Materials such as phosphorous pentoxide may be used in cases in which an aggressive drying agent is required. Note that samples removed from a vacuum oven must ideally be allowed to cool in a desiccator prior to analysis; otherwise, moisture can be absorbed while the sample is cooling to room temperature.

If a sample, solid or liquid, is known to be a mixture, solvent extraction may be used to selectively remove or extract a material. Solvent extraction is also a convenient method to enrich a particular component — either in the extracted phase or in the residual phase (what is left behind). Many of the comments provided earlier apply here. Obviously, it is necessary to find a solvent that has a high affinity for the component(s) to be extracted and a low affinity for the remaining material. It is not a good idea to select a solvent that is midway in its ability to extract an ingredient. If affinity exists for both phases, then there is a risk of establishing a stable emulsion.

Obviously, it is not always possible to overcome this problem, especially if the system contains surfactant materials. In such cases, in which aqueous systems are involved, the addition of a polar ionic substance, such as salt or sodium sulfate, may be necessary to help break (salt-out) an emulsion. Once a material has been extracted, it may be measured directly either in the extraction medium or on the isolated material after removing the extraction solvent. If the solvent is removed, make sure that all traces of solvent are eliminated prior to analysis, and always document what solvents have been used.

Examples of the practical use of solvent extraction are the use of water to remove water-soluble components from organic mixtures (the remaining organic phase is analyzed); the use of methanol with mineral oils or polymers to remove polar additives; the use of compound-selective solvents on powdered mixtures; and the use of Freon for the extraction and measurement of hydrocarbons and organics from soil and environmental water samples. In this last example, standard methods exist for the quantitative estimation of the extracted materials.

4.5.3 *Partial Evaporation or Distillation*

If an organic solution or mixture of solvents is known to contain components of different volatility, partial evaporation or selective distillation can provide preferential enrichment of one or more components. The collection of spectra at different stages of the evaporation process can provide an insight into the nature of the key components. Careful use of spectral subtraction can help in the numerical extraction of chemical components from the different fractions. Successive subtractions may be used, with moderate success, in multicomponent situations. The main obstacle in this case is spectral variations due to mutual interaction of components in solution. Obviously, if nonvolatile components are present, this approach provides a practical separation of some of the main components — by complete evaporation.

Normally, the analyses mentioned previously are performed on the residual materials. However, there is nothing to prevent measurements being made on the separated fractions or on condensate materials. In the case of samples containing water, be aware that the condensate will be enriched with water. An interesting approach to the controlled evaporation of materials is to use the combined techniques of thermogravimetric analysis and infrared. Although this technique is typically used for the determination of thermal decomposition products, it may also be used for controlled and reproducible selective removal of volatile components, with on-line monitoring of the volatile components.

4.5.4 *Dialysis*

Dialysis is not a technique widely used in analytical laboratories. However, it is a very convenient and useful method for separating mixtures of low and high-molecular-weight (e.g., polymers) compounds. It is used in the oil industry for the separation of polymeric additives from base oil components and low-molecular-weight additive in blended lubricants. For this separation, the blended lubricant is placed within a latex rubber membrane. The membrane is placed within a Soxhlet extractor, and hexane or a light petroleum distillate is used as the extraction solvent. After several hours of reflux, the higher-molecular-weight materials are retained in the membrane as a hexane solution. The application is not limited to lubricating oils and may be applied to any polymer-based system in which materials may be extracted by a hydrocarbon-based solvent on the basis of molecular weight. Note that the use of a hydrocarbon solvent is required to swell the membrane and to generate the required molecular porosity. Other membranes may be available for other chemical systems and for the use of other solvents.

4.5.5 *Chromatography: Adsorption and Ion Exchange*

The use of instrumental chromatographic methods was described earlier in the role of hyphenated or hybridized techniques when combined with infrared spectroscopy. In this case, the chromatographic front-end acts as a sophisticated sample preparation system for isolation of specific chemical species in a mixture. In cases in which a particular component needs to be removed or separated, it is not necessary to resort to an instrumental method. In such cases, the sample may be passed through a simple column containing the solid-separation phase. A convenient approach for liquids is to prepare small columns of adsorbent or ion-exchange resin in a Pasteur or dropper pipette. This is ideal as a method for sample cleanup or for selectively removing contaminants or specific chemical components.

In such cases, the eluted material can be analyzed directly and can be compared to the sample prior to elution. If a specific additive or chemical compound is removed by the solid-phase material, then its spectrum may be generated by subtraction of the spectra recorded before and after the column separation. Alternatively, the separated component may be eluted from the column by the use of an appropriate solvent.

This approach is not limited to liquids, and solid-phase adsorbents may be used for selective removal of components from gas streams. In cases in which molecular sieve compounds are used, the separated components may be studied following thermal desorption of the adsorbed component(s). A variant of this approach can be used for the analysis of particulates or dust in gas streams (or in the environment). In this case, the sample is drawn through a suitable membrane. Membranes made from materials such as PVC have sufficiently good infrared transmission so that separated components may be measured directly on the filter material. Membrane separations may also be applied to liquid systems. One convenient approach is to use a porous silver membrane and to measure any separated components (particulates or other insoluble matter) retained on the surface of the membrane directly via a reflection measurement.

4.5.6 *Pyrolysis-based Measurements*

The use of TGA as a controlled technique for monitoring the thermal breakdown of materials was mentioned earlier. A crude version of this approach is pyrolysis, in which the sample is heated in a Pyrex tube and the volatile thermal degradation products are collected as a condensate (also known as the pyrolysate) and analyzed. This is ideal for studying polymers, especially physically hard or heavily filled polymers, which are difficult to handle by the normal sample preparation procedures. It can be helpful to study both the pyrolysate (the condensate) and the pyrolysis residue. In this case, the residue can provide information on filler materials. Note that not all materials will form a condensate, and so it is also beneficial to analyze all gaseous components from the thermal degradation. Examples of materials that do not appear in the condensate are hydrogen chloride (from PVC), hydrogen cyanide (from urethanes and other nitrogenous polymers), and hydrogen sulfide (from sulfur-based polymers).

4.5.7 *Abrasion Sample Preparation*

Very hard surfaces can be studied by the abrasion of the surface with a sheet of abrasive material, such as silicon carbide or carborundum paper. At this point, a number of different methods may be used to analyze the abraded material. In essence, any solid sampling technique that is capable of handling fine powders — KBr pellet, diffuse reflectance, ATR, photoacoustic, etc. — may be used to study the material. An interesting variant is to use diffuse reflectance to study the abrasive for the residual material.

4.6 Recommended Spectrum Acquisition, Presentation Format and Data Manipulation Procedures

4.6.1 *Recommended Spectrum Acquisition Parameters*

It is difficult to be specific about the actual instrumental parameters for spectrum acquisition because these are dependent on both the instrument type and the method of sample preparation. The following are guidelines for average situations.

Normal sample, modern FTIR spectrometer: 1–4 minute scan time, with spectral resolution of 2–4 cm^{-1}, providing a signal-to-noise ratio (S/N) of between 1000:1 and 10,000:1. If a spectrum is to be used for qualitative purposes only, then a S/N of approximately 1000:1 to 2000:1 is normally adequate and is typically obtained in well under 1 minute; often within a few seconds. For quantitative applications, or applications involving spectral subtraction or resolution enhancement techniques, higher signal-to-noise performance is required. Note that scanning for more than 256 scans may approach a point diminishing returns in terms of spectral quality and S/N improvement (limited by the square root of n law, and the performance of the interferometer). Note that for qualitative work, 8 cm^{-1} resolution is often sufficient, and this is accompanied by a two-fold gain in S/N (from 4 cm^{-1}).

Normal sample, older- style dispersive spectrometer: 3- to 10-min scan, with a slit program (integrated scan mode) equivalent to 3–5 cm^{-1} (average value), providing S/N of 500:1 or better.

Low transmission sample (microsample or optically opaque material): 2- to 8-min scan, with a spectral resolution of 4–8 cm^{-1}, providing a S/N of 500:1 or better. The use of a MCT detector is suggested for samples with an overall transmission of 5%T or less, especially in the absence of an optimizing accessory (such as a beam condenser for microsamples). When using a MCT detector, it is important that the potential for nonlinearity is appreciated. Also, it is important to be aware of the occurrence of detector saturation, which will lead to extreme nonlinearity, and spectral distortion.

Irrespective of the final selection of data acquisition parameters, full documentation of the experimental details should be recorded with all spectra, especially if regulated procedures are followed. Also, as a general rule, it is recommended that good laboratory practice guidelines are adopted throughout.

4.6.2 *Recommended Spectral Range*

Unless special circumstances dictate otherwise, all mid infrared spectra are normally recorded within the traditional mid infrared spectral range, nominally defined as the region extending from 4000 cm^{-1} (2.5 microns, 2500 nm) to a minimum of 400 cm^{-1} (25 microns, 25,000 nm). It is expected that if special windows are used, or a restricted range detector is employed, the spectral range may be restricted to a region less than the nominal 4000-400 cm^{-1}. With a modern FTIR instrument, it is normal that the entire spectral range is obtained, and for some instruments the upper limit may be set beyond the traditional limit. The exceptions would be if a detector or a component optic constrains the spectral range or if special filter method (optical or digital) is imposed to improve the S/N performance for a specific application. If the range is too restricted, it must be appreciated that important, diagnostic spectral information may be lost, relative to either qualitative or quantitative applications.

4.6.3 *Recommended Data Presentation Formats*

The normal intensity scale obtained for a raw measurement is %T, usually scaled from 0 to 100%. If an abbreviated or expanded scale is used, it is important to ensure that the scale limits are presented, preferably with immediate scale markings shown. Note that for all numerical operations on spectral data, it is necessary to perform the operations on the absorbance form of the spectrum, including baseline corrections, spectral subtractions, and any numerical scaling of data. An important exception is the numerical and/or digital treatment of noise. Noise is essentially linear over the entire scale in transmittance (or %T) and is logarithmic in absorbance. All processing that assumes a linear distribution of noise, such as digital smoothing, must be applied to the transmittance form of the data.

The standard linear wavenumber scale is normally used for mid IR data; Fundamental vibration group frequencies are naturally associated with the wavenumber form of the scale. Wavelength-based scales are not normally used, except for the NIR. For mid IR instruments, options to provide a linear display in wavelength are usually available as an output function for the graphics. The raw data are seldom transformed and are usually retained linear in the frequency scale (cm^{-1}).

4.6.4 *Spectral Data Manipulation*

At all times, it is strongly recommended that raw spectral data are retained. If practical, the storage of single-beam data, or raw significance of the data is to be retained. For appearance purposes, some limited digital smoothing and nominal baseline corrections are acceptable if they improve overall readability of the final spectrum. If there are any doubts in terms of acceptability, it is recommended that original raw data are kept at hand to support any data that have been numerically manipulated (irrespective of the reasons for the manipulation). All documentation concerning data manipulation should be retained with the spectra. If spectral subtraction is used, it is important to retain all original raw spectral data including the component being subtracted.

4.6.5 *Standardization*

In the production of infrared spectral data, one of the greatest problems is a lack of standardization. This lack of standardization relates to sampling; the format of data; the actual recording of the spectrum; and the final digital representation used for storage and archiving. For spectral archiving, there have been attempts in recent years to gain consensus between the instrument vendors for the production of a standard for data transfer. The result of this has been the JCAMP-DX spectrum data format under the auspices of the IUPAC.[32] This documented ASCII format has been implemented for most modern instruments, although variation in the interpretation of the standard has led to some incompatibilities. A group involving the American Instrument Association and the FTIR vendors worked on a standard for spectral data transfer based on a *NetCDF* protocol.[33] This is not expected to replace JCAMP-DX, but it is expected to become an important standard for data transfer via the Internet and through various laboratory information management systems architectures.

However, the "physical" transfer of spectra between instruments is only one step in the complex chain of the standardization in spectra. The ideal is that a given sample provides a constant spectrum for a given physical state and a defined set of recording and sampling conditions. In the past, it was considered adequate to run a simple calibration standard, such as polystyrene. This is often sufficient as a simple validation of an instrument's performance relative to a prerecorded norm. However, it is not adequate for, and does not constitute, instrument standardization. Standardization implies a unified control of parameters, such as spectral resolution and band shape; actual spectral line position (wavelength calibration), and photometric recording accuracy; and all things that can impact these parameters in a practical measurement.

Currently, there is a focus on instrument standardization relative to the use of a fixed and reproducible method of sampling. Initially, the first implementations were for near infrared analysis, in which the analysis was primarily quantitative and in which there was a need to preserve the integrity of calibrations between different instruments (usually of the same design).[34] The technology and concepts exist for extending the approach to other spectral regions such as the mid infrared.

The next, and weakest link in the chain, is the sample itself and the method selected for sampling. A homogeneous sample, such as a gas/vapor or clear liquid, can be relatively straightforward as long as factors such as pressure and temperature are addressed. Solids present a different problem because morphology and form become an overriding issue. Some sampling techniques, such as photoacoustic detection/sampling, come close providing spectra that are independent of sample morphology. However, the more readily accessible techniques, especially the common KBr pellet, oil mull, and diffuse reflectance techniques, for powders, and grindable solids, which are very dependent on particle size, are particularly problematic.

There are surface-oriented sampling techniques such as internal reflectance (ATR) and external (specular) reflectance, both providing data that are influenced in one way or another by the sample's refractive index (at the measurement wavelength); the refractive index of the sampling medium (in particular, ATR); and the polarization of the infrared beam (in particular, external reflectance). For ideal cases, the mathematical relationship between the recorded spectrum and an idealized absorption (or transmission) spectrum is understood. In these situations, a correction algorithm can be applied to generate a pseudo-absorbance spectrum. This assumes that the sampling technique is well understood; that the nature of the differences between the recorded spectrum and the "absorbance" spectrum are appreciated; and that a standardized approach to the sample is adopted.

In general, there has not been a universal adoption of standardization regimes for infrared spectroscopic data. In some ways, this has limited the potential of the technique for generalized methods of analysis. For example, it has been stated by the US Food and Drug Administration that the reluctance to accept infrared spectroscopic methods (both near and mid infrared) stems from the inability to produce standardized spectral data. The goal is the ability to produce an exact and very reproducible standard spectrum for a given material by a prescribed method of sample preparation, independent of operator or instrument (model or vendor). Many of the factors that impact meeting that goal are now understood, and it is anticipated that future generations of instrumentation, with integrated sampling and data manipulation, may lead to a resolution of this problem for the majority of samples. In the meantime, readers are encouraged to utilize the standard procedures for sample preparation as defined in texts, such as the ASTM's *Annual Book of Standard's*.[35]

4.7 Summary

This chapter has focused on the issues of sampling as they pertain to routine sampling for qualitative and quantitative analysis. There are many routine specialized approaches to sampling that may be adopted for research applications. These are typically experiment dependent, and their success may be somewhat limited to the application for which they are customized. For this reason, such examples have been generally excluded from this text. The number of accessory companies has grown in the past 10–15 years, and the number of new sampling devices is growing steadily. It is beneficial to contact these companies to obtain the latest literature, and in this way, stay current with the latest developments in infrared sampling.

References

1. J. M. Hollas, *Modern Spectroscopy* (Wiley, Chichester, UK, 1996).
2. A. L. Smith, *Infrared Spectroscopy, Practical Handbook of Spectroscopy.* (J. W. Robinson, Ed.), (CRC Press, Boca Raton, FL, 1991).
3. N. B. Colthup, L. H. Daly, and S. E. Wiberley, *Introduction to Infrared and Raman Spectroscopy.* 3rd edn (Academic Press, Boston, 1990).
4. K. Feinstein, *Guide to Spectroscopic Identification of Organic Compounds* (CRC Press, Boca Raton, FL, 1995).
5. R. M. Silverstein, G. C. Bassler, and T. C. Morrill, *Spectrometric Identification of Organic Compounds.* 5th edn (Wiley, New York, 1991).
6. D. A. Burns, E. W. Ciurczak, (Eds.) *Handbook of Near Infrared Analysis,* 3rd edn. (CRC, Taylor and Francis, Boca Raton, FL, 2008).
7. A. L. Smith, *Applied Infrared Spectroscopy: Fundamentals, Techniques, and Problem Solving,* (Wiley-Interscience, New York, 1979).
8. B. Schrader, (Ed.) *Infrared and Raman Spectroscopy: Methods and Applications* (VCH Publishers, New York, 1995).
9. J. R. Ferraro, K. Krisnan, (Eds.) *Practical Fourier Transform Infrared Spectroscopy: Industrial and Chemical Analysis* (Academic Press, San Diego, 1989).
10. P. R. Griffiths and J. A. de Haseth, *Fourier Transform Infrared Spectroscopy* 2nd edn. (Wiley, New York, 2007).
11. P. B. Coleman, (Ed.) *Practical Sampling Techniques for Infrared Analysis* (CRC Press, Boca Raton, FL, 1993).
12. T. J. Porro and S. C. Pattacini, Sample handling for mid infrared spectroscopy, Part I: Solid and liquid sampling. *Spectroscopy* **8**(7) (1993) 40–47.
13. T. J. Porro and S. C. Pattacini, Sample handling for mid infrared spectroscopy, Part II: Specialized techniques. *Spectroscopy* **8**(8) (1993) 39–44.
14. C. Pouchert, *Aldrich Library of FTIR Spectra,* **3**: *Vapors.* (Aldrich Chemical Company, Milwaukee, WI, 1989).
15. *Solvents Condensed Phase, Gas Phase, and Mass Spectra, Book IV: Sprouse Collection of Infrared Spectra* (Sprouse Scientific Systems, Charlotte, NC, 1988).
16. K. Ohta and I. Ishida, Comparison among several numerical integration methods for Kramers-Kronig transformation. *Appl. Spectrosc.* **42** (1988) 952–957.
17. S. V. Compton, D. A. C. Compton, and R. G. Messerschmidt, Analysis of samples using infrared emission spectroscopy. *Spectroscopy* **6**(6) (1991) 35–39.
18. N. J. Harrick, *Internal Reflection Spectroscopy.* (Harrick Scientific, Ossining, NY 1987).

19. F. M. Mirabella, Jr., (Ed.) *Internal Reflection Spectroscopy: Theory and Applications* (M. Dekker, New York, 1993).
20. S. R. Culler, *Diffuse reflectance spectroscopy: Sampling techniques for qualitative/quantitative analysis of solids.* In *Practical Sampling Techniques: Techniques for Infrared Analysis.* (P. B. Coleman, Ed.), (CRC Press, Boca Raton, FL, 1993).
21. J. F. McClelland, Photoacoustic spectroscopy, *Anal. Chem.* **55**(1) (1983) 89A–105A.
22. J. F. McClelland, R. W. Jones, S. Luo, L. M. Seaverson, *A practical guide to FTIR photoacoustic spectroscopy.* In *Practical Sampling Techniques: Techniques for Infrared Analysis* (P. B. Coleman, Ed.), (CRC Press, Boca Raton, FL., 1993).
23. M. A. Harthcock, and R. G. Messerschmidt, (Eds.) (1988). *Infrared Microspectroscopy Theory and Applications* (Marcel Dekker, New York, 1993).
24. H. J. Humecki, (Ed.) *Practical Guide to Infrared Microspectroscopy,* (Marcel Dekker, New York, 1995).
25. R. White, *Chromatography/Fourier Transform Infrared Spectroscopy and its Applications.* (Marcel Dekker, New York, 1991).
26. W. Herres, *Capillary Gas Chromatography-Fourier Transform Infrared Spectroscopy, Theory, and Applications* (Huthig, New York, 1987).
27. C. D. Craver, *Halogenated Hydrocarbons* (A Coblentz Society special collection of infrared spectra. Coblentz Society, Kirkwood, MO, 1977).
28. C. D. Craver, *Infrared Spectra of Regulated and Major Industrial Chemicals* (A Coblentz Society special collection of infrared spectra. Coblentz Society, Kirkwood, MO, 1983).
29. C. Pouchert, *Aldrich Library of Infrared spectra,* 3rd *Ed.* (Aldrich Chemical Company, Milwaukee, WI, 1981).
30. *Solvents by Cylindrical Internal Reflectance, Book II: Sprouse Collection of Infrared Spectra.* (Sprouse Scientific Systems, Charlotte, NC, 1987).
31. *Infrared Spectra Handbook of Common Organic Solvents. Infrared Spectra Handbooks* (Bio-Rad, Sadtler Research Laboratories Division, Philadelphia, 1983).
32. R. S. Mc Donald and P. A. Wilks, Jr., JCAMP-DX: A Standard form for exchange of infrared spectra in computer readable form. *Appl. Spectrosc.* **42**(1) (1988) 151–162.
33. R. S. Lysakowski, NetCDF — A defacto standard framework for analytical data exchange, storage, and retrieval, Computerized Data Standards: Databases, Data Exchange, and Information Systems (R. S. Lysakowski and C. E. Graggs, Eds.), ASTM STP-1214, 57–74. (American Society for testing and Materials, Philadelphia, 1994).
34. J. Workman, Jr. and J. Coates, Multivariate calibration transfer: The importance of standardizing instrumentation. *Spectroscopy* **8**(9) (1993) 36–42.
35. ASTM International Methods and Practices for a Molecular Spectroscopy. In *ASTM Annual Book of Standards,* Vol. 03.06 (ASTM International, West Conshohocken, PA. 19428–2959, 2014).

Chapter 5

Data Acquisition for Infrared Spectra

Contents

5.1 Recommended Spectrum Acquisition, Presentation Format and Data Manipulation Procedures

5.1.1 *Recommended Spectrum Acquisition Parameters*

See Chapter 4 and Section 4.6 for more detailed descriptions of this topic for collecting infrared spectra. The following are guidelines for typical data acquisition situations using MIR spectroscopy. Depending upon the resolution setting of the infrared spectrophotometer, the time required for data acquisition will vary. Higher resolution requires more spectra to be collected to attain the required signal to noise. Infrared spectra may be collected as ATR (Attenuated Total Reflectance), transmittance, mulls, gels, films deposited on infrared transparent windows, gasses, liquids, caste films, DRIFTS (Diffuse Reflectance Infrared Fourier Transform Spectroscopy), open path, and other sampling methods. Each of these has optimized settings for both resolution and scan repetitions to obtain standard or acceptable spectra. See the various Chapters 3 and 4 in this text on Instrumentation and Sampling, respectively, for detailed information regarding these topics.

Commercially available interferometer instruments have multiple resolution settings, generally as 0.05, 0.1, 0.25, 0.5, 1, 2, 4, 8, 16, 32, and 64 cm^{-1}. As the resolution setting is decreased from one setting to the next, the number of data points recorded doubles. For example, a 4 cm^{-1} setting may require 8192 points, and a 2 cm^{-1} setting 16384 points.

Noise-reduction using co-added scans follows the square root of n reduction where the signal to noise is increased by a factor k equal to the square root of the number of replicate co-added scans. So 4 scans has 1/2 the noise and twice the signal to noise as compared to a single scan. Likewise 100 scans would have 1/10 the noise and 10 times the signal to noise as a single scan. In this way the number of scans is estimated for each type of spectrum measured.

For most mid infrared spectral measurements a resolution setting of 2 to 8 cm^{-1} and co-added scan numbers of 32–128 are common. For example, for most common samples using infrared the following set of parameters would be acceptable (Table 5.1).

Note that increased replicate scanning for more than 256 scans may approach a point diminishing returns in terms of spectral quality and S/N improvement (limited by the square root of n law for random variation, or white noise). MIR spectrometers may drift during a duration of several minutes, and structured noise not averaged by repetitive measurements may become a part of spectra measured for extended

Table 5.1 Typical instrument settings for liquids and solids measurements using infrared spectrophotometry.

Measurement parameter:	Typical setting:
Spectral Range:	4000 cm^{-1} to 500 cm^{-1}
Data Points:	1816
Infrared Source:	Globar
Infrared Detector:	DTGS KBr
Beamsplitter Material:	KBr
Autogain Velocity:	1.5825
Apodization Setting:	Happ-Genzel
Zero filling factor:	1 level
Aperture:	25
Resolution Setting:	4 cm^{-1}
Replicate Scans:	32 or 64 co-added scans
Sampling Methods:	KBr pellet
	Cast film on KBr disc
	ATR (ZnSe) 45°, 10 bounce
	Liquid cell

periods. Note that for gas samples much higher resolution is a requirement and would generally necessitate additional scan times and replicate measurements to attain optimal signal to noise.

Irrespective of the final selection of data acquisition parameters, full documentation of the experimental details should be recorded with all spectra, especially if regulated procedures are followed. Also, as a general rule, it is recommended that good laboratory practice guidelines are adopted throughout, which include laboratory notebooks, either paper or electronic and recorded measurement parameters, analyst, results, and corroborating signatures.[1]

5.1.2 *Recommended Spectral Range*

Unless special circumstances dictate otherwise, infrared spectra are recorded from $4000\,cm^{-1}$ to $400\,cm^{-1}$ (the ASTM International definition). A typical scan may include $4000\,cm^{-1}$ to $500\,cm^{-1}$ since data from $500\,cm^{-1}$ to $400\,cm^{-1}$ and below are most useful for Si-O absorption bands,[2] geological samples and some pharmaceutical molecule functional groups.

Typical research grade commercial infrared spectrophotometers will have a full scan range of $10\,cm^{-1}$ to above $4000\,cm^{-1}$. This wavenumber restriction is due to the number of octaves that may be covered with various optical systems, detectors, and infrared sources. Some very special research instruments may scan from $10\,cm^{-1}$ to above $28000\,cm^{-1}$ (i.e., Terahertz (THz or Far IR) through the visible region (Vis)) yielding full molecular and electronic spectra. Note that the pathlength for measurement changes with each region of the spectrum so that instruments are generally set-up to measure spectral regions with differing sample preparation or sample holders.

With modern instruments, it is normal to measure the entire individual spectral range with each measurement. When calibration or data transfer is required then data may be truncated (or windowed) so that one spectrometer collects the identical spectral range and data points as another spectrometer. Such data management allows easier compatibility of databases and calibrations developed using spectral search libraries for qualitative analysis. Note: information on calibrating instruments may be found in the index or the special chapters on data analysis (Chapter 6 and others) within this text volume.

5.1.3 *Recommended Presentation Formats*

The normal intensity scale obtained for spectral measurements are in %Reflectance or %Transmittance, usually scaled from 0 to 100%. If an abbreviated or expanded scale is used, it is important to ensure that the scale limits are presented, preferably with immediate scale markings shown. Note that for all numerical operations on spectral data, it is necessary to perform the operations on the absorbance form of the spectrum, including baseline corrections, spectral subtractions, and any numerical scaling of data. An important exception is the numerical and/or digital treatment of noise. Noise is essentially linear over the entire scale in reflectance (%R), or transmittance (%T) and is logarithmic in the absorbance scale. All processing that assumes a linear distribution of noise, such as digital smoothing, and digital signal processing must be applied to the reflectance or transmittance form of the data to perform a proper correction, and then converted to absorbance scale for quantitative or qualitative analysis.

The standard wavenumber scale is normally used for Mid IR data, representing the Fundamental vibration group frequencies and fingerprint frequencies associated with molecular vibrations. However, it is entirely appropriate to represent mid-infrared spectra in wavelength space as microns (μm), which has been traditionally used in some fields as 2.5 μm to 50 μm.

5.1.4 *Spectral Data Manipulation*

At all times, it is strongly recommended that the ratioed replicate spectral data are retained, along with the replicate spectra of the reference sample and dark background. Recall the formula representing Beer's law for spectroscopic measurements as shown in the following description of making a spectroscopic measurement. To convert the light reflected or transmitted from a sample (I) ratioed to the incident energy (I_0) to Absorbance (A), as a linear estimate of spectral response to concentration, we use the following equations when making spectrometer measurements. If only an internal reference is used for spectral collection, then the absorbance spectrum with respect to wavelength is computed as:

$$A = -\log_{10}\left(\frac{I}{I_0}\right) = \log_{10}\left(\frac{I_0}{I}\right) = \log_{10}\left[\left(\frac{\mathrm{R_I} - \mathrm{D_{R_I}}}{\mathrm{S} - \mathrm{D_S}}\right)\right] \tag{5.1}$$

If an external standard material is used to calibrate the internal reference material, then the absorbance spectrum with respect to wavelength is computed as:

$$A = -\log_{10}\left(\frac{I}{I_0}\right) = \log_{10}\left(\frac{I_0}{I}\right) = \log_{10}\left[\left(\frac{\mathrm{R_I} + \mathrm{R_\Delta} - \mathrm{D_{R_I}}}{\mathrm{S} - \mathrm{D_S}}\right)\right] \tag{5.2}$$

where

$$\mathrm{R_\Delta} = (\mathrm{R_E} - \mathrm{D_{RE}}) - (\mathrm{R_I} - \mathrm{D_{RI}}) \tag{5.3a}$$

Alternatively, Equation 5.3b (ratio of terms) may be substituted for Equation 5.3a; note that either equation for R_Δ is acceptable.

$$R_\Delta = (R_E - D_{RE})/(R_I - D_{RI}) \tag{5.3b}$$

and where S is the sample measurement; D_S is the Dark measurement for the Sample; R_I is the internal Reference Measurement; D_{RI} is the Dark Measurement for the Internal Reference Sample; R_E is the measurement of the external standard reference material for calibration; D_{RE} is the Dark Measurement for the external standard reference material; and R_Δ is the correction factor between the dark corrected R_E and R_I. The final spectrum is simplified to the ratio of (Internal Reference corrected by External Reference minus Dark) divided by (Sample-Dark). Note that Dark is a measurement where no energy from the source is allowed to the detector (it is blocked). The Dark measurement represents the dark current changes in the instrument during a measurement that relate to electronic noise where the detection electronics, computational electronics, and control electronics perform basic measurement functions with no imputed energy reaching the detector (i.e., the source is turned off or blocked using a shutter closure during the Dark measurement).

Retaining the individual replicate scan measurements allows improved diagnostics for determining the cause of error while taking repetitive measurements. For appearance purposes, some limited digital smoothing and nominal baseline corrections are acceptable if they improve overall quality of the final spectra for multivariate product calibration. If there are any doubts in terms of acceptability for data processing, it is recommended that original raw data are kept at hand to support any data that have been numerically manipulated (irrespective of the reasons for the manipulation). All documentation concerning data manipulation should be retained with the spectra in a digital format.

5.1.5 *Data Formatting*

A standard data format is essential for the utilization of mid infrared spectral data. Data from various instruments is routinely combined for quantitative calibration transfer and utilization of spectral libraries for qualitative searches for the purpose of identification.

In general, there has not been a universal adoption of standardization regimes for mid infrared spectroscopic data. In some ways, this has limited the potential of the technique for generalized methods of analysis. The goal for future spectrophotometry is to give users the ability to produce an exact and very reproducible standard spectrum for a given material by a prescribed method of sample preparation,

Table 5.2 Typical Spectroscopic Data file formats and extensions.

Data format name	Extension
AB SCIEX Data Explorer	(.pkm)
ASCII Table	(.csv)
Bruker	(.XML)
Excel	(.xls)
Finnegan	(.asc)
JCAMP-DX	(.dx)
MatLab	(.mat)
mzData	(.XML)
Perkin Elmer	(.sp)
Plain Text	(.txt)
Spectra-Calc (Archival)	(fixed point .spc)
Thermo Galactic	(.spc)
XLM Marked-up text	(.XML)

independent of operator or instrument (model or vendor). Many of the factors that impact meeting that goal are better understood, and it is anticipated that in some future generations of instrumentation many samples will be measured using specified metrical constraints, such that the same spectrum is obtained from different instruments. In the meantime, users are encouraged to refer to the standard procedures for sample preparation as defined in texts, such as the ASTM's *Annual Book of Standard's*.[3–4] (See Section 7.4.4, Chapter 7 of this volume for ASTM International standards related to infrared spectroscopy). Each instrument manufacturer may have from 1–10 (or more) data formats including older versions of their software. A list of the most typical spectroscopy software formats is given in Table 5.2.

5.1.6 *Summary*

This chapter has focused on the issues of sampling as they pertain to routine sampling for qualitative and quantitative mid infrared analysis. There are many routine specialized approaches to sampling that may be adopted for research applications. These are typically experiment dependent, and their success may be somewhat limited to the application for which they are customized. For this reason, such examples have been generally excluded from this text.

References

1. Guidance for Industry Good Laboratory Practices Questions and Answers, U.S. Department of Health and Human Services Food and Drug administration Office of Regulatory affairs, June 1981 (Minor editorial and formatting changes made December 1999 & July 2007).

2. G. Socrates, *Infrared Characteristic Group Frequencies*, 2nd edn. (John Wiley & Sons, 1997).
3. ASTM Practices for IR Spectroscopy, *Annual Book of ASTM Stardards*, Volume 03.06, 2014. ASTM international, 100 Barr Harbor Drive, PO Box C700, West Conshohocken, PA, 19428–2959 USA.
4. ASTM (American Society for Testing and Materials) Method E1655-05 (2012). "Standard Practices for Infrared Multivariate Quantitative Analysis", ASTM International, 100 Barr Harbor Drive, PO Box C700, West Conshohocken, PA (2012).

Chapter 6

Infrared Data Analysis and Presentation

Contents

Note: Acknowledgement: Portions of this chapter have resulted from previous collaborative work with Prof. Bruce Kowalski, Rasmus Bro, and Paul Mobley.[1–3] A special acknowledgement is extended to Dr. Clara Craver for her contributions to section 6.4 of this chapter.[4]

6.1 General Data Processing for Infrared Spectroscopy

6.1.1 *Introduction*

Data processing used in spectroscopy is often referred to in modern times as 'chemometrics'. This term chemometrics is defined as the application of mathematical and statistical methods applied to improve chemical measurement processes, and extract more useful chemical information from chemical and physical measurement data. At the time of writing of this chapter chemometrics may be used to refer to digital signal processing, spectral data manipulation and transforms, pattern recognition, imaging algorithms, quantitative and qualitative analysis algorithms and even some statistical methods such as statistical process control. A more formal definition may limit chemometrics to multivariate techniques, but then each spectrum is automatically multi-channel or multivariate in nature. Note to the reader that this chapter discusses multiple options for processing data, whereas Chapters 6 in volumes 2 and 3 discuss more details regarding the mechanics of data processing mathematics as well as the graphical display of the various data processing methods. The reader is referred to those chapters and the index for more details.

6.1.2 *Survey of Methods used for Infrared Data Processing*

Factor analysis has been used to determine both the presence and quantity of antioxidants in Extended Thermoplastic Rubber (ETPR) using FTIR spectroscopy.[5] A method is presented for data compression using B-splines with application to second-order FTIR spectra.[6] The B-splines compression method was selected for its suitability to modeling smooth curved data. The method implies that the primary goals of data compression are to reduce the overall file size and to reduce the computation required to analyze the compressed data. The data reduction for the method presented was from 16 bits per uncompressed data element to 0.16 bit per compressed data element. A comparison of a principal component model generated using uncompressed and compressed data demonstrated that the models are comparable, even with the 100 times reduction in data size.

MLR, PCR, and PLS are for quantitative analysis of latex in paper coatings using FTIR Spectroscopy and an ATR sample element. The styrene-butadiene latex on the surface of coated papers was quantified. The range of coating material varied between 5 and 25 parts, where 1 part is equivalent to grams of coating per 100 grams of mineral pigments. A relative error of 0.15 parts was found in the latex at $4\,cm^{-1}$ resolution.[7]

A method is described for the determination of vulcanized rubber additives using FTIR and partial least squares (PLS) regression for calibration. The method was reported to produce satisfactory performance for prediction of the individual industrial grade additives. Parameters measured include: stearic acid; N-phenyl-N'-isopropyl-p-phenylenediamine; aliphatic and aromatic processing oils; 1,2-dihydro-2,2 4-trimethyl-quinoline polymer; hindered bisphenol (NKF); Nickel dimethyldithiocarbamate; trimethylolpropane trimethacrylate; and N-(cyclohexylthio)phthalimide. The batches were prepared as standard solutions using carbon tetrachloride.[8]

The effects of spectral resolution on the performance of classical least squares (CLS) calibration in open-air long-path FTIR spectrometry is investigated. The combination of multivariate calibration using FTIR and open-air long-path FTIR spectrometry technology for the measurement of volatile organic compounds (VOCs) in the atmosphere is described.[9]

FTIR and attenuated total reflectance (ATR) combined with multivariate calibration applied to the quantification of individual sugar concentrations as glucose, maltose, maltotriose, and maltodexrin during a starch hydrolysis process. The solutions studied contained dry matter concentrations in the 25 to 30 percent range; glucose and maltose were predicted within desired accuracy; maltotriose and maltodextrin were not predicted within required accuracy. Predictions for glucose and maltose were tested for repeatability and reproducibility and were found to be acceptable.[10,11]

Multivariate calibration to the analysis of physiological samples using mid IR measurement data is demonstrated. Human plasma analysis is performed for protein, glucose, cholesterol, triglycerides, and urea. The measuring instrument utilizes mid infrared spectral data with an ATR cell. Partial least squares (PLS) regression was used to compute the calibration models.[12] This work was repeated using the NIR spectral region.[13] The use of Short Wave NIR (SW-NIR) resulted in mean-squared prediction errors (RMSEP) of 1.4% for total protein, 6.5% for total cholesterol, and 13.8% for triglycerides. Multivariate calibration methods were compared for the determination of multiple constituents in blood serum using mid IR spectra. PLS-ANN was applied for the determination of total proteins, glucose, and urea in human blood serum samples. The PLS-ANN calibration method did not yield significantly better prediction performance as compared with PLS and PCR.[14]

Partial least squares (PLS) and artificial neural networks (ANN) are compared as methods for multivariate calibrations using infrared data. The prediction results from both methods were equivalent; however, the computational time required for the PLS modeling approach is considerably shorter.[15]

The measurement of CO spectra in the range of 2.38 to 5100 mg/L (ppm) range is demonstrated. The spectra were measured at a resolution of $4\,cm^{-1}$ in the mid IR (2186 to $2001\,cm^{-1}$) region. The FTIR instrument was used in combination with a partial least squares (PLS1) calibration to compute a CO model from the teaching set. The instrument response (as absorbance) versus concentration of CO was found to be highly nonlinear. A linearization routine was applied to the spectroscopic data prior to computing the calibration model. It was found that the integrated absorbance areas, rather than the absorbance height values, were more useful for modeling. A fifth-order polynomial was used for the final model to calculate concentration from absorbance areas. The result was an improved calibration model with increased concentration prediction accuracy over the wide concentration range of CO.[16]

Mid IR spectra was used to rapidly assess pulpwood quality using samples of wood taken from Tasmanian Eucalyptus globulus. Measures of pulp wood quality included: yield, soda charge at kappa No. 15, total lignin, and hot water and alkali solubles. A PLS model was developed based on IR response information at 34 frequencies; the authors report success.[17]

PLS is used with vapor phase IR spectra of mixtures of o-, m-, and p-xylene to perform quantitative multicomponent analysis. The errors in prediction for mixtures of the xylenes was 1% for each component over a temperature range of 35° to 65°C. When calibration spectra were all measured at a single temperature (35°C), rather than over the expected temperature range for the unknown samples, the error was increased by a factor of 5.[18]

A modified least squares procedure was used in combination with FTIR spectra of coal samples to derive improved practices for smoothing, band resolution, peak finding, and data compression for this application.[19] A dissertation by P. Bhandare describes the use of IR absorption spectroscopy and chemometrics for the determination of a variety of constituents in blood tissue.[20] A technique termed 'Self-modeling multivariate mixture analysis' was presented to extract pure component information from complex sample mixtures, when spectra for the pure components within the mixture are unavailable. In the first experiment, differences between adjacent nonidentical monolayers of polymer laminate were distinguished. Spatial resolution was achieved for distinguishing between 2–3 micron thick inner layer, from a 4-layer polymer laminate. A second experiment produced pure component spectra from a KBr pellet made using a mixture of three compounds.[21]

ANN and PLS combined with IR spectra were used to classify spectra of pesticides as either organophosphorus or non-organophosphorus types. Classification by using these chemometric techniques resulted in a success rate approximately equal to results obtained from interpretation by spectroscopic experts. The differences between the methods tested lies primarily in the computational time required for optimizing the calibration model.[22]

6.1.3 *IR Special Data Processing Techniques*

Window Factor Analysis (WFA) is tested as a general method for chromatographic and spectroscopic data. A method is described using three approaches to WFA and compare these approaches as general data analysis techniques. The method showed that WFA with mean-centered sequential scores is sensitive to peak shapes; mean-centered sequential loadings WFA is a method of choice when low variability exists in the non-sequential direction. WFA with no mean-centering (as is) is a poor performer when heteroscedastic noise is present.[23]

6.1.3.1 *IR Expert Calibration Systems*

An expert system was developed for multivariate calibrations using PCR or PLS. A rule-based system was reported that develops calibrations automatically for FTIR spectroscopic data and have provided a commercial software package.[24] An expert calibration system has been proposed which includes a three-phase development strategy. This expert system would allow the use of MLR, PCA, PCR, PLS, LWR, and NNs in an automated fashion.[25]

6.1.3.2 *Vapor Phase Measurements*

A dissertation demonstrated the details of applying multivariate regression methods to multicomponent vapor phase infrared spectra.[26]

6.1.3.3 *Cluster/Discriminant Analysis*

A method is described for classifying human white and gray matter brain tissue using IR spectroscopy combined with a variety of multivariate techniques. The authors report that brain tissues identified by standard pathological methods as heavily, moderately, or minimally affected by Alzheimer's disease (AD), can be successfully classified at better than 90 percent accuracy using an IR-based classification method. Control tissues were distinguishable from known AD tissues with a success rate of 100 percent using the method.[27]

Diffuse reflectance Fourier transform spectroscopy (DRIFTS) is used to differentiate a series of cellulosic fabric samples, including dyed, undyed, and cotton based. Band assignments are discussed relative to previous work, and visual interpretation of spectra explored. All spectra were qualitatively and quantitatively measured with good differentiating results. The chemometric methods applied included principal component analysis (PCA) and soft independent modeling of class analogies (SIMCA). Discrimination of sample types was successful based on dye, fabric type, and textile processing techniques.[28]

An infrared method (700 to 3600 cm^{-1}) is demonstrated to examine the information content of sub-sections of this spectral region. The method demonstrated the redundancy within selected spectral regions and provided a method for determining information content, and comparing the merits of pre-specified wavenumber regions for their overall effective information content.[29]

DRIFTS is used to determine the type of fruit present in samples of jam. The experiment had as its purpose the differentiation between strawberry containing jams and non-strawberry containing jams. A group mean distance classification method was used following PCA or PLS decomposition of the data. Success in classification was near 100%. The use of ATR sample presentation resulted in a 91% classification success rate. The method was strongly influenced by the total sugar content of the jams tested, with the largest effect caused by the 'normal' and 'reduced' sugar content jams.[30]

Mixtures of crystalline and amorphous silica are classified using IR spectroscopy with multivariate analysis. The PLS multivariate approach was used and a paper reports the method can provide a low cost, rapid and robust method for quantitative analysis of complex silica dust mixtures.[31]

Samples of quartz crystals were analyzed using infrared spectroscopy, principal components analysis (PCA), and quantitative multivariate analysis. The calibration models were used to classify and quantify the sample population. The absorption spectra were measured from 3300 to 3600 cm^{-1} with corresponding reference chemical analysis performed using inductively coupled argon plasma spectrometry (ICP).[32]

FTIR spectroscopy and chemometrics were used to perform discriminant analysis for classification of fruit purees into three predetermined group classifications: apple, raspberry, and strawberry. The groups were classified with 100% success. A separate discriminant analysis was performed to determine whether fresh or freeze-thawed fruit samples could be distinguished. This experiment met with a 98.3% success rate for strawberry samples and 75% success for raspberry samples. A third experiment attempted to discriminate the fruit from each group into two levels of ripeness. Results for level of ripeness classification resulted in 92.5% correct group for raspberry, but strawberry samples were not distinguishable. Experiment four was a test as to the classification of apple samples according to added (or none) sulfur dioxide; producing 90% correct classification. A set of apple samples was tested for a fifth experiment as to variety classification; 86% of samples tested were correctly identified as Bramley or not.[33]

ANN and PLS regression for computerized studies of interpretive IR spectra were compared. Classification by skilled interpreters was superior in all cases and experiments performed using the particular chemometric method proposed by the authors were compared. ANN and PLS were nearly equivalent, with the exception that ANNs exhibit a higher score for individual band recognition.[34]

FTIR and factor analysis (FA) were used to evaluate the potential for measuring the hydration states of four different natural and synthetic nucleic acids.[35] PCA with IR spectral data were used to classify Saale River sediments from German origin.[36]

6.1.3.4 *Spectral Searching Algorithms*

The Adequate Peak Search (APS) algorithm as used for interpretation of infrared spectra is described. The method is evaluated for selectivity, trueness, ruggedness, and false results. The APS method is compared to 'Classical' interpretation of infrared spectra; the concepts of Limit of Detection, Optimization, and Automation are discussed relative to APS. The detection limit for APS was reported to be four to seven times lower than that attained by a panel of experts.[37]

6.1.3.5 *Calibration Transfer*

Calibration transfer methods involve the computation of calibration models using data from a primary scanning or interference filter spectrometers (i.e., the parent instrument), and transferring the calibration with a set of samples measured to a second (i.e., a child or secondary) instrument.[38] With this calibration transfer method bias and/or slope adjustment is often required to minimize the residuals from the child instrument and to adjust the model for the transfer procedure. This practice increases overall error of analysis on the child instrument, especially when a broad range of samples are measured. Calibrations are generally developed using 50–500 samples that define the dimensions of the vectors for the analytical line in a calibration model; often a statistically *insignificant* set of 10–20 samples is used to adjust the bias and slope. Based on statistical criteria,[39] slope adjustments should never be made, and bias adjustments are appropriate only after considerable testing. However, once a bias and or slope is made for any model, a separate test or validation set must be used to determine the prediction accuracy.

A second approach to calibration transfer requires sets of specially prepared reference samples (or a single special sample) measured on a primary, master, or reference instrument (on which the calibration model was developed) and a secondary, child, or transfer, instrument. The spectra of the same samples from each instrument are used to adjust the spectra of the second instrument to "look" virtually the same as those obtained from the original parent or primary instrument.[40,41] The problem with this method (the *Golden or Master Instrument* approach) is that the primary or reference instrument as well as the reference samples will change, thus common samples measured with both instruments over time will not be related to the original model produced on the primary instrument. Likewise, if the primary instrument is repaired, it may only approximate its original spectral performance. Thus, over time, libraries of sample spectra will have no direct relationship to any existing physical instrument or prepared samples.[42]

A third approach to calibration transfer uses chemometric techniques to reduce the impact, within a model, of normal instrument-to-instrument variation on analytical precision.[43–45] Calibration methods that include inter-instrument variations have the quality of producing models less sensitive to differences between spectrophotometers; these models make the requirements of instrument standardization potentially less important if equal performance can be achieved by using these techniques. Such modeling techniques decompose full spectral information into smaller spectral regions that are less subject to between instrument variations than other regions of the spectrum. The disadvantage of these techniques is that areas of the spectrum most subject to intra- and inter-instrument variation often contain valuable sample concentration information. Such information should be included within a calibration for optimum analytical performance and signal-to-noise; thus the method cannot be relied upon as a generalized approach to resolving the calibration transfer problem.

A fourth series of approaches to the calibration transfer problem involves the use of internal standards such as polystyrene to adjust dispersive instruments for wavelength accuracy or precision. The limitation for this technique is that it does not address issues of photometric throughput (accuracy), variability in bandpass, differences in stray light characteristics, and detector non-linearity. In addition, standards such as polystyrene provide wavelength or frequency calibration over limited spectral regions, defined by a few absorption bands. This approach does not ensure that the entire range of the instrument is calibrated or has fully linear dispersion.

6.1.3.6 *Instrument Standardization*

A more global approach to standardize spectrophotometers (make instruments more alike) has been developed. If instruments are alike and remain stable enough, calibration transfer no longer becomes an analytical performance issue. A method was developed to account for both the major and minor manufacturing and/or time-based variations between instruments.[46] Although individual instrument noise characteristics are known to cause difficulty in calibration model transfer, it is generally considered that the major variable in transferring calibrations between instruments is the difference in wavelength calibration (or wavelength accuracy or registration) between spectrophotometers.[47–49] This is considered to be true for both filter and continuous scanning instruments. Other problems requiring technology improvements include: the stability of internal standards, detector linearity, lineshape, spectral artifacts, dynamic range, linearity in wavenumber axis, and photometric axis registration.

One comprehensive approach to instrument characterization, leading to instrument standardization and calibration transferability involves five steps, a key part being the standardization of the instrument. This latter factor is addressed to ensure that spectra produced from different instruments of the same design are essentially identical. The proposed five step approach leading to transferability includes: 1) characterization and standardization of the instrument function, 2) stray light characterization and correction, 3) sample cell photometric corrections, 4) continuous background referencing for lamp, fiber, and detector effects, and 5) chemometric tools for optimizing model performance.[50]

A method for standardizing second-order instruments is presented using GC/MS and LC/UV data. This method proposes the use of the Gauss-Newman method and a common standard sample to reduce variation between sample measurements using second-order methods. A critical ingredient to the success of this method is the design of the common standard sample.[51] Another method for instrument standardization describes a correction factor that can be calculated to accommodate for non-linear instrument responses in calibration transfer, most often due to stray light.[40] The correction factor is termed the additive background term and is applied with the method delineated in reference 45.

6.1.3.7 *Continuum Regression (CR)*

Three excellent references describing the background for CR are given[52–54] that demonstrate the use of continuum regression "for the identification of finite impulse response (FIR) dynamic models." In CR a model is selected for a particular application using the selection of a minimum value selected over a response surface comprised of the abscissa (x-axis) as a power of the singular value decomposition (SVD) values for PCR, PLS, and MLR; an ordinate (y-axis) as the number of factors (or latent variables) in the model; and the z-axis as the Prediction Error Sum of Squares (PRESS) values corresponding to the x, y surface. In this way an optimum model can be selected based not only on the number of factors, but also on the type of modeling algorithm.

6.2 Data Preprocessing

6.2.1 *Introduction*

To begin this discussion we bring a caution to the reader and the user of data preprocessing methods. Note the following quotation,[55] "[Some chemometricians][56] have emphasized that data 'consist of numbers, of course. But these numbers are fed into the computer, not produced by it. These are numbers to be treated with considerable respect, never to be tampered with, nor subjected to a numerical process whose character you do not completely understand. You are well advised to acquire reverence for data that is rather different from the "sporty" attitude which is sometimes allowable, or even commendable, in other numerical tasks.' Yet by and large within chemometrics, preprocessing often seems to be carried out with little understanding of its fundamental effect on the structure of the data."

The intent of this early section of Chapter 6 is to identify published research as examples of typical data preprocessing methods and data transformations for infrared spectroscopy using chemometrics. Note that, in general, data transformations are applied to each sample measurement in a data set and each sample would be transformed similarly whether as a single sample or as a group of samples. Preprocessing methods alter each sample according to the characteristics of the entire data set, i.e., mean centering or autoscaling. Adding or subtracting a single sample from the data set would affect the preprocessing of each sample. In addition, 'pretreatment methods' may refer to either preprocessing methods or data transformations.

6.2.2 *Methods Which Utilize Noise Estimates*

The development of chemometric methods which use noise estimates require more precise experimental design work during the calibration step. Of particular interest to spectroscopists is the fact that many chemometric and statistical methods assume homoscedastic noise (i.e., a constant noise level across the spectrum); when in fact most spectroscopic methods produce data which contain heteroscedastic noise (i.e., errors varying in amplitude or magnitude across the spectrum). One paper describes the root causality of heteroscedastic noise in infrared (IR) absorbance data as well as several potential problems that result from this condition.[57] In this work a local windowing method is presented which estimates the actual pseudorank (loosely defined as the number of factors required to model the physically or chemically significant part of the data). Another paper proposes a modified Box-Cox power transform to convert heteroscedastic errors into homoscedastic errors. The other benefits of the method being the enhancement of information contained in small peaks compared to large ones and the preservation of strong linear correlations between signal amplitude.[58] A "balanced scaling" method is proposed, which takes a data matrix, X, and scales it in a three step algorithm using a matrix of error estimates, s, of the same size as X. It is suggested that by using this balanced scaling prior to performing PCA and also applying a method of factorization, which makes use of noise estimates, a more chemically relevant and meaningful set of factors is computed.[59]

6.2.3 *Spectral Enhancement and De-noising*

Research has also been conducted in the areas of spectral enhancement (narrowing the bandwidth of spectral bands) and noise reduction using numerical methods. A method called the data reflection algorithm (DRA) has been used in the spectral enhancement of Fourier Transform Ion Cyclotron Resonance (FT-ICR) mass spectroscopy and nuclear magnetic resonance (NMR) spectroscopy data. The method involved augmenting the time reflected data with the original data to form a centerburst spectrum.[60] This reference paper has asserted that the method is analogous to the well understood Michelson interferometer. In Michelson interferometry the centerburst interferogram is comprised of the signal detected as a function of pathlength differences between light beam components. With the DRA method, the centerburst time domain spectrum contains the phase information of the original data. The authors reported higher resolving power compared to the original magnitude-mode FT spectrum and a higher signal-to-noise ratio (S/N).

Research describes a method of minimizing noise in a spectrum using the Fourier transformation, circulent matrices, and singular value decomposition (SVD). One paper describes the algorithm and noted that little distortion of the signal waveform occurred during noise reduction. A second paper addresses difficulties in applying the method to continuum spectra and alleviated the problem of extensive computation time.[61,62] The problem of spike noise and proposed robust summation using missing-point fitting as the optimal method for compensating for spike noise are discussed.[63,64]

6.2.4 *Comparison of Preprocessing Methods*

Summary work to compare the application of different preprocessing methods has been described. These comparisons apply the various preprocessing methods on a single or small number of data sets to determine which methods performed optimally. One comparison paper not involving infrared spectroscopy is useful for summarizing the preprocessing methods in use. In this paper eight different preprocessing methods were applied to several data sets collected using high-performance liquid chromatography coupled with diode array detection (HPLC-DAD).[65] The effects of closure, normalization, and ratioing on principal components analysis (PCA) of a simulated data set have been explored. It was found that in some cases the rank of the data increased following preprocessing. In some data sets, "the rank of the ... data set [was] increased by the constant length and maximum value transformations, even though a degree of freedom [was] removed from the description of each object in the data set."[55,56] This problem is especially significant as one of the main goals of preprocessing is minimization of noise and instrumental effects so that the analyst can get at the chemical information. In pursuing this, preprocessing would ideally lower the rank of the data.

6.2.5 *Derivatives*

Derivatives are a commonly applied preprocessing technique in spectroscopy. The general rule of thumb is that first derivatives are useful in removing baseline offsets and that the second derivative corrects for baseline offsets and sloping baselines. See Volume 3, Chapter 6 for detailed explanation of derivative methods. From a calculation standpoint, Savitzky and Golay[66] delivered a landmark paper in 1962 that described what has become a standard method for smoothing spectra and calculating derivatives. Some years later, Gorry[67] extended this work to overcome the problem of truncating ends of the spectrum and also simplified computation for all derivatives. The issues associated with using second derivatives in curve fitting have been examined. The advantage of curve fitting is elimination of unnecessary parameters (factors), which reduces the likelihood of overfitting the data, as well as some improvement in the extraction of chemical information from the raw data. However, the negative trade-off for curve fitting is the potential for blurring profiles of small, overlapped peaks during the smoothing step.[68] Work by other authors describes the issues associated with spectral curve fitting, noting that the selection of window size is critical for optimizing the smoothing step and for preventing the blurring of small peaks.[69]

6.2.6 *Data Transformations*

Data transformations are important to the field of spectroscopy. A short list of these include the Hadamard transform, wavelet transform, and the Fourier Transform (FT). Fourier deconvolution,[70] using a Cauchy-Gauss production function, is found to enhance or optimize the line shape of infrared spectra.[71] A tutorial describing the mathematics behind FT and the use of interferometric FT methods in IR and ultraviolet (UV) spectroscopy is available.[72] FT may also be used to reduce the size of the data for storage and calibration speed.[73] A clear description of the Hadamard transform for use in spectroscopy and analytical chemistry has been published and compared to FT.[74]

Wavelet transformation is important for spectroscopists.[75,76] Tutorials are given in references 77 and 78. Source code for wavelet transforms in other computer languages is also available at several world wide web sites, including MatLab.[79]

6.2.7 *The Traditional Spectroscopic Data Preprocessing Methods*

Details of derivative computations are given in Chapter 6 of Volume 3 and the reader is also referred to the index for specific preprocessing method topics. To continue, the most common linearization of reflectance (R) or transmission data (T) to absorbance is given by:

$$A = -\log_{10} R_l \quad \text{and} \quad A = -\log_{10} T_l \tag{6.1}$$

where A is absorbance expressed as reflectance or transmission when R or $T = I/I_o$.

Absorbance ratio has been used for NIR[80] and is the ratio of absorbance at one wavelength divided by the absorbance at a second wavelength. This ratio method is common to infrared spectroscopists as using it results in a term which compensates for baseline offset. The ratio of transmission data is also useful for normalizing or reducing baseline offset in spectroscopic measurements. A typical absorbance ratio is given as:

$$\frac{A_{\lambda_1}}{A_{\lambda_2}} \tag{6.2}$$

Standard derivative nomenclature for first through fourth order derivatives as follows.

$$\text{First derivative} = f'(x) = \frac{dy}{dx} \tag{6.3}$$

$$\text{Second derivative} = f''(x) = \frac{d^2y}{dx^2} \tag{6.4}$$

$$\text{Third derivative} = f'''(x) = \frac{d^3y}{dx^3} \tag{6.5}$$

$$\text{Fourth derivative} = f''''(x) = \frac{d^4y}{dx^4} \tag{6.6}$$

The use of the moving-averaged-segment convolution (MASC) method for computing derivatives in spectroscopy, brings about the expression below describing a first derivative term as:

$$A_{\lambda_2} - A_{\lambda_1} = A_{\lambda_{1+\Delta\lambda}} - A_{\lambda_{1-\Delta\lambda}} \tag{6.7}$$

where the derivative is given as the difference in absorbance values at 2 different wavelengths; the position of each wavelength being determined as the gap ($\Delta\lambda$) distance + or − from a center wavelength (λ_1). Ratios of derivative terms have been applied to some spectroscopic data (NIR), and have been empirically shown to produce lower prediction errors in special cases. The expression shown is the first derivative ratio where two different center wavelengths have been used.

$$\frac{A_{\lambda_2} - A_{\lambda_1}}{A_{\lambda_4} - A_{\lambda_3}} = \frac{A_{\lambda_{1+\Delta\lambda}} - A_{\lambda_{1-\Delta\lambda}}}{A_{\lambda_{2+\Delta\lambda}} - A_{\lambda_{2-\Delta\lambda}}} \tag{6.8}$$

The MASC form of second derivative is shown below.[81,82] In this case, the second derivative term is defined by the sum of absorbances at two wavelengths ($\lambda_{1+\Delta\lambda}$, and $\lambda_{1-\Delta\lambda}$), minus 2 times the absorbance at a center wavelength (λ_1). In this case the second derivative gap size is designated as $\Delta\lambda$ (delta wavelength). The second derivative preprocessing step is quite effective in removing slope and offset variations in spectral measurement baselines. It also "assists" the calibration mathematics in defining spectral regions where small response changes can be useful in calibration modeling. Without the use of derivatives, these regions would not be beneficial for use in calibration.

$$A_{\lambda_1} + A_{\lambda_3} - 2A_{\lambda_2} = A_{\lambda_{1-\Delta\lambda}} + A_{\lambda_{1+\Delta\lambda}} - 2A_{\lambda_1} \tag{6.9}$$

Second derivative ratio preprocessing has also been demonstrated in the Norris regression method. Empirically this data preprocessing technique has proved useful for some applications. The relationship immediately below designates a second derivative ratio with two different center wavelengths.

$$\frac{A_{\lambda_1} + A_{\lambda_3} - 2A_{\lambda_2}}{A_{\lambda_4} + A_{\lambda_6} - 2A_{\lambda_5}} = \frac{A_{\lambda_{1-\Delta\lambda}} + A_{\lambda_{1+\Delta\lambda}} - 2A_{\lambda_1}}{A_{\lambda_{2-\Delta\lambda}} + A_{\lambda_{2+\Delta\lambda}} - 2A_{\lambda_2}} \tag{6.10}$$

Other derivative forms and formats are used to express computations of derivative spectra. These are a general form of the derivatives used in derivative processing software for spectrometers. These include the first derivative spectrum as a central difference method:

$$A'_{\lambda} = \frac{A_{\lambda+\Delta} - A_{\lambda}}{\lambda_1 - \lambda_2} = \frac{A_{\lambda+\Delta} - A_{\lambda}}{\Delta\lambda} \tag{6.11}$$

where Δ is the number of wavelength spaces between A_{λ} and $A_{\lambda+\Delta}$.

A central difference method for computing the first derivative spectrum is used to mitigate for the typical shifts resulting from computing the derivative of spectral data using other equation forms. For this nomenclature the $\lambda + \Delta$ represents a center wavelength plus a wavelength gap, represented by (Δ). Therefore $\lambda - \Delta$ represents the central wavelength minus the designated wavelength gap.

$$A'_{\lambda} = \frac{A_{\lambda+\Delta} - A_{\lambda-\Delta}}{2(\Delta\lambda)} \tag{6.12}$$

The method for computing the second derivative spectrum computation is given as:

$$A''_{\lambda} = \frac{(A_{\lambda+\Delta} - A_{\lambda}) - (A_{\lambda} - A_{\lambda-\Delta})}{(\Delta\lambda)^2} = \frac{(A_{\lambda+\Delta} - 2A_{\lambda} + A_{\lambda-\Delta})}{(\Delta\lambda)^2} \tag{6.13}$$

where Δ is the number of wavelength spaces (gap) between A_{λ} and $A_{\lambda+\Delta}$ and $A_{\lambda-\Delta}$.

The central difference method for computing the third derivative spectrum computation is given as

$$A'''_{\lambda} = \frac{(A_{\lambda+2\Delta} - 2A_{\lambda+\Delta} + A_{\lambda}) - (A_{\lambda} - 2A_{\lambda-\Delta} + A_{\lambda-2\Delta})}{(\Delta\lambda)^3},$$

and then

$$A'''_{\lambda} = \frac{(A_{\lambda+2\Delta} - 2A_{\lambda+\Delta} + 2A_{\lambda-\Delta} - A_{\lambda-2\Delta})}{(\Delta\lambda)^3} \tag{6.14}$$

The method for computing the fourth derivative spectrum computation is given as

$$A''''_{\lambda} = \frac{(A_{\lambda+2\Delta} - 3A_{\lambda+\Delta} + 3A_{\lambda} - A_{\lambda-\Delta}) - (A_{\lambda+\Delta} - 3A_{\lambda} + 3A_{\lambda-\Delta} - A_{\lambda-2\Delta})}{(\Delta\lambda)^4},$$

and then

$$A''''_{\lambda} = \frac{(A_{\lambda+2\Delta} - 4A_{\lambda+\Delta} + 6A_{\lambda} - 4A_{\lambda-\Delta} + A_{\lambda-2\Delta})}{(\Delta\lambda)^4} \tag{6.15}$$

For each of these the gap size is designated as Δ (delta) and the computational wavelength is λ (lambda).

The traditional Kubelka-Munk transform expression, where R is the reflectance value, is shown below. This expression is used for samples exhibiting scatter, and has proven beneficial for some measurements.

$$\frac{(1 - R_{\lambda_1})^2}{2R_{\lambda_1}} \tag{6.16}$$

The log[Kubelka-Munk] expression (below) has been demonstrated and may empirically prove useful for some scattering sample data sets.

$$\log_{10}\left[\frac{(1 - R_{\lambda_1})^2}{2R_{\lambda_1}}\right] \tag{6.17}$$

6.2.8 *Noise Compensation in Modeling*

(This is a general discussion on the rationale for adding noise prior to the modeling step.) Spectral interferences resulting from different spectral response between instruments are complex and result from the convolution of the measured sample optical properties (e.g. refractive index, temperature, composition, etc.), and the total instrument function. The sample in effect becomes an active optical element of the instrument. Noise Compensation[83–85] has been used by measuring calibration samples on different instruments while varying the measuring parameters (wavelength, temperature, sample presentation, etc.) for each instrument. If the type and magnitude of interferences was well understood, one could manually add these effects into a data set prior to modeling. Adding different types and magnitudes of instrument interferences (than those that exist in reality) reduces the quantity and quality of response data useful for calibration (thus reducing the model performance). The more exact the characterization of interference noise added during the modeling step, the more optimum the calibration performance becomes across instruments.

The generalized concept has existed for some time. U.S. Patent No. 4,944,589 broadly describes "A method for reducing the susceptibility to interference of the measured value from a measuring instrument.... During the calibration procedure, a variation of one or several properties of the measuring instrument, *or of representations thereof in the processing of the measured values* (italics mine), is introduced intentionally, said variation having the same order of magnitude as or being substantially greater than expected variations, when one or several test objects are being measured, without such information being supplied to the instrument." Claim 4 states, "The method as claimed in claim 2, the spectrometer consisting of a grating monochromator, wherein the wavelength is shifted by changing the angular position of the grating of said monochromator or by changing correspondingly said numerical representation of the angular position when calculating the wavelength, relative to the positions or values, for the main numbers of the measurements of the calibration test objects." In the body of this patent text it is stated, "Practical tests have shown that the interference thus intentionally introduced has implied a sharply increased insusceptibility to the much smaller variations in the measuring instrument parameter or parameters, which are normally expected during subsequent measurements of test objects of the same type as are used in the calibration procedure."[83]

Extensions of this method to include errors associated with variation in operators, sample presentation, sample non-homogeneity, and temperature (of sample and instrument) are discussed (and taught) in references.[84–86] Reference 86 adapts these techniques for octane measurement in gasoline (with the explicit addition of the terms: "transmittance shifts, wavelength shifts, absorbance-baseline shifts, and absorbance-baseline tilts").

Reference 87 provides a basis for the statement that it is important to use regions in the teaching set spectral data (for a calibration) where the relationship applies as $\beta_i \frac{\delta_{s_i}}{\delta_{p_i}} = 0$ when p_i is wavelength shift (or another variable property of the instrument response data). It is implicit within the statement $\frac{\delta_{\beta_i}}{\delta_{\lambda_i}} = 0$ from reference 87, that $\beta_i \frac{\delta_{s_i}}{\delta_{\lambda_i}} = 0$ is the desired condition for spectral data with respect to *the robustness condition* in calibration equations; relative to wavelength changes between (and within) instruments. Reference 86 extends this explanation to other interference properties related to octane analysis and other multivariate algorithms; although the details of explanation are not important for implementation of the method. For these regression coefficient relationships, β_i is the regression coefficient at wavelength i; s_i is the "true" spectrum response at wavelength i; and p_i is the interference property associated with wavelength variation and photometric variation between (and within) instruments at wavelength i.

6.2.9 *Trends in Data Pretreatment*

In calibration the expected benefits of data pretreatment are somewhat different from the goals of pretreatment prior to exploratory or classification methods. For example, with PCA the goal tends toward removing instrumental artifacts so that the chemical information in the spectra becomes more apparent and is maximized. In calibration, reference values as well as spectra may need to be pretreated and the desire for predictive ability means that the data may be altered so that the spectral features in the usual sense are lost, but predictive ability is improved.

A tutorial entitled, "A Primer on Multivariate Calibrations" explores calibration in general but has a good-sized discussion of data pretreatment. Several comparison papers have been written in which several methods were applied to a data set to determine the relative subsequent effect on predictive ability in calibration.[88]

A tutorial has been written for dealing with data sets exhibiting heteroscedastic noise (nonconstant variance across the spectrum). For certain cases the authors recommended the use of generalized least squares (GLS) and the variance function estimation (VFE) method. This tutorial provides a foundation for dealing with situations in which the use of data preprocessing or transformations do not provide enough flexibility.[89]

The effect of data preprocessing methods on the robustness and parsimony of multivariate calibration models has been reported. Proper data preprocessing eliminates non-relevant or non-chemically related sources of variance and nonlinearities such that more parsimonious (i.e., fewer factors and higher accuracy) calibration models are constructed. With proper preprocessing, the bias decreases more sharply as parameters are added. In addition, once the optimal number of parameters is exceeded, the error associated with measurement noise increases more sharply. The reduction in bias through proper data preprocessing results in a less complex model. Although predictive ability may not necessarily be enhanced in the short term, a more parsimonious calibration model is expected to be more robust over time.[90]

In one paper the importance of scaling the concentration values in partial least squares 2 (PLS2) is discussed. PLS2 differs from PLS1 in that it performs calibration of multiple analytes simultaneously. The prediction of minor components is improved and prediction is made more robust with proper scaling of the reference values.[91]

6.2.10 *Digital Filtering*

As with the exploration and classification literature, much work has been done in calibration using digital filters as a preprocessing technique. The similarities and difference between multivariate calibration methods and digital filters used in quantitative analysis has been discussed. The mathematics and statistics of the Kalman innovation filter has been compared to calibration methods such as principal components regression (PCR) and classical least squares (CLS).[92]

Gaussian-shaped bandpass digital filters using Fourier filter techniques have been applied to remove the absorbance associated with the spectral background in bovine plasma in which glucose was the analyte of interest. Digital filtering coupled with

PLS regression was used and digital filtering was found to improve predictive ability, minimize the importance of selecting a spectral range, and reduce the number of factors required to build the optimal PLS model.[93]

A narrow bandpass digital filter was applied to a shortened FTIR interferogram to isolate the spectral information related to a single analyte. The authors noted that since these filtered interferogram segments are analyte specific, univariate analysis may be used for quantitative analysis. The proposed method was found to perform similarly in predictive ability compared to conventional spectral calibration techniques, but was more efficient in terms of data storage and computation time.[94]

In an application using a wavelet transformation, quantitative analysis of 14 minerals contained in rock samples was measured using FTIR spectroscopy. The wavelet transform was used to separate mineralogical information contained in the spectrum from measurement noise. For this method wavelet coefficients that varied too much in replicate measured samples were ignored on the assumption those coefficients contained measurement errors. Samples which varied too little were also ignored as they were believed to be non-significant for discriminant or quantitative analysis. Using the remaining wavelet coefficients as the data set, an affine estimator was constructed for quantitative analysis; the method outperformed nonnegative least squares in many cases.[95]

6.2.11 *Mean Centering*

Mean centering is one of the most commonly used preprocessing methods. The use of error propagation formulas was applied to propose a theory for the effect of mean centering; this theory was tested using a simulation study. The observed results were that mean centering results in larger prediction errors when the data, (i) varies linearly with concentration, (ii) has a poorly defined baseline, and (iii) has no closure in the concentration levels.[96]

6.2.12 *Diffuse Reflectance and Scatter Correction*

Published research has examined multiple methods for correcting the light scattering effects due to different sizes and shapes of particles. There are several approaches to using multiplicative scatter correction (MSC) and the standard normal variate (SNV) transformation. In this paper eight different near infrared reflectance and transmittance data sets were then used to investigate the effect on predictive ability using first derivatives, second derivatives, SNV, and three variations of MSC as pretreatment methods.[97] Other classic papers describe the use of SNV, de-trending (DT), and multiplicative scatter correction (MSC). The basic problem set includes particle size variation, scattering effects, and multi-collinearity. The proposed approach to reduce these effects is to apply the standard normal variate (SNV) transformation at each wavelength to remove the multiplicative interferences of scatter and particle size; and de-trending (DT) to reduce the effects of curvilinearity in the baseline as well as baseline shifts.[98,99] Original discussions of these techniques are contained in three references[100–102] Note that piece-wise multiplicative scatter correction (PMSC) is the nonlinear version of MSC.[103]

6.2.13 *Miscellaneous Preprocessing Techniques*

Optimized scaling as OS-1 and OS-2, which combines the preprocessing step with the calibration step has been described. Two variations of the method are most common. OS-1 was devised for the situation where all the constituent concentrations are known. OS-2 was developed for the more common case where the reference values for only one constituent is used in calibration. Both OS-1 and OS-2 have been applied to several data sets and the results were compared to those obtained using the raw data, mean centered data, and data pretreated using MSC. When OS-2 is compared to MSC and several other preprocessing methods it has been found that OS-2 outperforms MSC in cases where the major constituents have large variations. In data sets where minor constituents were being predicted, MSC provided better predictive ability than OS-2.[104,105]

Adaptive Kalman filter have been used for quantifying analytes in the presence of uncharacterized interferents. Kalman filtering has been used previously in similar applications, but requires the overlap-free region is known beforehand. A proposed method determines the overlap-free region based on statistics using the residuals.[106]

6.3 Multi-Way Data Methods

6.3.1 *Introduction*

There are three main areas of interest in multi-way analysis: Curve-resolution, exploratory analysis, and calibration. Common models and review applications in these three areas will be discussed.

6.3.2 *Multi-way Data*

In standard multivariate data analysis data are arranged in a two-way structure; a table or a matrix. A typical example is a table where each row corresponds to a sample and each column to the absorbance at a particular wavelength. The two-way structure explicitly implies that for every sample the absorbance is determined at every wavelength and vice-versa, thus the data can be described using two indices, one defining the sample number and one defining the wavelength number. This arrangement is intimately connected to the data processing techniques subsequently used for analysis of the data (e.g., multivariate methods such as Regression, PCA, PLS, etc.). However, for a wide variety of data types a more appropriate structure would be a three-way table or array. An example could be a situation where for every sample for GC-IR data the absorbance is determined for several wavenumbers for different retention times. In such a case every data element can be logically indexed by three indices, one identifying the sample number, one the absorbance value at different wavenumbers, and one the retention time. Such data types are prime examples of problems that have been successfully exploited using multi-way analysis. However, consider a situation where spectral data are acquired on a sample at different chemical or physical circumstances, e.g., an IR spectrum measured at several different temperatures (or pH, or additive concentration or other experimental conditions that affects the analyte response in different relative proportions) on the same sample. Such data could also be arranged in three-way structure, indexed by sample, temperature, and wavenumber. Clearly, three-way data occur frequently, but are often not recognized as such, due to lack of awareness.[107–109]

6.3.3 *Important Terminology*

Some authors advocate using array notation,[110] others prefer notation adapted from tensor algebra[111,112] and others use modified matrix algebra notation.[113] It is unfortunate there is yet no consensus on the notation but it reflects that multi-way analysis is still in early stages. The terms mode, way and order are used more or less interchangeably though a distinction is sometimes made between the geometrical dimension of the array/cube — the number of ways — and the number of independent ways — which is the order/mode.[114–116] An ordinary two-way covariance matrix is thus a one-mode two-way array, because the variables are identical in the two ways. In reference 117, a very intuitive notation is used to describe different data structures depending on the number of modes referring to objects and the number of modes referring to variables in some sense.

6.3.4 *Multi-Way Algorithms*

Most multi-way models have their origin in the late sixties psychometrics society, and the interest in chemometrics was initiated much later. In many chemometrics research groups there is now a fruitful cooperation between psycho- and chemometricians. This cooperation resulted in several cooperative conferences. The main algorithms used for two-way analysis are PCA and PLS (partial least squares) regression. These methods have also been successfully applied to multi-way data, by rearranging (unfolding) the data to two-way data. As this approach is mostly justified by two-way methods simply being more accessible and as two-way methods have already been reviewed in detail we will only discuss methods that specifically use the multi-way structure of the data in the modeling. A paper describes the relationship among the most common three-way models and orders the models according to their complexity. He shows that the unfold PCA/Tucker1 model is shown to be the most general model in the sense that other models can be seen as constrained versions of unfold PCA. The most restricted and simple model is shown to be the trilinear model as it uses the fewest parameters.[116]

6.3.5 *Parafac/Gram Algorithm*

PCA is a bilinear method, which can be easily recognized by stating the mathematical model as:

$$x_{ij} = \sum_{f=1}^{F} a_{if} b_{jf} \tag{6.18}$$

where i designates the ith row and j the jth column of the data matrix $\mathbf{X}$. A natural extension to three-way data immediately follows as:

$$x_{ijk} = \sum_{f=1}^{F} a_{if} b_{jf} c_{kf} \tag{6.19}$$

This trilinear model forms the basis of PARAFAC (parallel factor analysis), CANDECOMP (canonical decomposition), and GRAM (generalized rank annihilation method). The PARAFAC[118] and the CANDECOMP[119] models are identical but developed in different contexts both in 1970.

6.3.6 *N-Mode Principal Component Analysis Algorithm*

During the sixties L. Tucker developed a series of three-way models, that are now known as the Tucker1, Tucker2, and Tucker3 models, also collectively called three-mode principal component analysis, or originally three-mode factor analysis.[120,121] These models are not unique as the trilinear models, but they have been extensively used primarily for exploratory analysis. The Tucker1 is essentially the well-known unfolding technique, where the data array is rearranged to a matrix which is then subjected to ordinary two-way PCA. The Tucker2 is a model of intermediate complexity compared with the Tucker1 and Tucker2 model. It will not be discussed much here as only few applications in chemistry have yet been seen. The Tucker3 model is a quadrilinear model in its three-way version as the model consists of four sets of parameters each set being conditionally linear. An alternating least squares algorithm called TUCKALS3 (or TUCKALS2 for the Tucker2 model) for estimating the model in a least squares sense with orthonormal loadings has been derived and described.[122–124]

6.4 Infrared Spectral Searching for Identification and Qualitative Analysis

Note: Computer Searching of Infrared Spectra (from personal interview with Clara D. Craver, D.Sc.)[4]

6.4.1 *State of the Art in Spectral Searching*

"Spectral searching is intended as a screening method to assist the analyst and is not an absolute identification technique.... Spectral searching is not intended to replace an expert in infrared spectroscopy. Spectral searching should not be used without suitable training," (Reference ASTM E 2310, *Standard Guide for Use of Spectral Searching by Curve Matching Algorithms with Data Recorded Using Mid Infrared Spectroscopy*). (Also note that ASTM International committees have published several practices for spectral searching including, E204, E1252, E1790, and E2310. See Chapter 7 and the index for more detailed descriptions of these ASTM Practices and similar techniques for qualitative analysis.

6.4.1.1 *Search Parameter Selection*

There are a wide variety of spectra and consideration of this fact should drive the choice of search options, especially for the initial search into any particular reference file. Additionally, the band shape and peak positions may be altered by the instrumentation used and by the physical state of the sample. Broadly, most spectra would be encompassed using the following description; spectra may have as few as 1–4 strong absorption peaks, and not uncommonly as many as 12 or more peaks whose intensity is the order of half of the strongest. Very often matching of significantly weaker peaks in the fingerprinting region is required for identification. These facts appear to call for two stepwise searches, especially in large spectral files.

There are series of molecules with weak to medium key group frequency peaks, which need a high weighting factor. Two of the regions where this higher weight could be especially important are "about" 3800-3000 cm^{-1} and 2700-1900 cm^{-1}. Note that one reason for not extending this special weighting routinely to lower frequency regions where weak bands may be critical, such as C=C, is that water vapor bands are most troublesome there. If water vapor is subtracted first, then weak bands around 1600 cm^{-1}, which are important in structure identification, could also be given high weight. Therefore, an important "keyed search option" could include one for "has water vapor and CO_2 been in balance or subtracted out" or "are water vapor or CO_2 still in the spectrum being searched".

There are classes of materials, e.g. homologous series, polymers, copolymers, minerals, gemstones, etc. for which there are about 1–6 peaks and only much weaker; these are useful for important identifying characteristics.

Absorption peaks may be sharp or broad. Band widths of principal peaks within about the top 5% of the peak's intensity may be as little as 2 cm^{-1} or as great as 10–30 cm^{-1} or more. Higher weight might be given to peaks that are a match for the reference spectra within $\pm 2\,cm^{-1}$, and at a lower weight for matching within $\pm 5\,cm^{-1}$ might be kept in the high ranking file.

These wide fluctuations suggest that clear separation of some categories of spectra is an optimal choice for long-term development of reference files. A good spectral search system needs to encompass strategies for all of these considerations.

6.4.2 *Identification of Polymers by Computer Search of Vibrational Spectra (A Summary)*

A successful analysis requires more than obtaining a single spectrum and running a computer search...

6.4.2.1 *Start with the Basics...*

It is the responsibility of the lab manager or designated chemist to find out and communicate to the chemist or technician performing the analysis all available, relevant information that could possibly be useful for identification of an unknown test material. The purpose of the test is to ascertain the composition of the test sample, not to test the qualifications of the analyst! The need to provide information directly to the person performing the problem solving is important and key to the success of any identification problem.

First of all, are there any safety concerns the analyst should be made aware of, such as flammability, sharpness, brittleness, inhalation concerns, or toxicity concerns? Other decisions needed: is primary consideration the time required or the certainty and completeness of results? This information may determine the extent of cleanup of the sample and number of searches possible. Is any possible answer known? Is the history or background of the sample known and to what extent of detail?

It is important that the person wanting a complex analysis be notified that the more information the analyst is given in advance, the better decision the analyst can make of how to proceed with their problem solving. The completeness and accuracy of this information affects the probability that the results provided are correct and complete and with the least unnecessary time being spent. This understanding is important. Other helpful information includes what is the source of the sample and what is needed in the results.

6.4.2.2 *Solving a Specific Problem: Polymers*

Many polymers or polymeric products cannot be correctly or fully identified especially by surface spectra only, as by ATR, or without inspection for particles, layers, streaks, or evidences of non-uniformity. Additionally polymeric materials may not be identifiable or may be wrongly identified by computer searching of the spectrum without removal of additives which can be present in significant concentrations, such as plasticizers or fillers. Laboratory treatment such as solvent extraction for partitioning of components with different solubility properties may be necessary.

Approaching the problem of identifying polymers by spectral search systems requires looking at all of the associated details of sampling, measuring the spectra, and seeking or creating optimum kind(s) or referencing libraries. After some of the options are considered, management decisions about how complex and varied the search programs may be allowed or are required. Multiple library systems and a choice of several search algorithms will probably prove optimum for a laboratory identifying or characterizing highly diverse polymers.

6.4.2.3 *Sampling and Processing*

In collection of samples for analysis, any visually different specimens should be kept separated for later independent analysis.

When there is a need for purification or fractionation:

- Multilayer films may be separated by mild solvent soaking
- Surface coatings may yield to mild solvents, e.g., hexane, chloroform, ethanol

- Powder layers are seldom optimum for measurement, thus try preparation by:
  - Pellet press
  - Dissolving to provide a cast film on KRS-5 or similar substrate window
- Pieces, pellets
  - Press into layers of useful thickness
  - Slice into film or cut one flat side for ATR
  - Dissolve and produce cast film on appropriate window
- Resinous, gummy, viscous, spreadable forms
  - Make film by pressing
  - Fractionate mixtures and make cast films of different solvent fractions
- Soluble polymers (PVAs, PEGs, etc.)
  - Cast film neat
  - Dilute and cast film
  - Fractionate by decreasing solubility of some components and precipitating them out of solution; measure precipitate and soluble materials
- Micro-sample
  - Carefully measure using microscope — can mask areas of sample to obtain multiple spectra for inhomogeneous samples

6.4.2.4 *Selection of Optimum Specimen Measuring Conditions*

- Transmittance
- Attenuated total reflection
- Reflection
- Photoacoustic spectra
- Diffuse Reflectance

Note that search requirements prefer that test spectrum and reference spectrum be measured using the identical technique

6.4.2.5 *Selection of Measurement Scan Conditions*

(a) Resolution: Generally $4\,cm^{-1}$ is adequate for polymers; consideration is needed to match resolution of sample spectrum to the reference spectra being searched. At $8\,cm^{-1}$ some significant spectral detail is lost. There are a few exceptions where $8\,cm^{-1}$ or $16\,cm^{-1}$ is acceptable, unless a process application requires the speed as a key priority over spectral information content.

(b) Scan Speed: Slower speeds are usually worth the time invested. Selection of scan speed is driven by energy losses from the sample size or area available, scattering by specimen, fillers, etc.

6.4.2.6 *Automated Search Strategy*

The strategy used for searching must take into consideration a number of search options and all the information provided with the sample.

Straightforward searching is needed:

1. To establish the complexity of the sample (details vary with sample type and origin)
2. Use as a routine peak search for ten (approximately) major IR absorption bands
3. If spectrum is strong (with some bands over 5 Au), use regions of no detected peaks and regions where bands are weaker (exclude strong band regions) for searches
4. If spectrum is weak (most bands well below 3 Au) search entire spectrum
5. To allow for possibility to identify polymers or blends, search for weak bands which correspond to reliable group frequency correlations, e.g., indications of carbonyls or aromatics, P=O, S=O, SiH, SiO, the bending frequencies for olefinic types, etc.

6.4.2.7 *Searching Strategy for General Sample Types and Problems*

To assist the analyst with search strategy options, questions related to the sample are important, these include:

1. Is the sample unblended as a pure incoming raw material?
   a. Check against known reference spectrum of this known material
   b. Subtract best hit if material is known to be in reference database
   c. Expand difference spectrum to approximately 10 X or higher depending upon QC requirements
   d. Evaluate the expanded difference spectrum for meaning
2. Is the sample a product — as a single polymer, a blend of polymers, a copolymer, and what are the potential additives suspected?
   a. Check test spectrum against reference library spectra of polymers
   b. Subtract closest pure polymer spectrum from test spectrum
   c. Expand difference spectrum (approximately 5–10 X) and repeat using search against reference database
   d. Repeat steps b and c for multiple iterations or until residual is in noise levels

3. Does the sample contain suspected additives? In polymers additives may have spectra similar to polymers. If they are suspected then fractionation is a necessity to make a definitive identification of all components in the mixture.

   a. Attempt fractionation via extraction, note that plasticizers, antioxidants, fillers, slip agents, adhesives, surfactants will require different fractionation techniques. *Note: Common solvents used for infrared analysis of polymers include chloroform, acetone, methanol, and hexane. Cast films are often made by dissolving or extracting a soluble solid or liquid sample in one of the above solvents; filtering the extract; evaporating most of the solvent in a suitable fume hood; and casting a thin film (by evaporation) onto an infrared-transparent window material. Note that each solvent has absorption bands as well and these bands must be considered during spectral interpretation activities.*

4. Is the test sample a new synthetic product? In this case there may be no comparative spectrum in the reference library. The steps used in Example 2 of this section may be useful for determining the optimum material identification.
5. Is the material completely new and unknown? In this case where no known reference spectrum is known to exist, one must refer to spectral interpretation using established techniques. Following the procedure used in Example 2 may also be helpful in this case.
6. The material is obviously degraded either by photo or thermal degradation (carbonized) and is no longer chemically similar to the original starting material. Following the procedure outlined in Example 2 may be helpful. Obtaining a reference database of polymer degradation products (by exposing known polymer samples to such conditions and then recording their spectra) is also particularly helpful.
7. Nothing is known about the sample. In this type of examination, following the procedures outlined in Example 2 is also quite useful. Other analytical techniques may be required when nothing is known about the sample and there are no similar compounds in the spectral reference library.

Note: If the search algorithm and peaks chosen for an initial search result in low match scores to a high quality polymer library, improved results may occur if subsequent searches for the test sample against the reference library are performed after excluding weak and medium peaks in spectral regions where common or strongly absorbing additives are likely to occur. Some judgment by the analyst may be required to select for removal the particularly well-known additives or modifiers. (An example of problematic modifiers as key components of polymers often causing difficulty for identification and interpretation include phthalates and trimellitate bands. These band regions should be excluded from several searches to identify basic polymer types and then specifically search these regions to identify their presence). It is also possible that selected absorption peaks could be eliminated over a specific range of intensities prior to initiating the search routine.

For copolymers it is probably essential to retain some key bands, even weak ones for the major polymeric structures such as OH, NH, C=O, aromatic ring, and olefinic bands that remain after polymerization reactions. This degree of sophistication in describing combinations of peaks to search, or subtract would require highly specific planning for the sets of bands that might be deleted or spared.

An obvious second stage of search is presented if one component in a mixture is identified and subtracted from the spectrum of the unknown sample. A new search on the remainder of the spectrum can be used to identify other component(s). Stepwise identification is possible.

For searching, consider:

1. Peak searching of about 10–15 major peaks
2. Include as "ranking match" information from regions where there are no peaks
3. Treatment of weak peaks is more subjective and may call for more parameter choices
4. Searches may be required for select regions of the spectrum only
5. Answers may be found without finding complete spectrum matching — the power of subtraction is important as a tool
6. Search routines may find closely related polymers (test vs. reference) but not exact matches. Fine structure details are known to be very specific in characterizing the properties of polymers — thus means of matching fine structure must be obtained and utilized
7. Guidelines using frequency characteristics for sets of peaks is important and should be used for sub-searches

6.4.3 *Example of Generalized ID Problem Solving Workflow*

1. Collect all available historical and chemical data relative to the test sample (optional at start)
2. If multicomponent sample is involved then perform appropriate fractionation (describe procedures per sample type)

   a. Sampling and Processing

3. Prepare sample and establish spectral measurement parameters
   a. Selection of Optimum Specimen Measuring Conditions
   b. Selection of Scan Conditions
4. Acquire spectrum and preprocess
   a. Water and Carbon Dioxide Removal and Baseline Correction
   b. Correct Spectrum for Measurement Conditions: ATR, PAS, DRIFTS, etc.
5. Perform spectral search against all pertinent libraries
   a. Customized Strategy for Computer Search Per Application
6. Perform results assessment and probability tests
   a. Probability Measures and Criteria Selected
7. Evaluate Search Results
   a. If Passes Probability Criteria for 2-5 Reference Spectra, Go To #(13)
   b. If Fails Probability Criteria, Go To #(8)
   c. If Too Many Comparisons Pass Criteria (I.E.>5), Go To #(10)
8. Create pre-filter based on presence or absence of "important" spectral features
   a. Repeat #(5) and #(6)
   b. If Passes Probability Criteria for 2-5 Reference Spectra, Go To #(13)
   c. If Fails Probability Criteria, Go To #(9)
   d. If Too Many Comparisons Pass Criteria (I.E.>5), Go To #(10)
9. If results are inconclusive, enter information and weight of spectral regions based on application knowledge, statistics, and results of general search and perform refined Search and evaluate closest matches
   a. If Passes Probability Criteria for 2-5 Reference Spectra, Go To #(13)
   b. If Fails Probability Criteria, Insufficient Information to Find Solution, Go To #(14)
   c. If Too Many Comparisons Pass Criteria (I.E.>5), Go To #(10)
10. Apply context filter (color, melting point, etc.) if multiple spectra have good matches
    a. Search Based on Indicator Variables Other Than Spectral Information
       I. If Excellent Results, Then Report Matches and Go To #(18)
       II. If no Matches. Then Go To #(10b)
    b. Search Database Again After Adding Verified Suspect Material Spectra to Reference Database, Go To #(6)

11. If match is good but name is meaningless, then cross check name with other chemical and commercial data bases
    a. Perform On-Line Web Search for Chemical Structure or Formulation Relative to Search Located Trade Name or Common Name
        I. When Information Found, Go To #(18)
        II. If No Additional Information Found, Go To #(18)
12. Are the best matches PASSING Statistical tests?
    a. If satisfied with ID, GO TO #(13)
    b. If not satisfied or if ID suspect, Go to #(11)
13. This is your best solution
    a. IF finished, go to #(18)
    b. IF Interpretation of structural information is desired, Go to #(15)
14. Insufficient Information exists to find solution, Go to #(15)
15. Search test spectrum using a Spectral Interpreter
    a. If Passes Probability Criteria For 2-5 Reference Spectra, Go To #(16)
    b. If Fails Probability Criteria, Go To #(17)
    c. If Too Many Comparisons Pass Criteria (I.E.$>$5), Go To #(17)
16. Find commonality (if any) between all matched reference spectra and GO TO #(17)
17. Search test spectrum against Model Functional Group and Fingerprint Spectra*
    a. If Passes Probability Criteria For 2-5 Reference Spectra, Find All Commonalities and Go To #(18)
    b. If All Searches Fail Probability Criteria, ID Has Failed, Go To #(18)
    c. If Too Many Comparisons Pass Criteria (I.E.$>$5), Find All Commonalities And Go To #(18)
18. Report Compiled Final Results, Make Recommendations — End

6.4.4 *Description of a General Analytical Forensic Screening Protocol*

The following information is typically collected by the analyst prior to performing spectral searches on reference library collections. This information effectively prunes the problem to a manageable search resulting in a few candidates/suspects prior to the search. The intellectual problem solving process is context dependent and requires standard forensic investigative techniques to eliminate the majority of compounds from the search in order to obtain a limited candidate list of potential suspects. The analyst will follow a more or less standard set of questions to narrow the field of suspects, for example:

1. What information is desired from this materials identification?
    a. What is in this sample (major component(s))?
    b. What is wrong with this sample (what are the major and minor contaminant(s))? (Including the spectrum of an uncontaminated control sample would be useful for comparison).
    c. What is different about this sample as compared to previous samples of the same or similar material? (Must include the spectrum of a control sample for comparison).
    d. How can I relate the chemistry from the spectral information to the overall product performance of the sample? (Must include the spectrum of a control sample for comparison).
    e. Is this sample what I think it is or have stated in my suspects list? (Including the spectrum of a control sample would be helpful for comparison).
    f. I want the software to perform an interpretation of the spectrum by functional groups.
    g. I want a basic report on commercial aspects of this material or these materials.
2. Is this the first known experience with this sample, or has it been seen before?
3. Do you have any additional analytical test data on this sample? For example: other spectra or chromatograms, melting point, boiling point, refractive index, etc.?
4. Where and when was the sample obtained?
5. What is the sample preparation for spectral measurement (Neat, powder, KBr pellet, cast film, polymer film, etc.)?
6. What is the spectral measurement method (ATR, Transmittance, DRIFTS, Specular reflectance, PAS, etc.)?
7. If this is a manufactured material has anything been changed in producing or sampling it since it was last measured?
8. What is the sample: Phase, Color, Shape, Hardness, Texture, Stickiness (Tack), Homogeneity, Physical or manufacturer markings, Optical density, Odor, Coating layer types? Is the material brittle, flammable, caustic, dusty, etc.? Does the analyst know the Source, Trade Name, Manufacturer, or whereabouts of ingredient label for materials in the sample? Is an MSDS sheet available for some or all of the components?

9. What chemical process was used to produce the sample? What are raw materials or catalysts used for the chemical reaction or manufacturing steps? What type of machinery is used to produce the material?
10. Is the potential for contamination based on standard materials found near machinery or used to lubricate, coat, or clean machinery during the manufacturing or treating process? Is the production of material near areas where food or beverages are ingested? Are samples stored or sold in such a way that standard handling might contaminate the sample with other materials in its environment? Was the sample placed in a clean container when obtained or was it brought in a contaminated container? What is a potential list of these contaminants?
11. What is the potential for degradation and what are potential suspects for this contamination? Has sample had excessive photo, high energy radiation, or heat exposure; or excessive mechanical pressure or chemical (acid or base) exposure?
12. Does the sample contain water soluble (polar) components in a mixture with organic soluble components (non-polar)? Samples in this grouping would include rubbers, adhesives, and some polymer systems. (These samples should be subjected to water and organic solvent extraction, with each extract being measured as a cast film using transmittance infrared spectroscopy. The individual components are more easily identified using this simple separation technique).
13. Can analyst list any ingredients contained within the mixture or list any perceived suspects? For example, ACME Gunk #710, Polyethylene, Floor grime with heat treated polymer and machine oil, etc. List all possible suspects.
14. List any other information you have on this sample.
15. Perform analysis, comparison, and validation of the sample using all available evidence.

6.4.4.1 *Steps to Follow in a Generalized Forensic Investigation*

1. Make sure the sample is preserved in its original form.
2. Supplement careful observations with experience and hunches about what to look for, and how to find it, in order to get to the bottom of what is going on.
3. The analyst has the responsibility of ensuring that as little of the input sample and data as possible is damaged or lost, thereby lessening its information content value as contributing to a meaningful reconstruction of what occurred.
4. Take careful notes of what was done and observed during the investigation.
5. Retain a record for a provable chain of custody.
6. Log all actions performed during the analysis.

7. Note as many pertinent facts as you can as you prepare, gather, and analyze evidence about the sample.
8. Only use a portion of the sample for analysis and carefully retain and label the rest. (For accurate analysis, do not modify the sample in any way).
9. Make certain the spectrum obtained is of the highest possible quality and taken using the proper set-up conditions for the instrument.
10. Profile the sample:
    a. Profiling Inputs: this is basically a collection of all evidence, including anything found on the scene (i.e. fibers, paint chips, etc.) and anything derived from the scene (for example, observations, notes, measurements, etc.).
    b. Decision Process Models: in this second step, evidence is arranged to locate any types of patterns, such as whether or not the sample is part of a series of samples from the same source or similar source, what do the samples have in common, etc.
    c. Sample History Assessment: now that the evidence has been organized, the sampling scene is reconstructed. Analysts use patterns to determine the sample composition and what could have happened to the sample and in what sequence.
    d. Sample Profile: the combined first three steps are used to create a sample profile incorporating the sample conditions, physical and chemical properties, and history of the sample. Also, the analyst uses this information to decide on the best way to measure the sample and compare the suspects based on their sample profiles.
    e. The Analysis: the sample profile is compared to the list of suspects for a match. The profile may be reassessed if no leads are found or if new information is learned.
    f. The Results: When a highly probable result candidate is found, the result is identified, and compared to the reference database. If the analyst has additional reason to conclude that the candidate is the accurate result, a result is posted.

6.4.5 *The Big Questions for Spectral Search Methods*

1. How to handle/solve complex mixture problems with an elegant working solution providing a meaningful match index with a well-defined statistical probability for success?
2. Basic question: Are first-order IR spectra underdetermined for solving many of the complex mixture problems (What is the limitation or threshold for solving such problems)?

3. Are all possible outcomes understood when applying expert search type algorithms? How do small spectral changes affect such results?
4. How does one solve the pure component peak set selection problem with complex mixtures?
5. Is it possible to ask a specific set of questions (up to 20) to enable a user answering them to reach a final solution to a complex mixture analysis problem?
6. The reference data base must be compatible with the unknown sample spectrum. Will these reference databases require additional search data such as melting point (m.p.), boiling point (b.p.), complete peak sets, etc.?
7. The confidence (or quality) of results obtained for a spectral search are limited by the quality of the sample and reference spectra. What are the parameters regulating this?
8. Are negative search results useful? This refers to identifying compounds that do not represent the unknown sample.
9. How can a useful knowledge base be made practicable and preferred by minimally trained users over automatic searching/matching techniques, e.g., increased interaction with the process (i.e., less automation) is generally less attractive to the minimally trained, general user base?
10. What are the key aspects or ingredients for a knowledge based expert match system?

6.4.6 *Other Important Considerations for Spectral Searches*

1. How much of the work of interpretation by experts can be automated or included in an automated software system? Note: This system could initially be set-up for specific types of compounds or problems by an expert analyst. It may be industry specific, or customized for each specific factory or laboratory, etc.
2. How would one describe the capabilities of a Spectrum Problem Solver? What would it be capable of? How much user interaction is required? What is the probability of successful identification and interpretation, and for what types of problems? What are the limitations of automating these functions?
3. What are the steps and procedures involved when identifying and interpreting vibrational spectra?

    a. What parameters other than spectral information are useful for identification and interpretation? For example, observations related to the sample might include:

        i. Melting or boiling point

   ii. Sample state and color
   iii. Source or origin of sample
   iv. Presence or absence of odor
   v. Solubility (simple solvent tests)
   vi. Refractive index (of liquids)

b. Mentally compose limited set of possible answers/functional groups/ compounds (mental list of potential hits)
c. Obtain a single or multiple high quality spectrum(a) of unknown
d. Obtain high quality reference spectrum(a)
e. Make first pass look at group frequencies and identify compound type (i.e., class)

   i. Group frequency *peak locations* (above 1500 cm^{-1})
   ii. Spectral regions where strongest bands occur
   iii. Spectral regions where no bands occur

f. Review fingerprint region to verify assumptions regarding sample classification, e.g., fingerprint region features (below 1500 cm^{-1})

   i. Band shape: broad (b), very broad (vb), sharp (s), very sharp (vs)
   ii. Band intensity: weak (w), medium (m), intense (i)
   iii. Regions with absence of bands
   iv. Locations for *sets of bands*

g. Identify individual functional groups by looking at peak/band properties of fingerprint region and the presence or absence of specific bands. What might a peak-picker function have?

   i. Identifies peak location, intensity, and width

      1. Position/location — Precise peak location identified (from 4000–500 cm^{-1})
      2. Intensity — assign numerical values over a range (e.g. 1–10)
      3. Widths — classified as sharp, average/medium, or broad

h. Look for inconsistencies and interpret possible contaminant bands or the characteristics of a classic mixture problem. Is it a simple mixture or a complex, multi-component mixture?
i. Apply search algorithms for spectral library comparisons. The limitations of this method is that it is limited to 2–5 component mixtures provided each of the individual components exceed approximately 1–2% absolute concentration. False positive matches can occur between very similar compounds due to band shifts. False negatives result as a function of spectral shifts and poor threshold selection for finding peaks. (N.B. Some of the best research efforts

to date generally result in approximately 25–30% false positives and false negatives for identification of the individual components in relatively complex mixtures.)

i. How can the 'Hit' or 'Match' index be derived such that there is a rigorous statistical or probabilistic meaning?
ii. 'Forward' search algorithms comparing unknowns to reference libraries are only highly effective for 'pure' compounds — not mixtures.
iii. When applying curve fitting or 'reverse' search algorithms (N.B. false assumption is that mixture spectra are a linear combination of reference spectra of the individual pure component spectra). Would it be possible to compensate for this non-linear additively and use pure component spectra to generate representative mixture spectra? (N.B. Solvent, temperature, humidity, and concentration are all influencers of spectral shapes and instrumental changes that affect spectral shape. These nonlinearities will often appear as peak location shifts, and occasionally as band shape changes).
iv. Using peak location tables or detailed windowed comparisons to reference spectra has proven only marginally useful for mixture problems.

6.4.7 *Major Factors Affecting General Spectral Search Methods (Spectral Quality)*

1. Baseline problems (e.g. offset, scaling, flatness, variable pathlength)
2. Sample properties:

   a. Neat
   b. Solution in solvent
   c. Solid (pure or mixture)
   d. Solid shape issues
   e. Hardness
   f. Liquid (pure or mixture)
   g. Viscosity
   h. Gas (pure or mixture)
   i. Stability
   j. Hazardous
   k. Contaminated (Contaminant at greater than 2%)
   l. Contaminated (Contaminant at less than 2%)
   m. Sample purity (constant versus reference library)
   n. Simple mixture (less than 5 components with more than 2% of each)

o. Complex mixture (more than 5 components with trace levels of some)
p. Sample too small (only trace quantity)
q. Sample too large (for sample compartment)
r. Must be measured nondestructively
s. Not representative of bulk material
t. Film (free)
u. Film (coated)
v. Film (composite)
w. Powder (crystalline)
x. Powder (amorphous)
y. Gel/paste
z. Wax
aa. Slurry

3. Sample presentation method:
   a. Transmittance
      i. Cast film
      ii. Solid film
      iii. Liquid cell
   b. Hydrate
   c. ATR
      i. Liquid
      ii. Slurry
      iii. Wax/paste
   d. DRIFTS
      i. Solid
      ii. Powder
      iii. Film/sheet
   e. PAS (photoacoustic spectroscopy)
      i. Solid
      ii. Slurry
      iii. Liquid
4. Sample spectrum quality issues
   a. Noise on spectrum
   b. Unknown contamination
   c. Sample thickness (pathlength) — Out of range peaks
   d. Absorbance linearity (Beer's law deviations) — uniformity of sample pathlength

e. Uniformity of sample across beam (transmittance measurements)
f. Homogeneity of sample
g. Interferences (H_2O, CO_2)
h. Instrument malfunction
i. Variable sample presentation (i.e., sample presentation errors)
j. Spectral resolution
k. Sample contact with optical element (ATR)

5. Reference spectrum quality issues

   a. No representative spectrum(a) in library
   b. Poor quality representative spectrum(a)
   c. Different measurement conditions for reference spectrum(a)
   d. Too many similar spectra
   e. Spurious bands (due to instrument or contaminants)
   f. Impurities
   g. Carbon dioxide bands
   h. Water vapor bands
   i. Mulling agent bands
   j. Halide salt artifacts (different window types)
   k. Water of hydration in sample
   l. Solvent bands
   m. Wavenumber range compatibility

6. Crystallinity differences due to sample presentation

   a. Annealed solid
   b. Crystalline melt
   c. Mull
   d. KBr pellet
   e. Cast film
   f. Liquid melt
   g. Neat solution

7. Spectral search issues

   a. Quality of unknown spectrum (i.e., baseline and X, Y correction)
   b. Quality of reference spectrum
   c. Preprocessing method selection
   d. Spectral window selection
   e. Algorithm type selection for search methods

f. Algorithm limitations

   i. All are susceptible to peak shifts and line shape changes
   ii. Current algorithms provide no probabilistic assessment for match confidence. Current methods have match or hit quality indices with no statistical meaning/interpretation
   iii. Methods do not accurately discern complex mixtures
   iv. Current methods are not intelligent — they rely mostly on user expertise
   v. Current methods do not allow 'weighting' of spectral regions or peaks depending upon the unique aspects of each problem solving exercise
   vi. Current methods do not provide for searching text variables to include independent variables, such as boiling point, melting point, refractive index, etc.

6.4.8 *Classic Spectral Search Algorithms and Terminology*

1. Spectral searching or comparison
2. Peak searching (set comparison)
3. Additional indicator or text variable comparisons (physical data or measurements)
4. Hit quality value ('similarity' value vs. library reference sample)

   a. Euclidean distance algorithm:

$$\mathrm{HQI} = \left\{ \sum_{i=1}^{n} (u_i - r_i)^2 \right\}^{\frac{1}{2}} \tag{6.20}$$

   b. First derivative Euclidean distance algorithm:

$$\mathrm{HQI} = \left\{ \sum_{i=1}^{n} (\delta_1 u_i - \delta_1 r_i)^2 \right\}^{\frac{1}{2}} \tag{6.21}$$

   c. Sum of differences as:

$$\mathrm{HQI} = \sum_{i=1}^{n} (u_i - r_i) \tag{6.22}$$

d. Correlation [row matrix (row vector) dot product] as:

$$HQI = (U \bullet R) = \sum_{i=1}^{n} U_i R_i \tag{6.23}$$

Hit quality index (table with hit quality values) with added probability statistic

5. Suggested new approaches, such as genetic algorithms, Hamming networks, pattern recognition, wavelets, and neural network learning systems are sometimes discussed but have not been commercially implemented.

6.4.8.1 *Standard Algorithms and Nomenclature used with Standard Commercial Software*

A spectrum in the full-spectrum format can also be considered as a point in a multidimensional space where each dimension in the space corresponds to a specific wavenumber location in the spectrum and the corresponding intensity is the distance along that dimension. In this representation, two points that are very close together in this hyper-space would be from the same or very similar molecules. A very common measure of closeness in a multidimensional space is the Euclidean distance between the two points. In the Euclidean distance calculation, x_i is the intensity at the ith dimension (wavenumber position) in the sample spectrum and r_i is the corresponding value for a reference spectrum from the spectral library. The squared difference is summed over all of the wavenumber positions and then the square root is calculated.

$$E_d = \sqrt{\Sigma (x_i - r_i)^2} \tag{6.24}$$

In a spectral search the points from the library spectra closest to the sample are determined and the distance values saved in increasing order. If the two spectra are "identical" the distance between them will be zero. However with real spectral data there is sufficient spectral noise and baseline drift that a perfect match rarely occurs even when spectra are obtained from the same material on the same instrument.

Another problem with real spectral data is the differences in peak intensity caused by changes in the sample pathlength or concentration. The use of a standardization or normalization technique is required to insure that all of the spectra share a common scale. One option is to scale all of the spectra between zero and a maximum value of one absorbance unit. The same normalization process must be used on the sample spectrum as well as each spectrum in the reference library. The distance algorithm shown below is used in Nicolet's software to compensate for intensity differences. Since the square root operation does not change the relative order of the distances we have eliminated it to reduce computation times and we normalize by the sum of the squares for the sample spectrum. This insures that all of the match values fall between zero and one. One somewhat arbitrary issue in spectral matching is defining a perfect match as zero distance or as a match value of 100%. We have chosen to subtract the normalized distance from one to give a perfect match value of one for identical spectra. We then multiply it by 100 to give a percent similarity value.

$$M_{SQ} = [1 - (N/D)]^{*}100; \text{ where } N = \Sigma(x_i - r_i)^2 \text{ and } D = \Sigma(x_i)^2 \quad (6.25)$$

If the unknown spectrum and the reference library contain perfect spectra with no contaminants or sampling differences, and if the library contains a spectrum of the sample compound, then almost any search algorithm will identify the unknown. However in a number of cases, we noticed that the search results were not satisfactory because the squared values strongly weighted differences in the large peaks relative to the smaller peaks. In these cases, the sum of the absolute differences gave better results.

$$M_{AB} = [1 - (N/D)]^{*}100; \text{ where } N = \Sigma|x_i - r_i| \text{ and } D = \Sigma|x_i| \quad (6.26)$$

With certain sample preparation techniques, spectral artifacts and scatter from particles in the sample can cause significant tilt to the spectral baseline. While this may actually be an important part of the spectral signature for a specific material it generally creates problems in spectral searching. The simplest way to correct for baseline variation and broad non-specific features in an automated fashion is to apply a first derivative computation. The "squared derivative" calculation is shown in the equation below where the comparison is made on the difference between adjacent points in the two spectra.

$$M_{SD} = [1 - (N/D)]^{*}100; \text{ where } N = \Sigma(y_i - s_i)^2 \text{ and } D = \Sigma(y_i)^2 \quad (6.27)$$

$$y_i = x_i - x_{i+1} \text{ and } s_i = r_i - r_{i+1}$$

The equivalent calculation for the "absolute derivative" is also included for completeness.

$$M_{AD} = [1 - (N/D)]^{*}100; \text{ where } N = \Sigma(y_i - s_i) \text{ and } D = \Sigma|y_i| \quad (6.28)$$

$$y_i = x_i - x_{i+1} \text{ and } s_i = r_i - r_{i+1}$$

A final search algorithm that is related to some of the chemometric techniques used in NIR spectroscopy is the correlation calculation shown below.

$$M_{CR} = N/D^{*}100; \text{ where } N = [(\Sigma y_{i*}s_i)^{*}(\Sigma y_{i*}s_i)] \quad (6.29)$$

$$D = [(\Sigma y_{i*}y_i)^{*}(\Sigma s_{i*}s_i)]; \quad y_i = x_i - x_{i+1} \text{ and } s_i = r_i - r_{i+1}$$

While each of these algorithms gives slightly different results, there is no clear "a priori" choice on which is better for a specific application. In general the correlation algorithm performs well and appears to work in a broad range of applications. There is some evidence that the absolute derivative works better for Raman spectra acquired with different instruments.

6.4.9 *Review Potential for Improving State-of-the-Art*

6.4.9.1 *Features of an Improved Spectral Interpretation System*

1. Spectral artifact removal (water and CO_2 in particular)
   a. Recommend improving these features using reference spectra at different resolutions for artifact removal or features to eliminate these regions from a spectral search
2. X-shift alignment function (maximum correlation, minimum distance)
3. Automated peak selection
4. Peak tables are generated for all reference and unknown spectra
5. Peak search against database including peak table for each reference spectrum
   a. Forward search — all peaks for the unknown sample are in the reference spectrum [the reference spectrum can have additional peaks (components)]
   b. Reverse search — all peaks for the reference standard are in the unknown spectrum [the unknown spectrum can have additional peaks (components)]
6. Peak threshold selection — the higher the threshold the fewer peaks are selected — this is important for identification purposes
7. Spectral similarity subtraction
8. Multi-level filter algorithm
9. Spectral similarity index ("Hit Quality") with interpretable and probabilistic meaning and resultant statistical reporting
10. Incorporation of derivatives and variations of them for library searching and for library data (all relevant preprocessing steps added)
11. Provision for 'weighting' of spectral regions or peaks depending upon the unique aspects of each problem solving exercise
12. Provision for searching text variables to include independent variables, such as boiling point, melting point, refractive index, absorptivity, etc.
13. Provision for saving unknown spectra into a reference database and later identifying their occurrence with comment line from past work on a similar sample (What is new here is flagging samples as separate from standard reference database and addition of large (100 word) comment section)

14. Provision to use cursor to click bands of highest priority for search and then enter weighting factor for each (e.g., 1x, 2x, etc.)

6.4.9.2 *Additional Chemical or Physical Properties of Compounds that can be used for Identification*

1. Color
2. Physical state at room temperature
3. Refractive index
4. Melting point
5. Boiling point
6. Freezing point
7. Cloud point
8. Flash point
9. Density/specific gravity
10. Dielectric constant
11. Conductivity
12. pH
13. UV cutoff
14. Resistivity
15. Compression modulus
16. Compression strength
17. Hardness
18. Thermal expansion
19. Specific heat
20. Stain affinity
21. Flame emission spectrum
22. Absorptivity (ε)
23. Critical temperature
24. Critical pressure
25. Critical volume
26. Hydrophobicity
27. Solubility
28. DSC (Differential scanning calorimetry)
29. Color (visible spectrum)
30. UV spectrum
31. Raman spectrum
32. Near infrared spectrum

33. Fluorescence spectrum
34. NMR spectrum
35. Mass spectrum
36. Vibrational Circular Dichroism spectrum
37. GC separation
38. HPLC separation
39. TLC
40. Chemical identification and classification tests

6.4.9.3 *Important Guidelines, Cautions, and General Rules for using Spectral Search Methods*

1. Previous attempts at Expert Systems for identifying individual components in mixtures have produced typical results such as:
   a. Sensitivity of 69–74% (identifying all components present that were present 26–31% false negatives (i.e., a determination that a component is *not* present when it in fact *is* present)
   b. Specificity of 98% (distinguishing between components accurately)
   c. False positives of 28% (identifying components as present that were not)
2. When using spectral searching approaches, every entry in a reference database will match the unknown sample to some degree (The question becomes, when is a hit index value considered a real/true hit?)
3. The spectroscopic rules and history of the sample are transparent to spectral searching approaches, i.e., dumb algorithms make mistakes
4. Chemical class is often an indicator of the identification for an unknown
5. The absolute HQI values are less important than the distance between successive HQI values
6. HQI is generally scaled from 0–1000, with >940 considered a 'reasonable' match — this should be made more rigorous and have an absolute probabilistic meaning for objective component determinations
7. Peak compare is important for all positions and intensities

6.4.9.4 *Compound Types, Groupings, and Classes*

1. Aliphatics
   a. Alkanes or paraffins (Saturated hydrocarbons)
      i. Alkyl groups

1. Unsubstituted
2. Substituted

ii. Cycloalkanes

1. Unsubstituted
2. Substituted

iii. Aliphatics can be

— Long chain
— Short chain
— Gem dimethyl
— Isopropyl group
— Tertiary butyl group
— Methyl on oxygen
— Methyl on nitrogen
— Methyl on iodine
— Methyl on silicone
— Methyl on phosphorus
— Methyl on sulfur
— Cyclic
— Cyclopropane

b. Alkenes or alkadienes (containing double [C=C] bonds)
c. Alkynes or acetylenes (containing triple [C≡C] bonds)

2. Aromatics

a. Colthup compound taxonomy:

1. Aliphatic
2. Triple bonds
3. Cumulated double bonds
4. Carbon-Carbon double bonds
5. Aromatic rings
6. Ethers and related C-O-C Groups
7. Alcohols, phenols, and related O-H Compounds
8. Peroxides
9. Aldehydes

10. Ketones
11. Esters
12. Carboxylic acids and salts
13. Anhydrides
14. Amides and related compounds
15. Carbamates
16. Ureas
17. Imides and related compounds
18. Chloro carbonyl compounds
19. Amines and amine salts
20. C=N compounds
21. N=O compounds
22. N=N compounds
23. Nitrogen heteroaromatics
24. Five-membered rings with two C=C bonds
25. Boron compounds
26. Phosphorus compounds
27. Sulfur compounds
28. Fluorine compounds
29. Chlorine, Br, and I compounds
30. Metal carbonyls
31. Arsenic compounds
32. Tin compounds
33. Inorganic ions

6.4.9.5 *Atmospheric Correction Issues*

Current methods do not adequately remove water vapor and CO_2 from spectra. It has been suggested that improvements at removing these artifacts are essential to improved spectral identification and interpretation. If this cannot be accomplished on schedule it is suggested that the moisture and carbon dioxide window regions be eliminated from any spectral search to mitigate their influence on the results.

New approaches to elimination of these regions includes the archiving of both water vapor and carbon dioxide at multiple resolutions in order to subtract these from the test sample. This is a separate area of investigation and is beyond the specific scope of this section.

6.4.9.6 *Spectrum Interpretation Issues*

Use the group frequencies (4000–1500 cm^{-1}) to raise specific questions about the identification of the sample of interest and then refer to the fingerprint region for verification. When using the group frequency region there are characteristic bands to look for to identify specific functional groups of atoms. These functional group bands are best to use for interpretation having the following characteristics.

1. Are reliable — always appear when functional group is present
2. Are relatively narrow bands ±25 cm^{-1}
3. Bands that yield an exact, constant cm^{-1} position
4. Bands that are intense
5. Bands that are isolated with rarely found interferences
6. Bands above 1500 cm^{-1}
7. Bands that have a characteristic and constant property other than frequency location, for example, very sharp (vs), very broad (vb), doublet, etc. Best examples are: carbonyl (>C=O), hydroxyl (-O-H), cyano (-C≡N). Other examples having more than one characteristic frequency are: methyl (-CH_3), aromatic (-C_6H_5), carboxylic (-COOH), amide (-$CONH_2$), -nitrate (-NO_2).

Fingerprint regions are utilized by forward and backward searches and comparisons to peak table sets, model compound sets, and specific compound peak tables.

6.4.10 *Probability Measures Associated with Hit Quality Index*

Note: Questions related to numbers of degrees of freedom or probability levels to use for these statistical tests are minor considerations that may be addressed later. For example once a problem is defined probability levels for such a test can be determined more precisely.

1. *Use of database management* such that pure compounds representing a specific compound class become multidimensional reference points of that class to the unknown spectrum. The pure compounds are specific individual compound spectra or averages of classes of compounds that are used as model spectra of the compound class in spectral space.
2. Euclidean distances in multidimensional space are given for each of the classes from the unknown sample. Distance metrics are given for the distance from the unknown sample and any model reference spectrum over the frequency window of interest by:

$$D = \{(\nu_{u1} - \nu_{r1})^2 + (\nu_{u2} - \nu_{r2})^2 + \cdots + (\nu_{un} - \nu_{rn})^2\}^{\frac{1}{2}} \qquad (6.30)$$

For example, the distance from functional groups and model compounds is listed for each unknown sample. So that distance from each group is given. If the distance is large then there is a low probability that the unknown sample contains the group or resembles the reference compound. Test groups used include: carbonyl (>C=O), hydroxyl (-O-H), cyano (-C≡N), methyl ($-CH_3$), aromatic ($-C_6H_5$), carboxylic (-COOH), amide ($-CONH_2$), -nitrate ($-NO_2$). Verification/validation is performed by measuring reference compounds as unknown samples and comparing reported statistics. F- and t-tests for distances from the rank ordered reference spectra are compared, and distances from known or model spectra are also compared; these statistics have been proposed and are valid tests for Euclidean distance metrics. The specific probability levels associated with these tests need to be determined using actual case studies.

3. *Measure correlation* over the same model compounds using:

$$\mathrm{r} = (U \bullet R) = \sum_{i=1}^{n} U_i R_i \tag{6.31}$$

4. A finite number of spectral points are selected at predetermined intervals (wavenumbers) and the correlations calculated. *A t-statistic is calculated for the correlation between the unknown spectral points and the reference spectral points* using Equation 6.32, noting Tables 6.1 and 6.2.

$$t = \frac{r\{n-2\}^{\frac{1}{2}}}{\{1-r^2\}^{\frac{1}{2}}} \tag{6.32}$$

Table 6.1 Students t-Table (For 5 random data points).

r	r^2	% Explained variation*	% Unexplained variation**	α	P	t
0.9911	0.982	98.2	1.8	0.0005	0.9995	12.924
0.986	0.972	97.2	2.8	0.0010	0.9990	10.215
0.974	0.949	94.9	5.1	0.0025	0.9975	7.453
0.958	0.918	91.8	8.2	0.005	0.9950	5.841
0.935	0.874	87.4	12.6	0.010	0.990	4.541
0.878	0.771	77.1	22.9	0.025	0.975	3.182
0.80	0.640	64	36	0.05	0.950	2.353
0.685	0.469	47	53	0.10	0.900	1.638
0.405	0.164	16.4	83.6	0.25	0.750	0.765

*$r^2 \times 100 =$ % Explained Variation.
**$[100 - (r^2 \times 100)] =$ % Unexplained Variation.

Table 6.2 Student's t-Table for $\gamma = \infty$, 25, and 5.

α	P	$t_{\alpha,\infty}$	$t_{\alpha,25}$	$t_{\alpha,5}$
0.001	0.999	3.300	3.768	12.924
0.005	0.990	2.576	2.807	5.841
0.010	0.980	2.326	2.500	4.541
0.025	0.950	1.960	2.069	3.182
0.050	0.900	1.645	1.714	2.353

Note that H_0: $\rho = 0$ and H_1: $\rho > 0$

5. For any measured correlation, compute the ***confidence limits for r*** as measured.

   a. *Compute the Z statistic* for example as correlation measured as 0.99 (P = 0.95) for 1000 data points.

$$\text{Lower limit: } Z_{LL} = 1.1513 \cdot \log_{10}\left(\frac{1+r}{1-r}\right) - t_{\alpha,\gamma}\left(\frac{1}{\{n-3\}^{\frac{1}{2}}}\right) \quad (6.33)$$

$$\text{Upper limit: } Z_{LL} = 1.1513 \cdot \log_{10}\left(\frac{1+r}{1-r}\right) + t_{\alpha,\gamma}\left(\frac{1}{\{n-3\}^{\frac{1}{2}}}\right) \quad (6.34)$$

   Z range is calculated as 2.58 to 2.71

   b. Compute the *measured correlation coefficient upper and lower confidence limits* (ρ_{LL} and ρ_{UL}) by solving:

$$\text{Lower r limit: } Z_{LL} = 1.1513 \cdot \log_{10}\left(\frac{1+\rho_{LL}}{1-\rho_{LL}}\right) \quad (6.35)$$

$$\text{Upper r limit: } Z_{UL} = 1.1513 \cdot \log_{10}\left(\frac{1+\rho_{UL}}{1-\rho_{UL}}\right) \quad (6.36)$$

This corresponds to an r range at $t_{0.025}$ ($P_{0.95}$) of 0.9885 to 0.9913.

6. To determine whether a measured correlation coefficient (e.g. $r_m = 0.95$) is statistically equivalent to 1 (i.e. 0.999) for n = 1000 we use:

$$Z_l = 1.1513 \cdot \log_{10}\left(\frac{1+0.999}{1-0.999}\right) = 3.800 \quad (6.37)$$

$$Z_m = 1.1513 \cdot \log_{10}\left(\frac{1+0.95}{1-0.95}\right) = 1.832 \quad (6.38)$$

$$\text{and } \sigma_{Z1-Zm} = \left\{\frac{1}{N_1-3} + \frac{1}{N_2-3}\right\}^{\frac{1}{2}} = 0.045 \quad (6.39)$$

Note that H_0: $\rho = 1$ and H_1: $\rho \neq 1$

$$\text{To test } H_0 \text{ we use: } z = \frac{Z_l - Z_m - (1 - r_m)}{\sigma_{Z1-Zm}} = \frac{3.800 - 1.832 - 0}{0.045} = 43.73 \quad (6.40)$$

We reject H_0 if $z > 1.96$, and accept H_1, thus 0.95 is not statistically equivalent to a perfect match score for this case.

7. *Recommended statistical match summary report*

   a. Class Distances (raw or normalized) for each compound class or functional group in rank order
   b. Repeat above for spectral search windows
   c. Compound distances (for samples in reference library) in rank order
   d. Correlation coefficient (unknown versus reference spectra) in rank order
   e. t-statistic for correlation scores with r^2, % Explained Variation, % Unexplained Variation, α, and P (also rank order). This is completed for searches against functional groups, negative peaks, peak tables, random or fixed points, model spectra, or full spectra
   f. Confidence limits for r (rank order)
   g. Determine if the measured correlation coefficient (r_m) is statistically equivalent to 1.0 or some other preset threshold (YES/NO)
   h. Compute test statistics for best hits and successive hits (e.g., top 5)

6.4.10.1 *Recommendations for Search Software Features*

1. Use of database management (simplify problem to optimize search using sample history and chemical/physical property information)
2. Search unknown against functional groups, negative peaks, peak tables, random or fixed points, model spectra, or full spectra
3. Spectral artifact removal (water and CO_2 in particular)
4. X-shift alignment function (maximum correlation, minimum distance)
5. Automated peak selection
6. Peak tables are generated for all reference and unknown spectra
7. Peak search against database including peak table for each reference spectrum

   a. Forward search — all peaks for the unknown sample are in the reference spectrum [the reference spectrum can have additional peaks (components)]
   b. Reverse search — all peaks for the reference standard are in the unknown spectrum [the unknown spectrum can have additional peaks (components)]

8. Peak threshold selection — the higher the threshold the fewer peaks are selected — this is important for identification purposes
9. Spectral similarity subtraction
10. Multi-level filter algorithm
11. Apply statistical testing for match reporting
    a. Class Distances (raw or normalized) for each compound class or functional group in rank order
    b. Repeat above for spectral search windows
    c. Compound distances (for samples in reference library) in rank order
    d. Correlation coefficient (unknown versus reference spectra) in rank order
    e. t-statistic for correlation scores with r^2, % Explained Variation, % Unexplained Variation, α, and P (also rank order). This is completed for searches against functional groups, negative peaks, peak tables, random or fixed points, model spectra, or full spectra
    f. Confidence limits for r (rank order)
    g. Determine if the measured correlation coefficient (r_m) is statistically equivalent to 1.0 or some other preset threshold (YES/NO)
    h. Compute test statistics for best hits and successive hits (e.g., top 5)
12. Incorporation of derivatives and variations of them for library searching and for library data (all relevant preprocessing steps added)
13. Provision for 'weighting' of spectral regions or peaks depending upon the unique aspects of each problem solving exercise
14. Provision for searching text variables to include independent variables, such as boiling point, melting point, refractive index, absorptivity, etc.
15. Provision for saving unknown spectra into a reference database and later identifying their occurrence with comment line from past work on a similar sample (What is new here is flagging samples as separate from standard reference database and addition of large (100 word) comment section)
16. Provision to use cursor to click bands of highest priority for search and then enter weighting factor for each (e.g., 1x, 2x, etc.)

6.4.10.2 *Common Spectral Matching Approaches*

The ASTM (American Society for Testing and Materials) has published a "Standard Practice for General Techniques for Qualitative Analysis" (Method E1252-98(2013)e1). The method describes techniques useful for qualitative evaluation of liquids, solids, and gases using the spectral measurement region of 4000 to 50 cm^{-1} (above 2500 nm).

6.5 Normalization of Spectra

Normalization generally consists in one of the following data processing methods: mean centering, variance scaling, or autoscaling peak height or area. For mean centering, the sum of all spectrum or chromatogram measurement values (i.e., data points) = 0; for variance scaling the variance of all spectrum or chromatogram measurement values = 1.0; and for autoscaling either the peak heights or peak areas for the measurement may be set to any value, most often = 1.0. Note that autoscaling may be accomplished by dividing each peak, or each data point of a spectrum, by either the peak height or peak area of a reference (or selected) peak. For demonstration, two unique examples are presented in sections 6.5.2 and 6.5.3 with peak height and peak area being made equivalent, respectively. Unlike the examples shown, often a reference material peak will be used for autoscaling.

6.5.1 *Spectrum Normalization*

When the spectrum vector is divided by the norm of the spectrum it is termed as spectrum normalization. With this normalization only the relative weights are maintained but no specific information regarding the absolute or spectrum intensity. The equation for this normalization process is as:

$$sn_i = \frac{s_i}{\left(\sum_{i=1}^{n} s_i^2\right)^{\frac{1}{2}}} \tag{6.41}$$

where sn_i are the normalized spectrum data points. If there are no negative spectrum values the normalized spectrum in the above equation is identical to that calculated using only the sum of the spectrum values as the denominator as given in the following equation:

$$sn_i = \frac{s_i}{\left(\sum_{i=1}^{n} s_i\right)} \tag{6.42}$$

These results for a small test example are shown in Table 6.3 and Figures 6.1 and 6.2. This type of normalization is used for qualitative analysis or for creating a normalized weighting spectrum function. Such weighting spectral functions are common for use in color or solar measurements for the UV-Vis-NIR spectral regions. (See, for example, Volume 2, Chapter 7 for Solar Measurement technique.) This normalization technique is used for qualitative analysis in IR and Raman where spectral matching techniques are highly influenced mathematically by the numerical amplitude or absorbance values. Column 1 is the wavenumber value for a spectrum; column 2 is the absorbance values for a spectrum; and column 3 are the normalized

Table 6.3 Test example of spectrum normalization as described in Equations 6.41 and 6.42.

cm^{-1}	Abs	NormAbs
4000	0.10	0.015
3750	0.10	0.015
3500	0.25	0.038
3250	0.10	0.015
3000	0.50	0.076
2750	0.10	0.015
2500	1.00	0.153
2250	0.10	0.015
2000	1.50	0.229
1750	0.10	0.015
1500	2.00	0.305
1250	0.10	0.015
1000	0.20	0.031
750	0.10	0.015
500	0.10	0.015
SUM	**6.35**	**1.00**

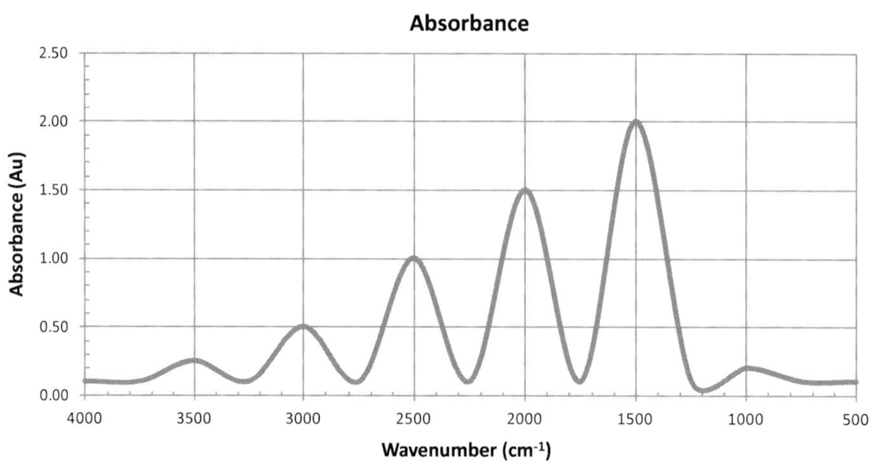

Fig. 6.1 Example spectrum from data in Table 6.3 (column 2) demonstrating an Absorbance spectrum. The relative amplitudes are as measured.

absorbance values for column 2. The absorbance and normalized absorbance spectra are shown in the Figures 6.1 and 6.2, respectively.

6.5.2 *Autoscaling or Normalization by Peak Height*

For autoscaling by peak height, the spectrum peak are measured as their height difference from the selected baseline points and then the peaks are divided by their individual heights. All peaks are then set to a maximum value of 1.0. This autoscaling

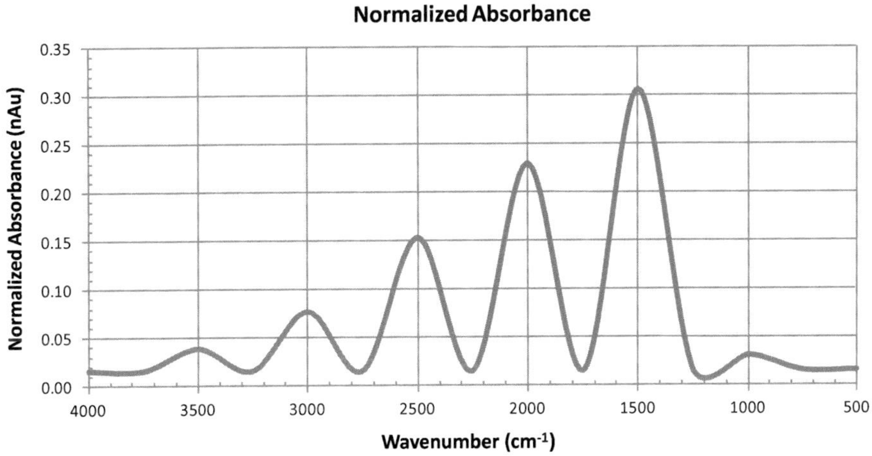

Fig. 6.2 Example spectrum from Table 6.3 (column 3) as normalized absorbance spectrum. Note the sum of a normalized spectrum is equal to 1.0, yet the relative weights remain identical to the absorbance spectrum (Figure 6.1).

Table 6.4 Data for example of autoscaling or normalization by peak height.

cm^{-1}	Abs	Peak height	Hi/Hb
4000	0.10	0	0
3750	0.10	0	0
3500	0.25	0.15	1.00
3250	0.10	0	0
3000	0.50	0.40	1.00
2750	0.10	0	0
2500	1.00	0.90	1.00
2250	0.10	0	0
2000	1.50	1.40	1.00
1750	0.10	0	0
1500	2.00	1.90	1.00
1250	0.10	0	0
1000	0.20	0.10	1.00
750	0.10	0	0
500	0.10	0	0
SUM	**6.35**	**4.85**	**6.00**

method may be used for qualitative analysis to remove the relative weighting of each peak height for functional or molecular identification purposes. The data and illustration of the method are shown in Table 6.4 and Figure 6.3. Alternatively, each peak may be divided by the peak height of the single maximum peak absorbance such that each height would be relative to that of the highest peak of 1.0. Height normalization for a single peak is also performed by dividing the amplitude at each data point by the peak height maximum amplitude.

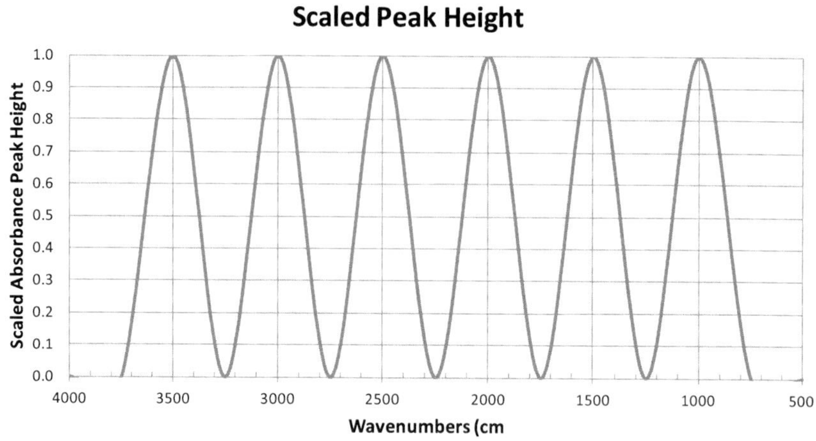

Fig. 6.3 Data from Table 6.4 (column 2 Absorbance data), which has been autoscaled/normalized by peak height as given in column 4. All peaks heights are set to 1.0. Peak positions are maintained and all peaks are divided by their respective peak heights (column 3). In a spectral search all autoscaled/normalized bands/peaks would receive equal weighting.

6.5.3 *Autoscaling or Normalization by Peak Area*

For autoscaling by peak area, the spectrum peak areas are measured as from the selected baseline points and then the peak areas are divided by their individual areas. All peak areas are then set to a maximum value of 1.0. This autoscaling method may be used for qualitative analysis to remove the relative weighting of each peak area for functional or molecular identification purposes. The data and illustration of the method are shown in Table 6.5 and Figure 6.4. Alternatively, each peak area may be divided by the peak area of a single reference band such that each area is corrected relative to that of the single reference band. If the maximum area band were used as the reference then the maximum band area would be 1.0 and all other bands would have areas less than 1.0. Area normalization for a single peak is also performed by dividing the amplitude at each data point by the peak area.

6.5.4 *Normalization for GC-IR*

Area normalization for GC-IR peaks within a chromatogram are used for quantitative analysis. The concentration (c_i) of a particular compound in a sample chromatogram is determined using an area (a) computation for each eluted peak. The following equation applies:

$$\%c_i = \frac{100 \cdot a_i}{a_t} \tag{6.43}$$

where $\%c_i$ is the percent concentration of the molecule from the chromatogram; a_i is the peak area of the component of interest in the chromatogram; and a_t is the sum of all peak areas in the chromatogram.

Table 6.5 Data for example of autoscaling or normalization by peak area.

cm^{-1}	Abs	Peak/Band area	Ai/Ab
4000	0.10	0	0.00
3750	0.10	0	0.00
3500	0.25	37.5	4.00E-03
3250	0.10	0	0.00
3000	0.50	100	4.00E-03
2750	0.10	0	0.00
2500	1.00	225	4.00E-03
2250	0.10	0	0.00
2000	1.50	350	4.00E-03
1750	0.10	0	0.00
1500	2.00	475	4.00E-03
1250	0.10	0	0.00
1000	0.20	25	4.00E-03
750	0.10	0	0.00
500	0.10	0	0.00
SUM	**6.35**	**1213**	**2.40E-02**

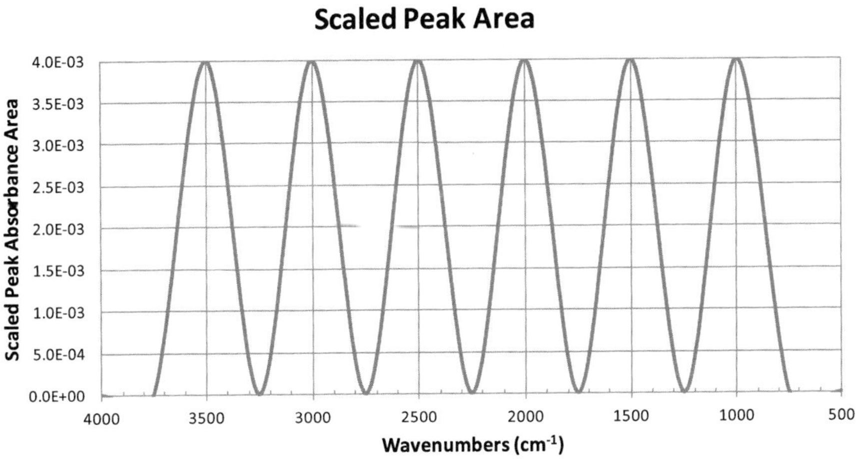

Fig. 6.4 Data taken from Table 6.5 (column 2), which is absorbance data that has been autoscaled/normalized by peak/band area (column 3). All peak areas are set to 1.0 by computing the peak areas at each peak and dividing each respective peak by its area. For each peak the area is identical. Thus any search technique using area calculations would treat all absorbance bands with equal area weighting.

Since the sensitivity of the infrared detection of each molecule is different, the method must be calibrated for each component for quantitative analysis using a regression of the known concentration of a material (X axis or abscissa) versus the instrument response (Y axis or ordinate). Once these response curves are known the peak area may be inserted into a prediction equation to determine actual concentration. The regression line is computed from measuring a standard mixture of materials with known concentrations and determining the regression line as illustrated in Table 6.6 and Figures 6.5 and 6.6. The final concentration for component B is determined using the following equation based on the regression results from data in the table. The initial concentration for component **B** from using only the chromatogram would be 30% of the material injected. An IR spectrum is taken of this material as it eluted to verify its composition. However, a corrected concentration value for component B is computed using the regression correction for detector response using the linear regression correction $Y = mX + B$, i.e., $X = (Y-B)/m$) in the form of Conc. = (Signal Area + 150)/108.25, as shown in Table 6.7 and Figure 6.6. For component B this would be computed as 29.1% of the material

Table 6.6 Data Table for GC-IR Example: Components A, B, C, D peak areas are given.

RT (min)	A	B	C	D	Sum	Area	% Conc (uncorrected)
0	0	0	0	0	0	0	0
0.5	0	0	0	0	0	0	0
1	0	0	0	0	0	0	0
1.5	0	0	0	0	0	0	0
2	4000	0	0	0	4000	2000	20
2.5	0	0	0	0	0	0	0
3	0	0	0	0	0	0	0
3.5	0	0	0	0	0	0	0
4	0	0	0	0	0	0	0
4.5	**0**	**6000**	**0**	**0**	**6000**	**3000**	**30**
5	0	0	0	0	0	0	0
5.5	0	0	0	0	0	0	0
6	0	0	0	0	0	0	0
6.5	0	0	0	0	0	0	0
7	0	0	7000	0	7000	3500	35
7.5	0	0	0	0	0	0	0
8	0	0	0	0	0	0	0
8.5	0	0	0	3000	3000	1500	15
9	0	0	0	0	0	0	0
9.5	0	0	0	0	0	0	0
10	0	0	0	0	0	0	0
Sum						**10000**	100

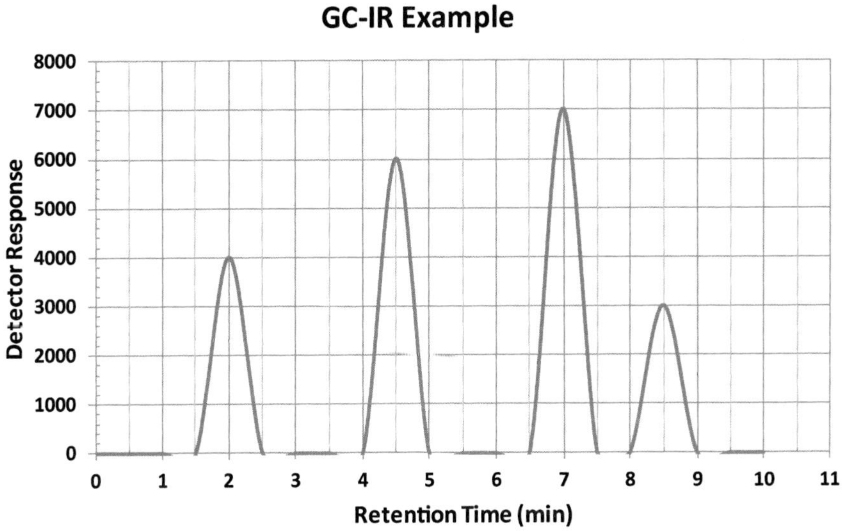

Fig. 6.5 Example GC-IR chromatogram peak area data from Table 6.6.

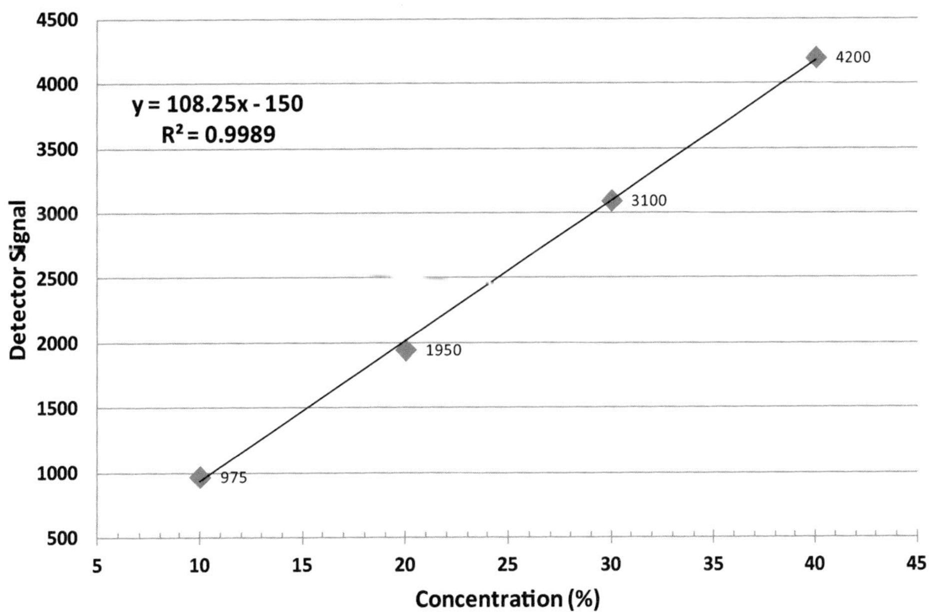

Fig. 6.6 Regression line for Concentration (x) in percent (%) versus Detector Signal (Peak Area) for component B in solution (see data in Tables 6.6 and 6.7).

Table 6.7 Data used for Component B calculations. RF = Area/Conc.

Conc. (%)*	Signal (area)	RF	Xest
10	975	97.50	10.4
20	1950	97.50	19.4
30	3100	103.33	30.0
40	4200	105.00	40.2
Test	3000		29.1

*Actual concentration (Note that Xest column is estimated from regression in Figure 6.6)

injected. Note: The units may be changed from % to ug/L or ppm, etc. as appropriate for any computation. Note that 1/RF x Measured Area = Corrected Concentration (in original concentration units).

6.6 Testing FTIR Instrument Performance

Chapter 6 of Volume 3 describes a series of instrument performance tests that are useful for any spectrophotometer and include a host of important measurements and procedures for recording performance as well as diagnosing instrument functionality and design limitations. The reader is referred to the main index in Volume 5 for multiple subjects relative to detailed instrument design, testing, and performance.

6.6.1 *Types of Tests*

Recommended tests for FTIR instruments include ASTM E1421,[125] which specifies Level zero and Level one tests. Level zero tests are comprised of routine checks of instrument performance to determine significant changes and record a history of data from each instrument to be used for later comparison purposes. Level one tests are used to provide quantitative data on instrument function and performance with the purpose of diagnosing any significant instrument performance issues.

6.6.2 *Sample for FTIR Evaluation*

The sample used to assess instrument performance for FTIR spectrophotometers is normally a 0.038 mm (i.e., 38 micron or 1.5 mil) thick matte-finish polystyrene film mounted on a card that readily is inserted into standard film holders for transmission measurements in most commercial FTIR instruments. The outer dimensions of a typical reference sample within the film holder are useful for mounting into most commercial instruments and have a clear aperture of 2.5 cm (1 inch) at the center of

Table 6.8 Energy spectrum test reported values.

Energy Spectrum Ratio	$4000\,cm^{-1}/2000\,cm^{-1}$	$2000\,cm^{-1}/1000\,cm^{-1}$
Aligned	0.200 to 0.230	0.700 to 0.720
Slight Misalignment	0.100 to 0.200	0.600 to 0.700
Misalignment	0.025 to 0.100	0.500 to 0.600

the card for transmission measurements. Chapter 4 of this volume illustrates sample holders for polymer film analysis.

6.6.3 *Measurements Required for Level 0 Testing*

6.6.3.1 *Energy Spectrum Test*

This test consists in ratioing different wavenumber energy measurements of the single-beam energy spectrum of an interferometer with no sample in the beam, and with the sample compartment purged of water vapor and carbon dioxide. The measurements consist of the energy at $4000\,cm^{-1}$ ratioed to the energy at $2000\,cm^{-1}$; and the energy at $2000\,cm^{-1}$ ratioed to the energy at $1000\,cm^{-1}$. The ratioed measurements are reported as Table 6.8. Note the measurement results in each column and the indication of interferometer alignment (column 1). The decrease in the ratios of the energy spectra over time or measurement replicates indicates interferometer misalignment. Note that the greatest indicator of misalignment is the largest deviations at the higher wavenumbers.

6.6.3.2 *One Hundred Percent Transmittance Test*

A ratio of 2 energy spectra is made with the resultant spectrum being a 100% transmittance line. This is indicated as a line at 100% Transmittance across the entire infrared spectral region from $5000\,cm^{-1}$ to $400\,cm^{-1}$. This line is an indication of interferometer performance, and detector and electronics performance. Compute RMS noise or peak-to-peak noise. These statistical parameters are computed for $100\,cm^{-1}$ intervals at 4000, 2000, 1000, and $500\,cm^{-1}$. RMS noise is computed as 100/standard deviation of the transmittance values over the computed range (see the index for other discussions of RMS noise computations). The peak to peak values are the high transmittance minus the low transmittance values over the measured range. For calculations of noise in spectra, see also Chapter 6 of Volume 3. Note the reported values for this test are computed as shown in Table 6.9. Large deviations above manufacturers specifications indicate instability or misalignment of interferometer, optics and detector, or source dynamics. Also electronics or purge issues may be present depending upon test range demonstrating the most noise.

Table 6.9 One hundred percent transmittance test reported values.

Test Range (cm^{-1})	RMS Noise	Peak-to-peak
4050–3950	—	—
2050–1950	—	—
1050–950	—	—
550–450	—	—

6.6.3.3 *Polystyrene Subtraction Test*

Subtract 2 polystyrene transmittance spectra, one from an early reference and one spectrum at least 2 hours later. (Note: This may be completed in absorbance or transmittance space.) The transmittance difference spectrum should show a line with relatively equivalent noise across the spectral region at 0% Transmittance. An absorbance difference spectrum should show a line at 0 Absorbance with roughly equivalent noise across the spectral region. For the absorbance difference spectrum a sharp peak near 3600 cm^{-1} and 1600 cm^{-1} indicates water vapor in the optical path and an insufficient purge of the instrument. An insufficient purge is also observed due to excess carbon dioxide (CO_2) at 2360 cm^{-1} (doublet band) and 667 cm^{-1} (sharp spike). Dramatic baseline shifts from 0 nominal position, or peak shape changes are indicative of instrument performance malfunctioning.

6.6.3.4 *Polystyrene Peak, Resolution, and Photometry Test*

These tests are sometimes substituted for the polystyrene subtraction test. The reader is referred to ASTM E1866[126] for more rigorous testing; this ASTM practice describes additional evaluation of an FTIR spectrophotometer. The use of peak position tests, resolution tests, and photometry tests to determine instrument fitness and performance are described. See these individual topics in the index for more information contained in this set of volumes.

6.6.4 *Measurements Required for Level 1 Testing*

6.6.4.1 *Energy Spectrum Test (Is performed as the level 0 test)*

6.6.4.2 *One Hundred Percent Line Test (Is performed as the level 0 test)*

6.6.4.3 *Stability Test*

The RMS noise and peak-to-peak noise is computed as in Table 6.9. Compare the RMS and peak-to-peak variation over a series of Transmittance spectra computed over time. The RMS variation at the highest wavenumbers indicate system stability.

Source flicker, interferometer alignment, and detector and electronic instability are causes for high noise measurements.

6.6.4.4 *Signal Averaging Test*

Subspectra are collected as 4 times the previous number of replicate scans starting with 1 scan, such as Table 6.10. The noise levels are computed over the ranges as shown in this table. Obtain a series of replicate scan-to-scan spectra in transmittance mode, and compute a subset of replicate scans and process as described. Do this for the following number of scans: 1, 4, 16, 64, 256, 1024, 4096, 16384, etc., up to the maximum measurement time of interest. Ratio each pair of sample and background spectra and calculate the RMS noise level using Equation [6.44] at the four 100 cm^{-1} wavenumber regions. The noise level should be reduced by a factor of 2 for each successive ratioed spectrum; for example, if 1 scan gave a noise level of 1, 4 scans would give 1/2, 16 would give 1/4, 64 would give 1/8 and so on until signal averaging fails. The percent noise level for each successive ratioed spectrum should be a factor of 2 lower; for example, 1, 1/2, 1/4, 1/8, 1/16, 1/32, 1/64, 1/128, etc. (see results reporting as in Table 6.10).

Calculate the mean and standard deviation (RMS) of the transmittance values at the four wavenumber positions for scan-to-scan (within replicate samples). This statistic is calculated for each wavenumber data point as:

$$s_i = \sqrt{\frac{\sum_{j=1}^{n} (T_{ij} - \bar{T}_i)^2}{n - 1}} \tag{6.44}$$

where (s_i) is the RMS or root mean square standard deviation for each set of scans in transmittance units (%T) repeatability for the scan-to-scan measurements at each wavenumber; and where (T_{ij}) are individual measurements of the transmittance for wavenumber i and replicate scan number j; ($\bar{T}_i$) are the mean measured values for the scan for each wavenumber; and n is the replicate number of spectra (i.e., 1 to n). The results are reported as follows. The RMS noise is computed at each wavenumber for the series of spectra (as s_i) and the total RMS for each spectral region is computed as the square root of the sum of the variances at each wavenumber (equivalent to s_i^2) divided by the total number of data points (wavenumbers). The worksheet shown as Table 6.10 is populated with the calculations for comparison purposes. If this test fails then the noise is too high for the instrument under test. Note: Each successive RMS value should decrease by 1/2 as one increases the number of replicates down the table.

Table 6.10 Table for reporting signal averaging test.

Number of Scans	RMS Test 4050–3950 cm^{-1}	RMS Test 2050–1950 cm^{-1}	RMS Test 1050–950 cm^{-1}	RMS Test 550–450 cm^{-1}
1	—	—	—	—
4	—	—	—	—
16	—	—	—	—
64	—	—	—	—
256	—	—	—	—
1024	—	—	—	—
4096	—	—	—	—
16384	—	—	—	—
65536	—	—	—	—

Table 6.11 Report wavenumber positions for the following peak positions (in transmittance).

Nominal peak positions	Measured peak positions
3082 cm^{-1}	—
3060 cm^{-1}	—
3028 cm^{-1}	—
2852 cm^{-1}	—
1945 cm^{-1}	—
1601 cm^{-1}	—
1028 cm^{-1}	—
907 cm^{-1}	—
540 cm^{-1}	—

6.6.4.5 *Polystyrene Test*

Measure a polystyrene single beam spectrum and ratio that to a reference single beam spectrum; the reference spectrum having no sample present. Report peak positions for the transmittance spectrum as given in Table 6.11, and Figure 6.7.

6.6.4.6 *Resolution Test*

A method for assessing resolution involving the determination of peak ratios using an absorbance spectrum is quite valuable for determining the resolution of a spectrophotometer. In the case of using a polystyrene absorbance spectrum, Figure 6.8, one may observe the following peaks and record a series of absorbance ratios to characterize resolution from one measurement time to the next Table 6.12; use and record a fixed resolution for this comparative test. (Note: A resolution

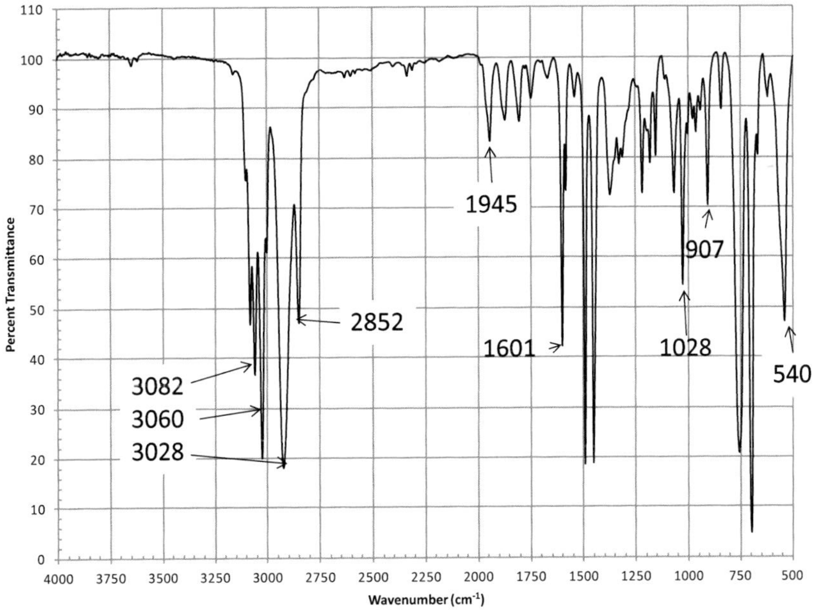

Fig. 6.7 Measured peak positions for Polystyrene (transmittance spectrum).

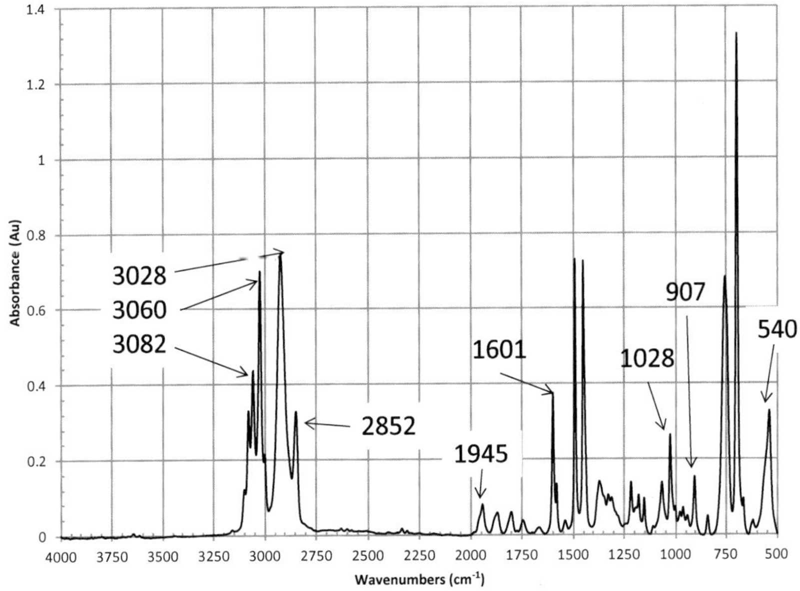

Fig. 6.8 Polystyrene film absorbance spectrum with important peak positions labeled.

Table 6.12 Peak ratios indicative of spectrophotometer resolution using polystyrene absorbance spectrum. Absorbance at wavenumbers specified in table are used for calculations.

Peak ratio (Abs at cm^{-1})	Measured ratio at time 1	Measured at time 2
3082/2849	—	—
1583/1154	—	—
(3082–3096)/(2849–2870)	—	—
(1601–1564)/(1583–1564)	—	—
(91028–994)/(1002–994)	—	—

Table 6.13 Transmittance peak differences using polystyrene to determine photometric variation, aka, photometric jitter.

Spectral bands (in cm^{-1})	%T RMS at time 1	%T RMS at time 2
3082–2984	—	—
3060–2984	—	—
3001–2984	—	—
2923–2870	—	—
2849–2870	—	—
1601–1564	—	—
1583–1564	—	—
1069–1104	—	—
1028–997	—	—
1002–994	—	—

setting of 4 cm^{-1} is a useful benchmark.) Changes in these ratios indicate a change in resolution. The user is looking for consistency in the ratioed values; neither increasing values nor decreasing values. Change indicates variation in resolution, which has serious consequences on results from quantitative and qualitative analysis.

6.6.4.7 *Photometric Jitter Test*

Measure and record transmittance spectra for polystyrene at 30 different times, e.g., these measurements may be successive or several minutes apart; then compute the RMS (standard deviation) of the transmittance at each of the peak center positions for the spectral bands given in Table 6.13. Any changes in these RMS %Transmittance values over time indicate a combination of noise effects and photometric jitter. One may compare overall noise over the regions to the peak center position noise in order to partition the amount of variation mostly due to noise and that due to photometric jitter at the peak positions.

6.7 IR Imaging

6.7.1 *Brief Description and Illustrations of Imaging Techniques*

Imaging techniques include multi-wavelength or multi-frequency scanning with a limited size detector while rastering or collecting data by gradually moving the detection sensor over a large spatial range; or it involves making imaging measurements using an array detection camera capable of real-time large area imaging using multi-wavelength or multi-frequency scanning. These basic methods are illustrated in Figures 6.9 and 6.10. Imaging allows the information from spatial dimensions to be combined with chemical information, as key requirements for understanding the interrelationship between structure, composition, and function. Imaging data begins with second-order or higher order data as illustrated in Figures 6.13 through 6.16. A series of illustrations for the different orders of spectroscopic data are shown and described in Figures 6.11 through 6.16. Note that simple spectroscopy where there is an amplitude at one wavelength is referred to as zero-order data (Figure 6.11); standard scan data taken with a scanning

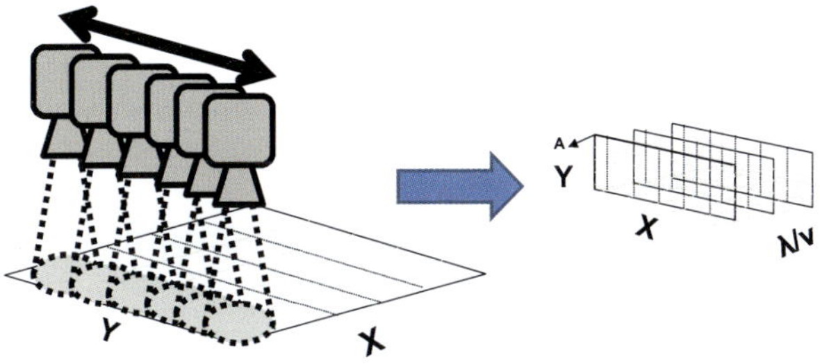

Fig. 6.9 Single multi wavelength camera rastering (motioning) across spatial area to create hyperspectral data cube (i.e., third-order data).

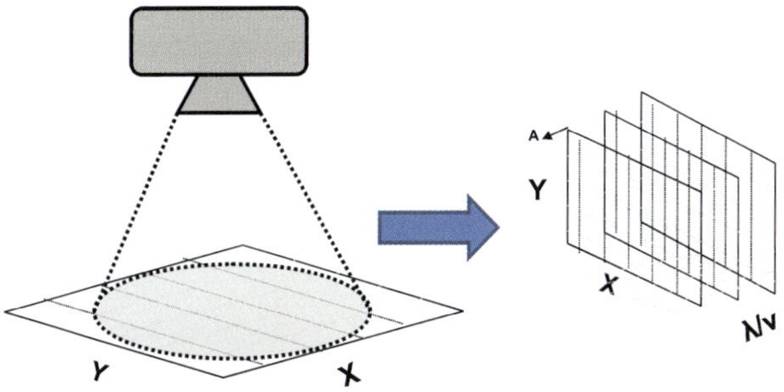

Fig. 6.10 Large area detection multi-wavelength camera measuring spatial area to create hyperspectral cube (i.e., third-order data).

Fig. 6.11 Illustration of zero-order data showing amplitude at a single wavelength or frequency.

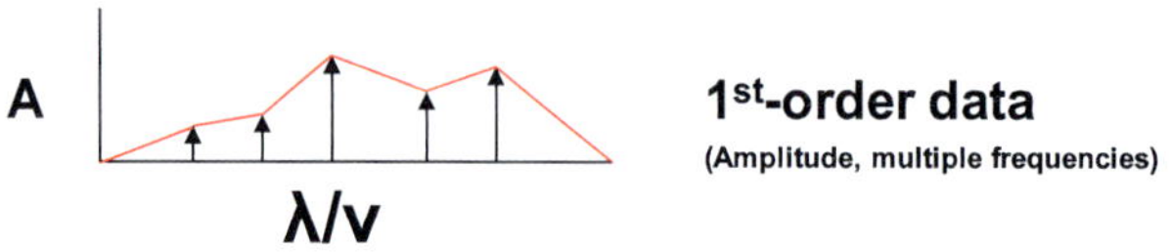

Fig. 6.12 Illustration of first-order data showing amplitude at multiple wavelengths or frequencies.

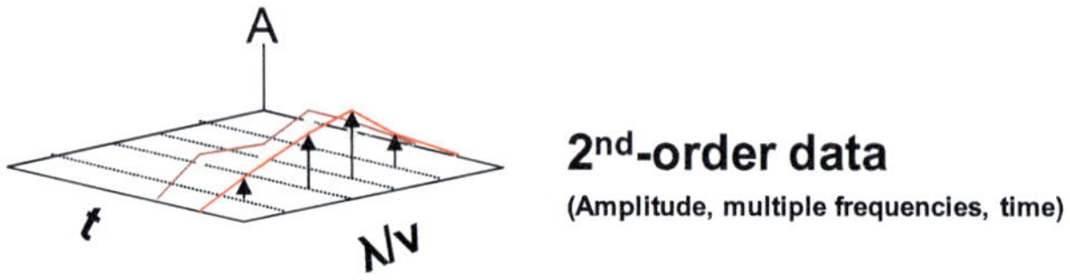

Fig. 6.13 Illustration of second-order data showing amplitude, at various wavelengths or frequencies, over different times slices (intervals).

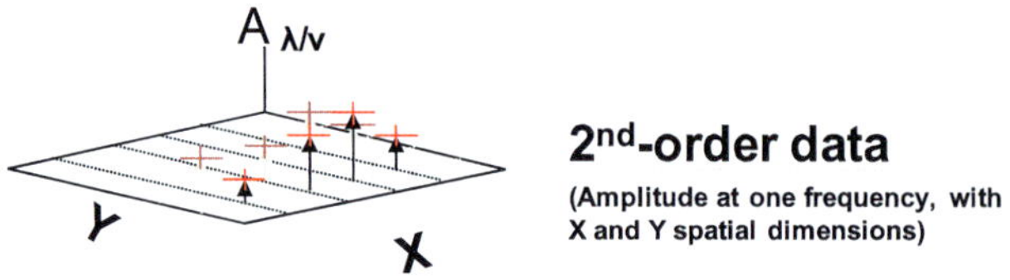

Fig. 6.14 Illustration of second-order data showing amplitude at a single wavelength or frequency versus X and Y spatial data.

spectrophotometer is referred to as first-order data (Figure 6.12). Higher order data is collected using imaging techniques as illustrated in Figures 6.13 through 6.16.

6.7.2 *Overview of Infrared Imaging Applications*

The following cited papers give brief descriptions and a survey of the of the various infrared imaging techniques. A descriptive article describing the field of fast FTIR imaging including theory and applications is presented by Koenig and Snively. This basic review describes instrumental aspects, imaging methods, data processing, and applications with 20 references.[127] An *Applied Spectroscopy* focal point article describing infrared spectroscopic imaging in terms of focal-plane arrays, remote

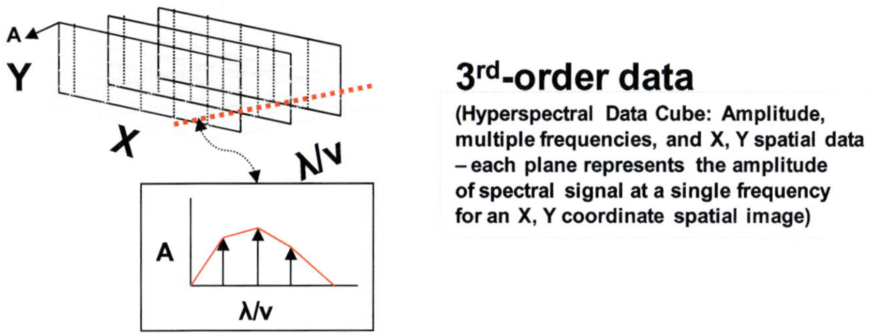

Fig. 6.15 Illustration of third-order data (i.e., hyperspectral data cube) showing amplitude at multiple wavelengths or frequencies along with X and Y spatial data.

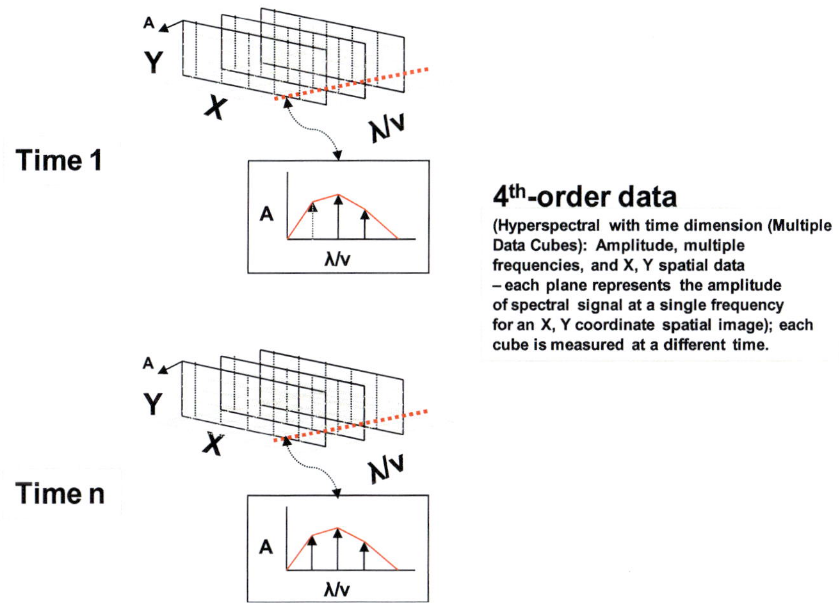

Fig. 6.16 Illustration of fourth-order data (i.e., hyperspectral data cube with time sequence dimension) showing amplitude at multiple wavelengths or frequencies along with X and Y spatial data at time 1 and at time n.

sensing and astronomy, terrestrial and medical applications, is described in detail with 35 references by Colarusso *et al.*[128] Villringer and Chance[129] published work entitled, "Noninvasive optical spectroscopy and imaging of human brain function". Crowley and Zimbelman, of the U.S. Geological Survey, Reston, VA[130] completed work entitled, "Mapping hydrothermally altered rocks on Mount Rainier, Washington, with Airborne Visible/Infrared Imaging Spectrometer (AVIRIS) data."

Ning *et al.*[131] of Oregon State Univ., Corvallis, OR published, "Five novel applications of imaging visible and short near infrared spectrophotometry and fluorometry in the plant sciences. Part I: Photographic and Histological Applications." Rowlands and Neville[132] of CAL Corporation, Ottawa, ON published, "Calcite and dolomite discrimination using airborne SWIR imaging spectrometer data." Otten[133] of Kestrel Corporation, Albuquerque, NM , "Measured performance of an airborne Fourier transform hyperspectral imager." Valdez *et al.*[134] of the Electrical and Computer Engineering University of New Mexico, Albuquerque, NM, USA published "Selection of spectral bands for interpretation of hyperspectral remotely sensed images." Silk and Schildkraut[135] of Block Engineering, Marlborough, MA completed work entitled, "Imaging Fourier transform spectroscopy for remote chemical sensing." Rowan *et al.*[136] of the U.S. Geological Survey, Reston, VA completed, "Analysis of airborne visible-infrared imaging spectrometer (AVIRIS) data of the Iron Hill, Colorado, carbonatite-alkalic igneous complex."

References

1. J. Workman, P. Mobley, B. Kowalski, R. Bro, Review of Chemometrics Applied to Spectroscopy, Part 1, *Applied Spectroscopy Reviews* **31**(1&2) (1996) 73–124.
2. P. Mobley, B. Kowalski, J. Workman, R. Bro, Review of Chemometrics Applied to Spectroscopy, Part 2, *Applied Spectroscopy Reviews* **31**(4) (1996) 347–368.
3. R. Bro, J. Workman, P. Mobley, and B. Kowalski, Review of Chemometrics Applied to Spectroscopy: 1985–1995, Part 3, *Applied Spectroscopy Reviews* **32**(3) (1997) 237–261.
4. Personal Communications with Dr. Clara Craver, A series of conversations over several years, beginning in 2000.
5. J. Thaiman, A. Debska, A. Eckard, J. A. Weaver, *Proceedings of the 42nd International Wire Cable Sym.* (1993) 547–51.
6. B. K. Alsberg, E. Nodland, and O. M. Kvalheim, Compression of nth-Order Data Arrays by B-Splines. Part 2: Application to Second-Order FTIR Spectra, *J. Chemometrics* **8** (1994) 127–145.
7. N. Dupuy, L. Duponchel, B. Amram, J. P. Huvenne, and P. Legrans, Quantitative Analysis of Latex in Paper Coating by ATR-FTIR Spectroscopy, *J. Chemometrics* **8** (1994) 333–347.
8. M. Blanco, J. Coello, H. Iturriaga, S. Maspoch, E. Bertran, Simultaneous Determination of Rubber Additives by FTIR Spectrophotometry with Multivariate Calibration, *Appl. Spectrosc.* **49**(6) (1995) 747–53.
9. T. L. Marshall, C. T. Chaffin, V. D. Makepeace, R. M. Hoffman, R. M. Hammaker, W. G. Fateley, P. Saarinen, and J. Kauppinen, Investigations of the effects of resolution on the performance of classical least-squares (CLS) spectral interpretation programs when applied to volatile organic compounds (VOCs) of interest in remote sensing using open-air long-path Fourier transform infrared (FTIR) spectrometry, *J. Mol. Struct.* **324** (1994) 19–28 (Invited paper, dedicated to Norman Jones).

10. V. Bellon-Maurel, C. Vallat, D. Goffinet, Quantitative Analysis of Individual Sugars during Starch Hydrolysis by FTIR/ATR Spectrometry. Part I: Multivariate Calibration Study — Repeatibility and Reproducibility, *Appl. Spectrosc.* **49**(5) (1995) 556–562.
11. V. Bellon-Maurel, C. Vallat, D. Goffinet, Quantitative Analysis of Individual Sugars during Starch Hydrolysis by FTIR/ATR Spectrometry. Part II: Influence of External Factors and Wavelength Parameters, *Appl. Spectrosc.* **49**(5) (1995) 563–568.
12. H. M. Heise and A. Bittner, Multivariate calibration for physiological samples using infrared spectra with choice of different intensity data, *J Mol Struct.* **348** (1995) 127–130.
13. A. Bittner, R. Marbach, H. M. Heise, Multivariate calibration for protein, cholesterol and triglycerides in human plasma using short-wave near infrared spectrometry. *J. Mol. Struct.* **349** (1995) 341–344.
14. P. Bhandare, Y. Medelson, E. Stohr, R. A. Peura, Comparison of Multivariate Calibration Techniques for Mid IR Absorption Spectrometric Determination of Blood Serum Constituents, *Appl. Spectrosc.* **48**(2) (1994) 271–273.
15. H. J. Luinge, J. H. van der Maas, T. Visser, Partial least squares regression as a multivariate tool for the interpretation of infrared spectra, *Chemom. Intell. Lab. Syst.* **28**(1) (1995) 129–138.
16. J. Bak, A. Larsen, Quantitative Gas Analysis with FTIR: A Method for CO Calibration Using Partial Least-Squares with Linearized Data, *Appl. Spectrosc.* **49**(4) (1995) 437–443.
17. A. J. Michell, Vibrational spectroscopy: A rapid means of estimating plantation pulpwood quality, *Appita J.* **47** (1994) 29–37.
18. D. Qin and P. R. Griffiths, Minimization of Quantitative Errors for Analysis of Vapor-Phase Infrared Spectra Measured at Different Temperatures by Partial Least Squares Regression, *J. Quant. Spectrosc. Radiat. Transfer* **52**(1) (1994) 51–58.
19. C. E. Alciaturi, T. Montero, C. De La Cruz, M. E. Escobar, The prediction of coal properties using compressed infrared data from osculating polynomials, *Analytica Chimica Acta* **340**(1–3) (1997) 233–240.
20. P. S. Bhandare, Doctoral Dissertation, Worchester Polytech. Inst. (MA, USA), Univ. Microfilms Int., Order No. DA9323307 (1993) 327 pp.
21. J. Guilment, S. Markel, W. Windig, Infrared Chemical Micro-Imaging Assisted by Interactive Self-Modeling Multivariate Analysis, *Appl. Spectrosc.* **48**(3) (1994) 320–326.
22. T. Visser, H-J Luinge, J. van der Maas, Evaluation of artificial neural networks and partial least squares regression for computerized interpretation of FTIR spectra, 9th international conference on fourier transform spectroscopy, August 1993, Calgary, Canada. Bellingham: SPIE — The International Society for Optical Engineering (SPIE proceedings series 2089) (1993) 232–233.
23. R. G. Brereton, S. P. Gurden, J. A. Groves, Use of eigenvalues for determining the number of components in window factor analysis of spectroscopic and chromatographic data, *Chemom. Intell. Lab. Syst.* **27**(1) (1995) 73–87.
24. R. E. Aries, D. P. Lidiard, R. A. Spragg, Expert system for developing multivariate calibrations, *Proc. SPIE* 2089, 9th International Conference on Fourier Transform Spectroscopy, 9th international conference on Fourier transform spectroscopy, August 1993, Calgary, Canada. Bellingham: SPIE — The International Society for Optical Engineering (SPIE proceedings series 2089) (1993) 338–339.
25. J. Workman, B. Kowalski, and P. Mobley, The Design of an Expert Calibration System for Spectroscopic Based Process Analytical Chemistry, In Proc. Instrument Society of America (ISA) Conference, Toronto, Canada, (1995) 97–106.

26. D. Qin, Doctoral Dissertation, Univ. Idaho, (ID, USA), Univ. Microfilms Int., Order No. DA9425983 (1994) 201 pp.
27. L. P. Choo, J. R. Mansfield, N. Pizzi, R. L. Somorjai, M. Jackson, W. C. Halliday, and H. H. Mantsch, Infrared spectra of human central nervous system tissue: Diagnosis of Alzheimer's disease by multivariate analysis, *Biospectroscopy* **1** (1995) 141–148.
28. C. Gilbert, S. Kokot, Discrimination of cellulosic fabrics by diffuse reflectance infrared Fourier transform spectroscopy and chemometrics, *Vibrational Spectroscopy* **9**(2) (1995) 161–167.
29. H. Schulz, K. Tetzlaff, and L. Vogel, Simple encoding of infrared spectra for pattern recognition. 1: Statistical examination of the effectiveness and information content by factor analysis, *J. Chemometr.* **9**(3) (1995) 143–168.
30. M. Defernez, R. H. Wilson, Mid infrared spectroscopy and chemometrics for determining the type of fruit used in jam, *J. Sci. Food Agric.* **67** (1995) 461–467.
31. E. Bye, Proc. 7th Int. Symp. on Inhaled Part. VII. Published by Permagon Press, Oxford, U.K. (1994) 519–25.
32. P. Zecchini, K. Yamni, B. Viard, D. Dothee, Proc. 48th IEEE Freq. Control Symp. (1994) 91–98.
33. M. Defernez, E. K. Kemsley, R. H. Wilson, Use of infrared-spectroscopy and chemometrics for the authentication of fruit purees, *Journal of Agricultural and Food Chemistry* **43** (1995) 109–113.
34. T. Visser, H. J. Luinge, J. H. van der Maas, Recognition of visual characteristics of infrared spectra by artificial neural networks and partial least squares regression, *Anal. Chim. Acta* **296** (1994) 141–154.
35. W. Pohle, E. B. Starikov, W. Scheiding, A. Rupprecht, Proc. 5th Int. Conf. Spectrosc. Biol. Mol., Published by Kluwer, Dordrecht, Neth. (1993) 47–50.
36. H. Hobert, U. Schulz, J. Einax, Characterization of Saale River Sediments by IR Spectroscopy Charakterisierung von Saale-Sedimenten durch Infrarotspektroskopie, *Acta Hydrochim. Hydrobiol.* **22**(2) (1994) 76–84.
37. W. G. De Ruig and J. M. Weseman, A New Approach to Confirmation by Infrared Spectrometry, *J. Chemometrics* **4** (1990) 61–77.
38. J. Coates, T. Davidson, L. McDermott, The Design and Application of Spectrometric Analyzers for the Chemical Process Industry, *Spectroscopy* **7**(9) (1992) 40–49.
39. The American Society for Testing and Materials (ASTM) International Practice ASTM E1655-05(2012) Standard Practices for Infrared Multivariate Quantitative Analysis, ASTM Annual Book of Standards, West Conshohocken, PA 19428–2959 USA, (2012) **03.06**.
40. J. S. Shenk, M. O. Westerhaus, New standardization and calibration procedures for NIRS analytical systems, *Crop Sci.* **31** (1991) 1694–1696.
41. J. S. Shenk, M. O. Westerhaus, W. C. Templeton, Calibration Transfer Between near Infrared Reflectance Spectrophotometers, *Crop Sci.* **25**(1) (1985) 159–161.
42. E. Bouveresse, D. L. Massart, and P. Dardenne, Modified Algorihm for Standardization of Near Infrared Spectrometric Instruments, *Anal. Chem.* **67** (1995) 1381–1389.
43. H. Mark, J. Workman, A New Approach to Generating Transferable Calibrations for Quantitative Near Infrared Spectroscopy. *Spectroscopy* **3**(11) (1988) 28–36.
44. K. G. Nordqvist, Method of Reducing the Susceptibility to Interference of a Measuring Instrument, U.S. Patent No. 4,944,589 (1990).
45. Y. Wang, D. J. Veltkamp, B. Kowalski, Multivariate instrument standardization, *Anal. Chem.* **63** (1991) 2750–2756.

46. A. Ganz, R. Hoult, D. Tracy, Standardizing and Calibrating a Spectrometric Instrument, U.S. Patent No. 5,303,165, (April 12, 1994).
47. J. S. Shenk, Near Infrared Reflectance Spectroscopy (NIRS): Analysis of Forage Quality, G. C. Marten, J. S. Shenk, F. E. Barton II, eds. (Agriculture Handbook #643, U.S. Department of Agriculture, Washington, D.C., (1985) 41–44.
48. T. Hirschfeld and E. Stark, In Analysis of Foods and Beverages (Academic Press, San Diego, CA, 1984).
49. H. Mark and J. Workman, Jr., Statistics in Spectroscopy, 1st edn. (Academic Press, Boston, MA, 1991).
50. J. Workman and J. Coates, Multivariate Calibration Transfer, The Importance of Standardizing Instrumentation, *Spectroscopy* **8**(9) (1993) 36–42.
51. Y. Wang and B. Kowalski, Standardization of Second-Order Instruments, *Anal. Chem.* **65** (1993) 1174–1180.
52. M. Stone and R. J. Brooks, Continuum Regression: Cross-Validated Sequentially Constructed Prediction Embracing Ordinary Least Squares, Partial Least Squares and Principal Components Regression, *Journal of the Royal Statistical Society. Series B (Methodological)* **52**(2) (1990) 237–269.
53. A. Lorber, L. E. Wangen, and B. R. Kowalski, A Theoretical Foundation for the PLS Algorithm, *The Journal of Chemometrics* **1**(1) (1987) 19–31.
54. B. M. Wise and N. L. Ricker, Identification of Finite Impulse Response Models with Continuum Regression, *J. Chemometr.* **7**(1) (1993) 1–14.
55. S. N. Deming, J. A. Palasota, J. M. Nocerino, The Geometry of Multivariate Object Preprocessing, *J. Chemometr.* **7** (1993) 393.
56. W. H. Press, B. P. Flannery, S. A. Teukolsky, W. T. Vetterling, *Numerical Recipes in Pascal: The Art of Scientific Computing*, (Cambridge University Press, New York, 1989).
57. J. Toft, O. M. Kvalheim, Eigenstructure Tracking Analysis for Revealing Noise Pattern and Local Rank in Instrumental Profiles: Application to Transmittance and Absorbance IR Spectroscopy, *Chemom. Intell. Lab. Syst.* **19** (1993) 65.
58. O. M. Kvalheim, F. Brakstad, Y. Liang, Preprocessing of Analytical Profiles in the Presence of Homoscedastic or Heteroscedastic Noise, *Anal. Chem.* **66** (1994) 43.
59. P. Paatero, U. Tapper, Analysis of Different Modes of Factor Analysis as Least Squares Fit Problems, *Chemom. Intell. Lab. Syst.* **18** (1993) 183.
60. M. V. Gorshkov, R. T. Kouzes, Data Reflection Algorithm for Spectral Enhancement in Fourier Transform ICR and NMR Spectroscopies, *Anal. Chem.* **67** (1995) 3412.
61. T. Iwata, J. Koshoubu, A New Method to Eliminate the Background Noise from a Line Spectrum, *Appl. Spec.* **48** (1994) 1453.
62. T. Iwata, J. Koshoubu, Minimization of Noise in Spectral Data, *Appl. Spec.* **50** (1996) 747.
63. H. Takeuchi, S. Hashimoto, I. Harada, Simple and Efficient Method to Eliminate Spike Noise from Spectra Recorder on Charge-Coupled Device Detectors, *Appl. Spec.* **47** (1993) 129.
64. W. Hill, Comment on a 'Simple and Efficient Method to Eliminate Spike Noise from Spectra Recorded on Charge-Coupled Device Detectors, *Appl. Spec.* **47** (1993) 2171.
65. F. C. Sanchez, P. J. Lewi, D. L. Massart, Effect of Different Preprocessing Methods for Principal Component Analysis Applied to the Composition of Mixtures: Detection of Impurities in HPLC-DAD, *Chemom. Intell. Lab. Syst.* **25** (1994) 157.
66. A. Savitzky, M. J. E. Golay, Smoothing and Differentiation of Data by Simplified Least Squares Procedures, *Anal. Chem.* **36** (1964) 1627.

67. P. A. Gorry, General Least-Squares Smoothing and Differentiation by the Convolution (Savitszky-Golay) Method, *Anal. Chem.* **62** (1990) 570.
68. F. Holler, D. H. Burns, J. B. Callis, Direct Use of Second Derivatives in Curve-Fitting Procedures, *Appl. Spec.*, **43** (1989) 877.
69. W. Windig, The Use of Second-Derivative Spectra for Pure-Variable Based Self-Modeling Mixture Analysis Techniques, *Chemom. Intell. Lab. Syst.* **23** (1994) 71.
70. J. K. Kauppinen, D. J. Moffatt, H. H. Mantsch, D. G. Cameron, Fourier Self-Deconvolutions: A Method for Resolving Intrinsically Overlapped Bands, *Appl. Spec.* **35** (1981) 271.
71. C. Chapados, M. Trudel, J. Miletic, Enhancing Infrared Spectra by Fourier Deconvolution using the Cauchy-Gauss Product Function, *Chemom. Intell. Lab. Syst.* **22** (1994) 209.
72. R. G. Brereton, Fourier Transforms: Use, Theory and Applications to Spectroscopic and Related Data, *Chemom. Intell. Lab. Syst.* **1** (1986) 17.
73. M. Devaux, D. Bertrand, P. Robert, J. Morat, Extraction of Near Infra-red Spectral Information by Fast Fourier Transform and Principal Component Analysis. Application to the Discrimination of Baking Quality of Wheat Flours, *J. Chemometr.* **1** (1987) 103.
74. P. J. Treado, M. D. Morris, A Thousand Points of Light: The Hadamard Transform in Chemical Analysis and Instrumentation, *Anal. Chem.* **61** (1989) 723A.
75. M. Bos, J. A. M. Vrielink, The Wavelet Transform for Pre-processing IR Spectra in the Identification of Mono- and Di-substituted Benzenes, *Chemom. Intell. Lab. Syst.* **23** (1994) 115.
76. B. Walczak, B. van den Bogaert, D. L. Massart, Application of Wavelet Packet Transform in Pattern Recognition of Near-IR Data, *Anal. Chem.* **68** (1996) 1742.
77. B. Vidakovic, P. Muller, Wavelets for Kids: A Tutorial, http://www.diku.dk/~jda/biosignal/kidsA.pdf
78. W. H. Press, B. P. Flannery, S. A. Teukolsky, W. T. Vetterling, *Numerical Recipes in Fortran: The Art of Scientific Computing*, (Cambridge University Press, New York 1992).
79. Mathworks wavelet toolbox, http://www.mathworks.com/wavelet.html
80. P. Williams, K. Norris, (Eds.) *Near Infrared Technology*, 2 edn. (Amer. Assoc. Cereal Chemists, St. Paul, MN. 2001).
81. W. F. McClure and A. M. C. Davies, More on derivatives: Part 1: Segments, gaps and ghosts, *NIR news* **4**(6) (1994) 12.
82. W. F. McClure, More on derivatives. Part 2: Band shifting and noise, *NIR news* **5**(1) (1995) 14–16.
83. K. G. Nordquist, Method of reducing the susceptibility to interference of a measuring instrument, U.S. Patent No. 4,944,589; Filed Sept. 22, 1988; Awarded July 31, 1990.
84. J. Workman, Advances in the use of Near Infrared Spectroscopy in Monogastric Research, *In Proceedings of the Georgia Nutrition Conference*, Atlanta, GA (1991) 37–47.
85. J. Workman, H. Andren, NIR Spectroscopy for At-Line Meat Product Measurements, *American Laboratory* **24**(18) (1992) R20–V20.
86. R. DiFoggio, M. Sadhukhan, Method for improving infrared analysis estimations by automatically compensating for instrument instabilities, U.S. Patent No. 5,397,899; Filed July 21, 1992; Awarded Mar. 14, 1995.
87. H. Mark, J. Workman, A New Approach to Generating Transferable Calibrations for Quantitative Near Infrared Spectroscopy, Spectrosc. **3**(11) (1988), 28–36.
88. E. V. Thomas, A Primer on Multivariate Calibration, *Anal. Chem.* **66** (1994) 795A.

89. M. Davidian, P. D. Haaland, Regression and Calibration with Nonconstant Error Variance, *Chemom. Intell. Lab. Syst.* **9** (1990) 231.
90. O. E. de Noord, The Influence of Data Preprocessing on the Robustness and Parsimony of Multivariate Calibration Models, *Chemom. Intell. Lab. Syst.* **23** (1994) 65.
91. L. Jancar, W. Wegscheider, Effect of Scaling Regimes on the Prediction of Analytical Results from Multivariate Calibration, *Anal. Chim. Acta* **248** (1991) 459.
92. C. L. Erickson, M. L. Lysaght, J. B. Callis, Relationship between Digital Filtering and Multivariate Regression in Quantitative Analysis, *Anal. Chem.* **64** (1992) 1155A.
93. G. W. Small, M. A. Arnold, L. A. Marquardt, Strategies for Coupling Digital Filtering with Partial Least-Squares Regression: Application to the Determination of Glucose in Plasma by Fourier Transform Near Infrared Spectroscopy, *Anal. Chem.* **65** (1993) 3279.
94. M. J. Mattu, G. W. Small, Quantitative Analysis of Bandpass-Filtered Fourier Transform Infrared Interferograms, *Anal. Chem.* **67** (1995) 2269.
95. P. B. Stark, M. M. Herron, A. Matteson, Empirically Minimax Affine Mineralogy Estimates from Fourier Transform Infrared Spectrometry Using a Decimated Wavelet Basis, *Appl. Spec.* **47** (1993) 1820.
96. M. B. Seasholtz, B. R. Kowalski, The Effect of Mean Centering on Prediction in Multivariate Calibration, *J. Chemometr.* **6** (1992) 103.
97. I. S. Helland, T. Naes, T. Isaksson, Related Versions of the Multiplicative Scatter Correction Method for Preprocessing Spectroscopic Data, *Chemom. Intell. Lab. Syst.* **29** (1995) 233.
98. R. J. Barnes, M. S. Dhanoa, S. J. Lister, Standard Normal Variate Transformation and De-trending of Near Infrared Diffuse Reflectance Spectra, *Appl. Spec.* **45** (1989) 772.
99. M. S. Dhanoa, S. J. Lister, R. J. Barnes, On the Scales Associated with Near Infrared Reflectance Difference Spectra, *Appl. Spec.* **49** (1995) 765.
100. T. Isaksson, T. Naes, The Effect of Multiplicative Scatter Correction (MSC) and Linearity Improvement in NIR Spectroscopy, *Appl. Spec.* **42** (1988) 1273.
101. P. Kubelka, F. Munk, Ein Beitrag zur Optik der Farbanstriche, *Z. Tech. Phys. (Leipzig)* **12** (1931) 593–601.
102. J. L. Sanderson, Calculation of the Color of Pigmented Plastics, *J. Opt. Soc. Am.* **32** (1942) 727-729.
103. T. Isaksson, B. R. Kowalski, Piece Wise Multiplicative Scatter Correction Applied to Near — Infrared Diffuse Transmittance Data from Meat Products, *Appl. Spec.* **47** (1993) 702.
104. T. V. Karstang, R. Manne, Optimized Scaling: A Novel Approach to Linear Calibration with Closed Data Sets, *Chemom. Intell. Lab. Syst.* **14** (1992) 165.
105. T. Isaksson, Z. Wang, B. R. Kowalski, Optimised Scaling (OS-2) Regression Applied to Near Infrared Diffuse Spectroscopy Data from Food Products, *J. Near Infrared Spec.* **1** (1993) 85.
106. A. A. Agbodjan, S. C. Rutan, Optimization of an Adaptive Kalman Filter Based on Information Theory, *Chemom. Intell. Lab. Syst.* **24** (1994) 137.
107. C. L. de Ligny, M. C. Spanjer, J. C. van Houwelingen, H.M. Weesie, Three-mode factor analysis of data on retention in normal-phase high-performance liquid chromatography, Journal of Chromatography, *J. Chromatography* **301**(3) (1984) 311–324.
108. A. K. Smilde, P. H. van der Graaf, D. A. Doornbos, T. Steerneman and A. Sleurink, Multivariate calibration of reversed-phase chromatographic systems. Some designs based on three-way data analysis, *Analytica Chimica Acta* **235** (1990) 41–51.
109. A. K. Smilde and D. A. Doornbos, Three-way methods for the calibration of chromatographic systems: comparing PARAFAC and three-way PLS, *J. Chemometr.* **5** (1991) 345–360.

110. S. Leurgans and R. T. Ross, Multilinear models in spectroscopy, *Stat. Sci.* **7** (1992) 289–319.
111. D. S. Burdick, An Introduction to Tensor Products with Applications to Multiway Data Analysis, *Chemom. Intell. Lab. Syst.* **28** (1995) 229–237.
112. E. Sanchez and B. R. Kowalski, Tensorial calibration: II. Second-order calibration, *J. Chemometr.* **2** (1988) 265–280.
113. S. Wold, P. Geladi, K. H. Esbensen, and J. Öhman, Multi-way principal components and PLS-analysis, *J. Chemometr.* **1** (1987) 41–56.
114. J. D. Carroll and P. Arabie, Multidimensional scaling, *Ann. Rev. Psychol.* **31** (1980) 607–649.
115. P. Geladi, Analysis of multi-way (multi-mode) data, *Chemom. Intell. Lab. Syst.* **7**(1–2) (1989) 11–30.
116. H. A. L. Kiers, Hierarchical relations among three-way methods, *Psychometrika* **56** (1991a) 449–470.
117. K. H. Esbensen, S. Wold, P. Geladi, Relationships between higher-order data array configurations and problem formulations in multivariate data analysis, *J. Chemometr.* **3** (1988) 33–48.
118. R. A. Harshman, Foundations of the PARAFAC procedure: Models and conditions for an "explanatory" multi-modal factor analysis. UCLA Working Papers in Phonetics (UMI Serials in Microform, No. 10,085) **16** (1970) 1–84.
119. J. D. Carroll and J. Chang, Analysis of individual differences in multidimensional scaling via an N-way generalization of 'Eckart–Young' decomposition, *Psychometrika* **35** (1970) 283–319.
120. N. M. Faber, L. M. C. Buydens, G. J. Kateman, Generalized rank annihilation method. II: Bias and variance in the estimated eigenvalues, *J. Chemometr.* **8**(2) (1994) 181–203.
121. N. M. Faber, L. M. C. Buydens, G. J. Kateman, Standard errors in the eigenvalues of a cross-product matrix: Theory and applications, *J. Chemometr.* **7**(6) (1993) 495–526.
122. B. E. Wilson, W. Lindberg, and B. R. Kowalski, Multicomponent quantitative analysis using second-order nonbilinear data: theory and simulations, *J. Am. Chem. Soc.* **111**(11) (1989) 3797–3804.
123. Y. Wang, O. S. Borgen, B. R. Kowalski, M. Gu, and F. Turecek, Advances in second-order calibration, *J. Chemometr.* **7**(2) (1993) 117–130.
124. E. Sanchez, B. R. Kowalski, Tensorial resolution: A direct trilinear decomposition, *J. Chemometr.* **4**(1) (1990) 29–45.
125. American Society of Testing and Materials International Practice ASTM ASTM E1421-99 (2009), "Standard Practice for Describing and Measuring Performance of Fourier Transform Mid Infrared (FT-MIR) Spectrometers: Level Zero and Level One Tests", ASTM International, 100 Barr Harbor Drive, PO Box C700, West Conshohocken, PA, 19428–2959 USA (2009).
126. American Society of Testing and Materials International Practice ASTM E1866-97 (2013), "Standard Guide for Establishing Spectrophotometer Performance Tests", ASTM International, 100 Barr Harbor Drive, PO Box C700, West Conshohocken, PA, 19428–2959 USA (2009).
127. J. L. Koenig and C. M. Snively, Fast FTIR Imaging: Theory and Applications, *Spectroscopy* **13**(11) (1998) 22–28.
128. P. Colarusso, L. H. Kidder, I. W. Levin, J. C. Fraser, J. F. Arens and E. N. Lewis, Infrared Spectroscopic Imaging: From Planetary to Cellular Systems, *Appl. Spectrosc.* **52**(3) (1998) 106A–120A.
129. A. Villringer, B. Chance, Noninvasive optical spectroscopy and imaging of human brain function, *Trends Neurosci.* **20**(10) (1997) 435–442.

130. J. K. Crowley and D. R. Zimbelman, Mapping hydrothermally altered rocks on Mount Rainier, Washington, with Airborne Visible/Infrared Imaging Spectrometer (AVIRIS) data, *Geology* **25**(6) (1997) 559–562.
131. L. Ning, A. M. Chozinski, A. Azarenko, L. S. Daley, W. J. Bowyer, T. Buban, G. E. Edwards, J. B. Callis and G.B. Strobel, Five novel applications of imaging visible and short near — infrared spectrophotometry and fluorimetry in the plant sciences. Part I: Photographic and Histological Applications, *Spectroscopy* **11**(9) (1996) 44.
132. N. Rowlands and R. Neville, Calcite and dolomite discrimination using airborne SWIR imaging spectrometer data. Proc. SPIE-Int. Soc. Opt. Eng. Vol. 2819: Imaging Spectrometry II (1996) 36–44.
133. L. John Otten, III, A. D. Meigs, R. G. Sellar and J. B. Rafert, Measured performance of an airborne Fourier transform hyperspectral imager, Proc. SPIE-Int. Soc. Opt. Eng., Vol. 2819: Imaging Spectrometry II (1996) 291–299.
134. P. F. Valdez, G. W. Donohoe, M. R. Descour and S. Motomatsu, Selection of spectral bands for interpretation of hyperspectral remotely sensed images, Proc. SPIE-Int. Soc. Opt. Eng., Vol. 2819: Imaging Spectrometry II, (1996) 195–203.
135. K. Silk and E. R. Schildkraut, Imaging Fourier transform spectroscopy for remote chemical sensing, Proc. SPIE-Int. Soc. Opt. Eng., Vol. 2763: Electro-Optical Technology for Remote Chemical Detection and Identification (1996) 169–177.
136. L. C. Rowan, T. L. Bowers, J. K. Crowley, C. Anton-Pacheco, P. Gumiel and M. J. Kingston, Analysis of airborne visible-infrared imaging spectrometer (AVIRIS) data of the Iron Hill, Colorado, carbonatite-alkalic igneous complex, *Econ. Geol.* **90**(7) (1995) 1966–82.

Chapter 7

Applications and Spectral Examples for Infrared Spectroscopy

Contents

7.1 Infrared History and Applications

7.1.1 *Introduction*

Infrared energy is the electromagnetic energy of molecular vibration. The energy band is defined for convenience as the near infrared (0.78 to 2.50 microns (i.e., 780 to 2500 nm); the infrared (or mid infrared) 2.50 to 25.0 microns (i.e., 4000 to 400 cm^{-1}); and the far infrared or terahertz region of 25.0 to 1000 microns (i.e., 400 to 10 cm^{-1}). Dominant mid infrared spectral features include the C-H (methyl, methylene, aromatic, methoxy, and carbonyl) fundamental stretching and bending molecular vibrations, O-H (hydroxyl) stretch fundamental vibrations, N-H (amine) stretching, C-F (fluorocarbon) stretching, -C≡N (nitrile) stretching, -C=O (carbonyl) stretch from esters, acetates, and amides, C-Cl stretch from chlorinated hydrocarbons, and $-NO_2$ from nitro- containing compounds.

Table 7.1 and Figures 7.1 and 7.2 demonstrate the major infrared bands, locations, and band characteristics. Additional detailed figures and tables identifying spectra-structure correlation for infrared bands are also included within this chapter. For example, refer to Figures 7.3 through 7.24, and Table 7.4.

Table 7.1 Summary of infrared band positions and strengths for major functional groups.

Functional group	Location (cm^{-1})	Band character	Comments
O-H	3350–3600	s, brd	Broad due to H bonding
N-H Stretching	3350	m	Asymmetric Stretching
N-H Stretching	3180	m	Symmetric Stretching
N-H Bending	3100	m	Overtone of Bending
N-H Scissoring and bending	1550–1630	m	**Amide II Band**
C-H	2800–3050	s	Range of C-H
Ar-H	3030	s, sharp	Aromatic C-H Stretching
CH_3 Stretching	2962	s	Asymmetric Stretching
CH_2 Stretching	2926	s	Asymmetric Stretching
CH_3 Stretching	2872	s	Symmetric Stretching
CH_2 Stretching	2853	s	Symmetric Stretching
CH_2 Scissoring	1460	m	Symmetric
CH_3 Bending	1455	m	Asymmetric Stretching
CH_3 Bending	1375	s	Symmetric Stretching
CH_2 Rocking	725	s	Symmetric Stretching
C=O Stretching	1715	s, sharp	Exact position depends on type of carbonyl
C=O Stretching	1630-1690	s, sharp	**Amide I Band**
C-O Stretching	1100	s	Ethers and Alcohols
C=C Stretching	1650	w, alkene m-s, aromatic	Alkenes
C≡C Stretching	2150	w, sharp	Alkynes
C≡N Stretching	2250	m, sharp	Nitriles

Note: The following abbreviations for Band Intensities: w = weak, m = medium, m-s = medium-strong, s = strong, υ = variable.

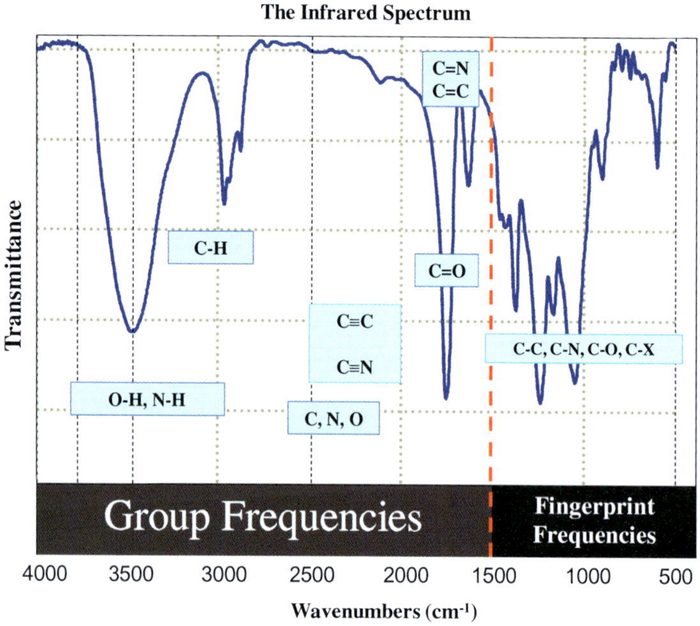

Fig. 7.1 MIR spectral chart: MIR fundamental group frequency and fingerprint frequency locations, showing a transmittance spectrum of cellulose acetate butyrate with typical NIR functional groups and band locations labeled (4000 to 500 cm^{-1}).

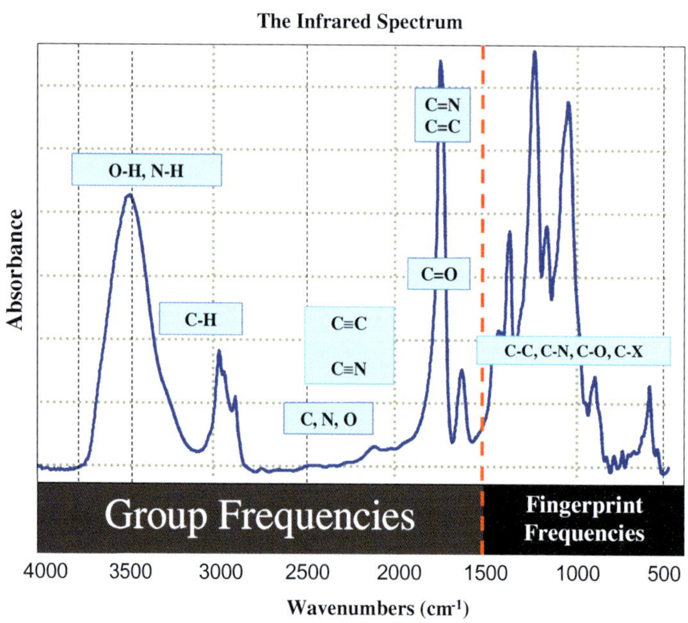

Fig. 7.2 MIR spectral chart: MIR fundamental group frequency and fingerprint frequency locations, showing an absorbance spectrum of cellulose acetate butyrate with typical NIR functional groups and band locations labeled (4000 to 500 cm^{-1}).

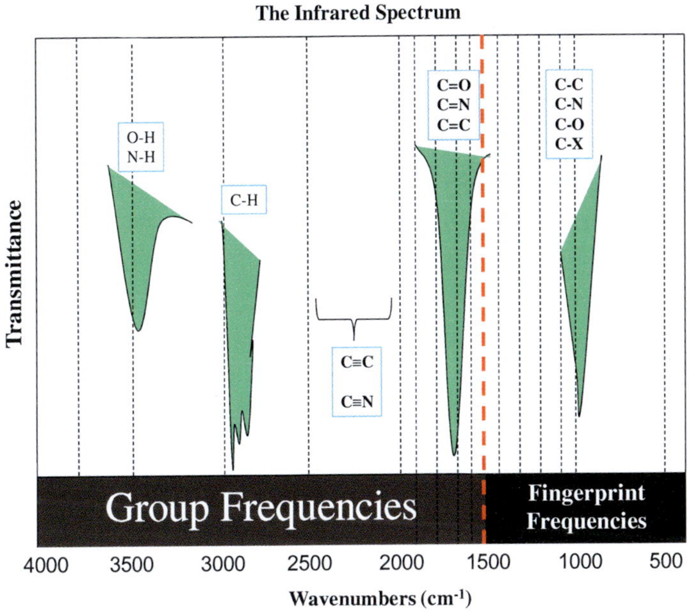

Fig. 7.3 MIR spectral chart: MIR fundamental group frequency and fingerprint frequency locations, showing transmittance band shapes and intensities with typical functional group bands are labeled (4000 to 500 cm^{-1}).

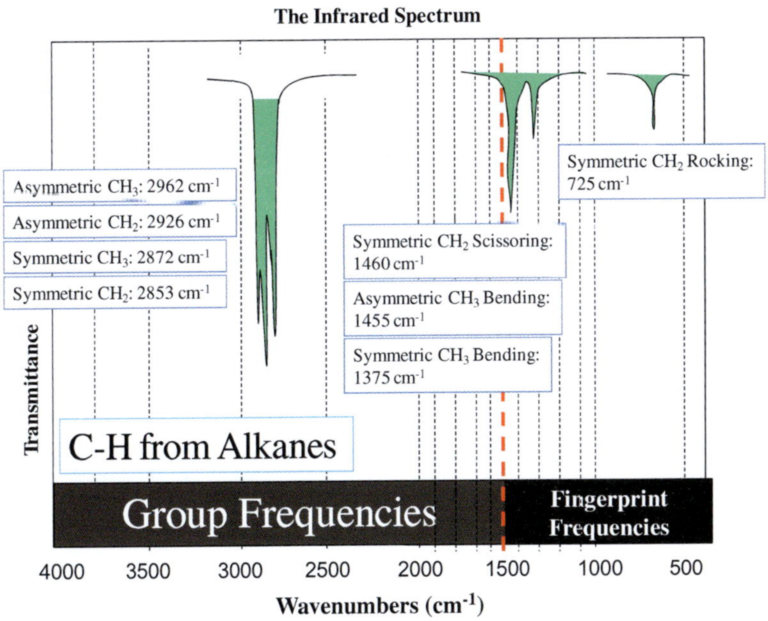

Fig. 7.4 MIR spectral chart: MIR fundamental group frequency and fingerprint frequency locations, showing transmittance C-H band shapes and intensities for n-Alkanes with typical functional group bands are labeled (4000 to 500 cm^{-1}).

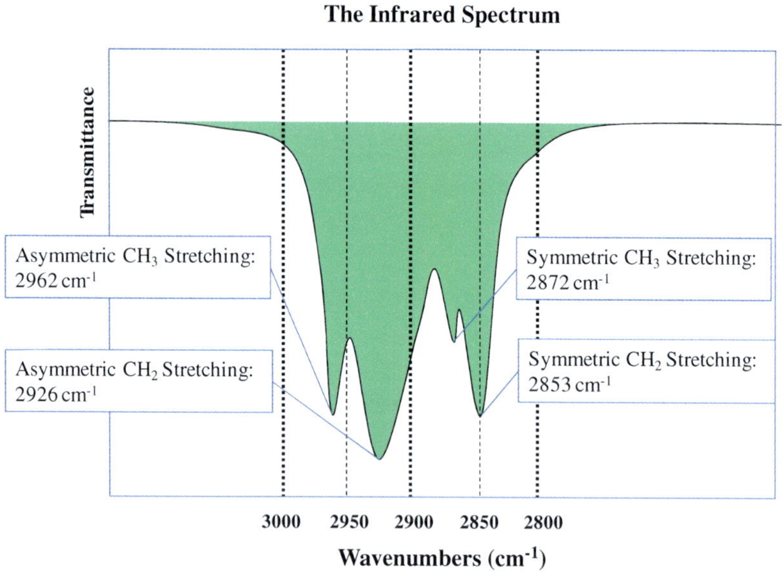

Fig. 7.5 MIR spectral chart: MIR fundamental group frequency and fingerprint frequency locations, showing transmittance band shapes and intensities for n-Alkanes C-H stretching for Methyl (CH_3) and Methylene (CH_2) functional group bands are labeled (4000 to 500 cm^{-1}).

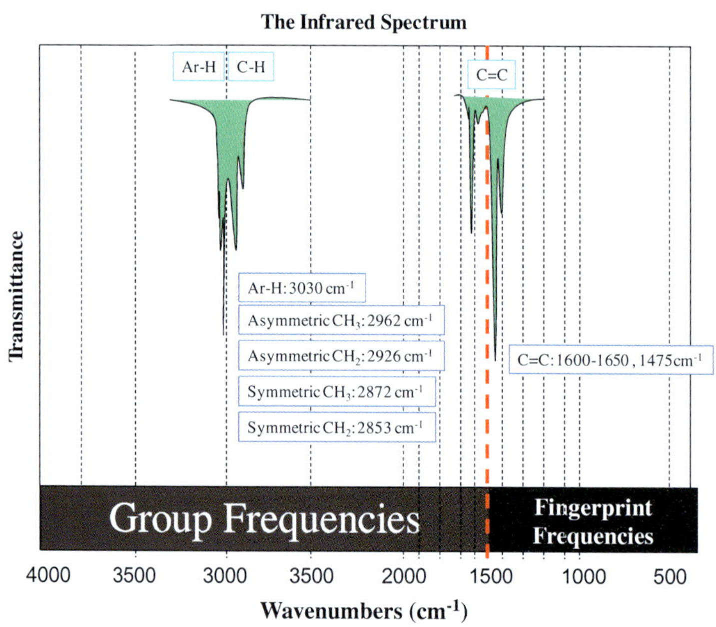

Fig. 7.6 MIR spectral chart: MIR fundamental group frequency and fingerprint frequency locations, showing transmittance band shapes and intensities for aromatic compounds with aliphatic attachments. Note aromatic (Ar) C-H, Aliphatic C-H, and Alkene (C=C) functional group bands are labeled (4000 to 500 cm^{-1}).

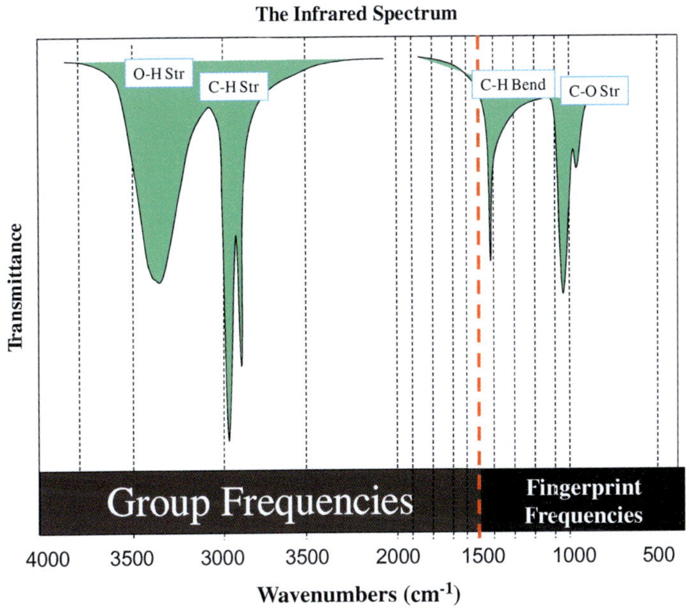

Fig. 7.7 MIR spectral chart: MIR fundamental group frequency and fingerprint frequency locations, showing transmittance band shapes and intensities for O-H Stretching, Aliphatic C-H Stretching, C-H Bending, and C-O Stretching functional group bands are labeled (4000 to 500 cm^{-1}).

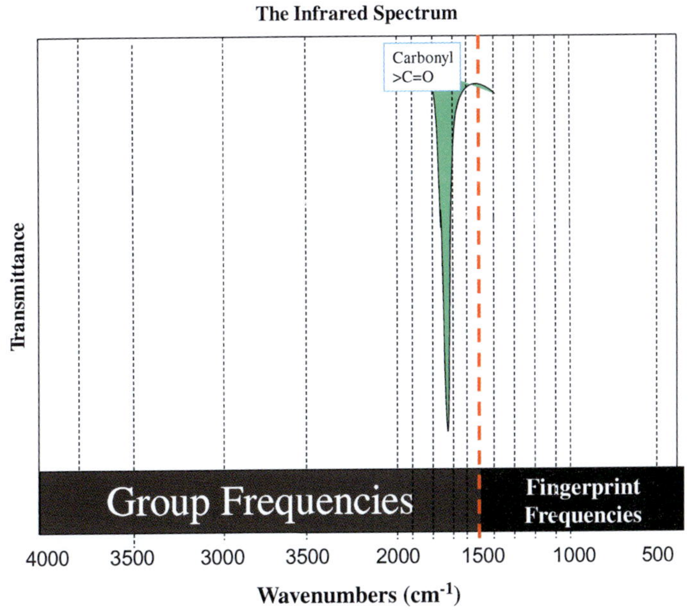

Fig. 7.8 MIR spectral chart: MIR fundamental group frequency and fingerprint frequency locations, showing transmittance band shape and intensity for Dimethyl Ketone Carbonyl (>C=O) Stretching functional group band is labeled near 1715 cm^{-1} (4000 to 500 cm^{-1}).

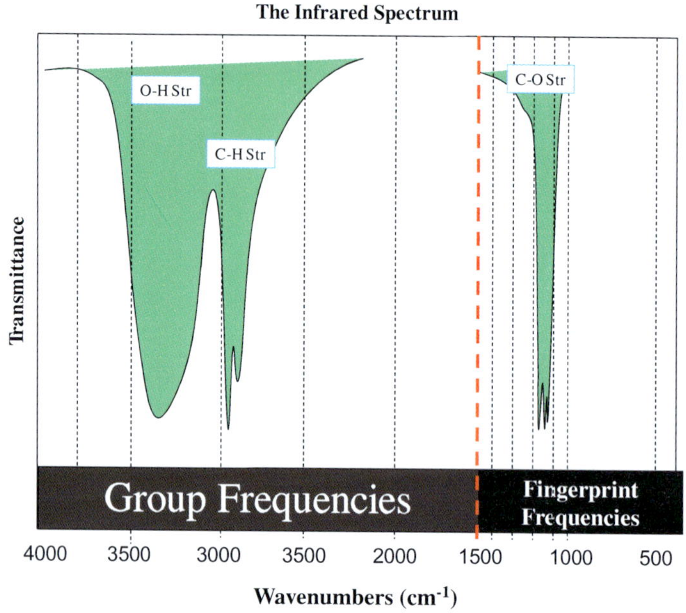

Fig. 7.9 MIR spectral chart: MIR fundamental group frequency and fingerprint frequency locations, showing transmittance band shapes and intensities for 2-Propanol O-H Stretching, Aliphatic C-H Stretching, and C-O Stretching functional group bands are labeled (4000 to 500 cm^{-1}); O-H band near 3340–3350 cm^{-1}.

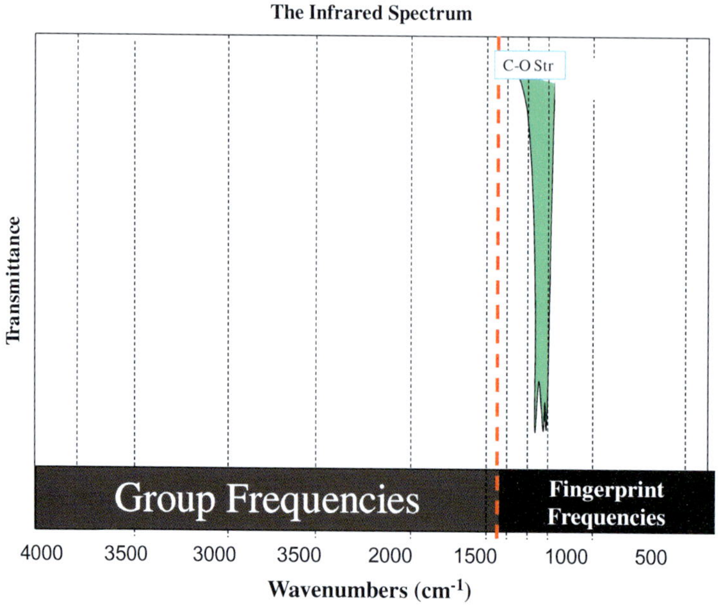

Fig. 7.10 MIR spectral chart: MIR fundamental group frequency and fingerprint frequency locations, showing transmittance band shape and intensity for 2-Propanol C-O Stretching (4000 to 500 cm^{-1}); Sharp C-O band near 1100–1170 cm^{-1}.

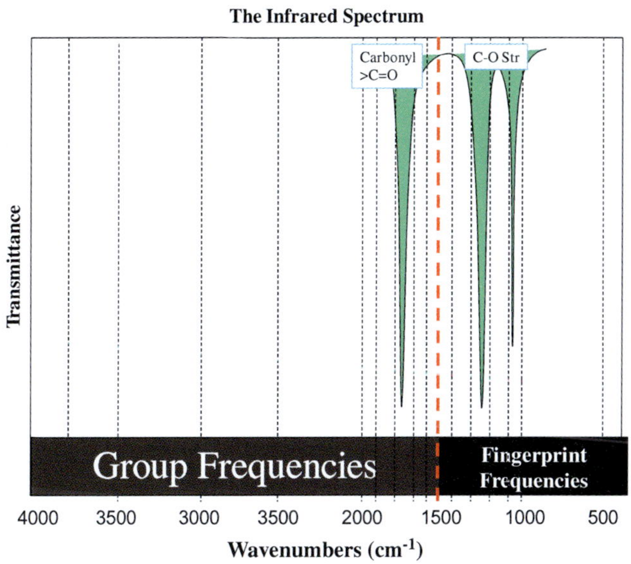

Fig. 7.11 MIR spectral chart: MIR fundamental group frequency and fingerprint frequency locations, showing transmittance band shapes and intensities for Methyl Acetate (Ester) Carbonyl (>C=O) Stretching, and C-O Stretching functional group bands are labeled (4000 to 500 cm^{-1}); C=O band near 1735–1720 cm^{-1}, C-O Bands near 1250 and 1050 cm^{-1}.

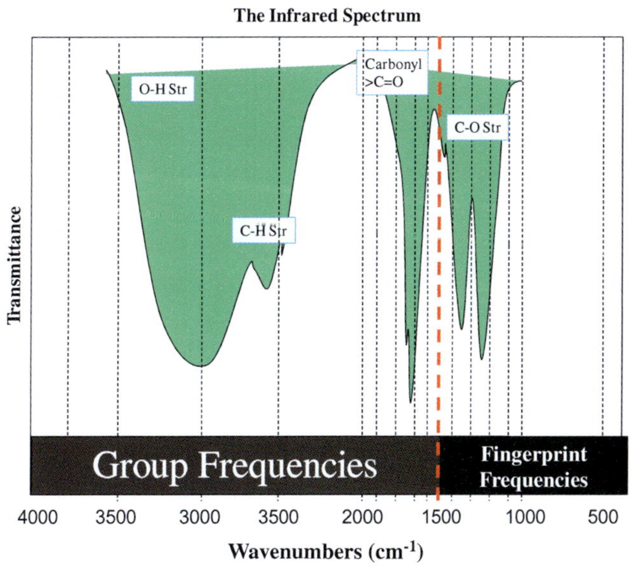

Fig. 7.12 MIR spectral chart: MIR fundamental group frequency and fingerprint frequency locations, showing transmittance band shapes and intensities for Ethanoic Acid (Carboxylic Acid) for O-H Stretching, Aliphatic C-H Stretching, Carbonyl (>C=O) Stretching, and C-O Stretching functional group bands are labeled (4000 to 500 cm^{-1}); O-H Stretching near 3050 cm^{-1}, C-H Stretching near 3200 to 3000 cm^{-1}, C=O Stretching near 1715 cm^{-1} and C-O stretching near 1280 cm^{-1}.

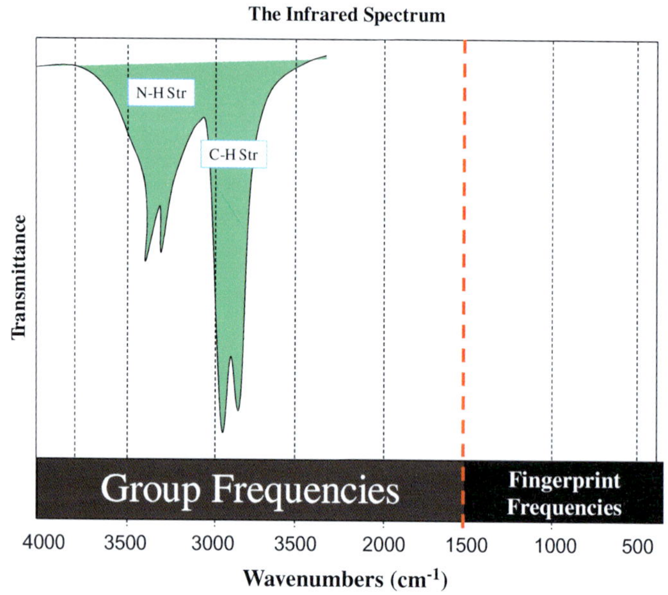

Fig. 7.13 MIR spectral chart: MIR fundamental group frequency and fingerprint frequency locations, showing transmittance band shapes and intensities for 1-Aminobutane (Primary Amine) for N-H Stretching, and Aliphatic C-H Stretching functional group bands are labeled (4000 to $500\,cm^{-1}$); N–H bands near 3400 to $3100\,cm^{-1}$, C-H bands between 3000 and $2800\,cm^{-1}$.

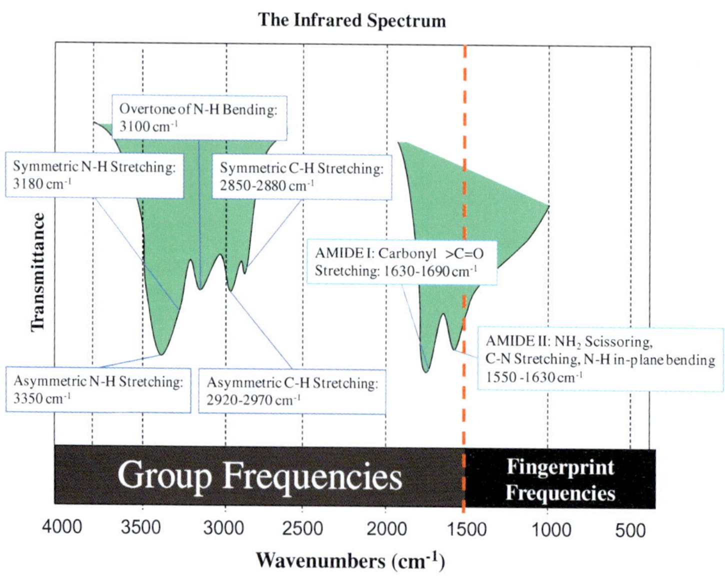

Fig. 7.14 MIR spectral chart: MIR fundamental group frequency and fingerprint frequency locations, showing transmittance band shapes and intensities for N-Methyl Acetamide (Secondary Amide) for N-H Stretching, and Aliphatic C-H Stretching, Carbonyl (>C=O) Stretching as Amide I band; and NH, C-N, and N-H comprising the typical Amide II Band. These functional group bands are labeled (4000 to $500\,cm^{-1}$).

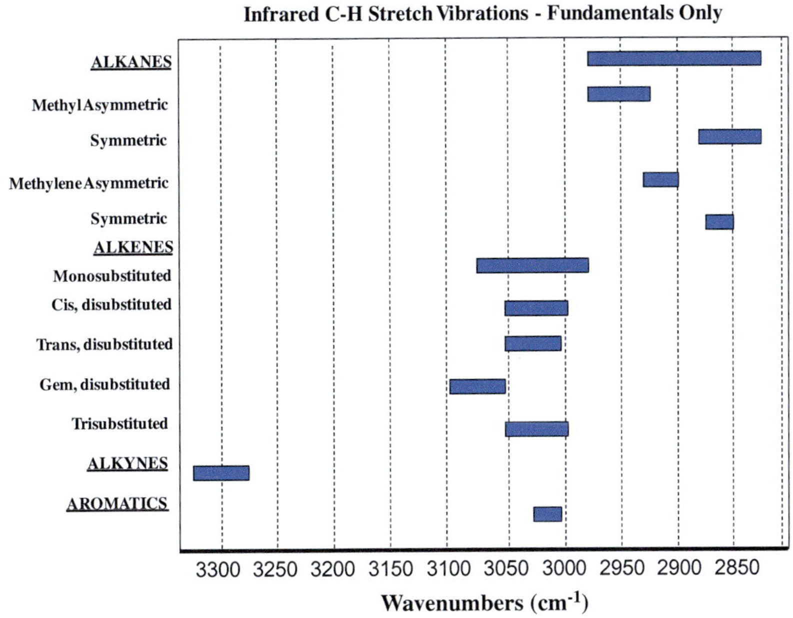

Fig. 7.15 IR Spectra-Structure Band Positions Chart — MIR Fundamental C-H absorptions indicated relative to wavenumbers (as 3300 to 2800 cm^{-1}).

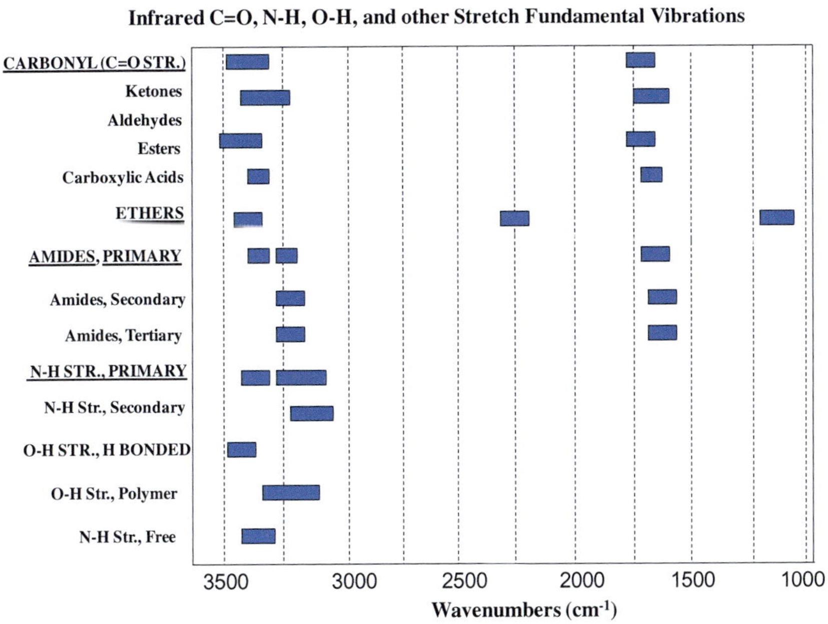

Fig. 7.16 IR Spectra-Structure Band Positions Chart: Fundamental C=O, N-H, and O-H absorptions indicated relative to wavenumbers (as 3500 to 1000 cm^{-1}).

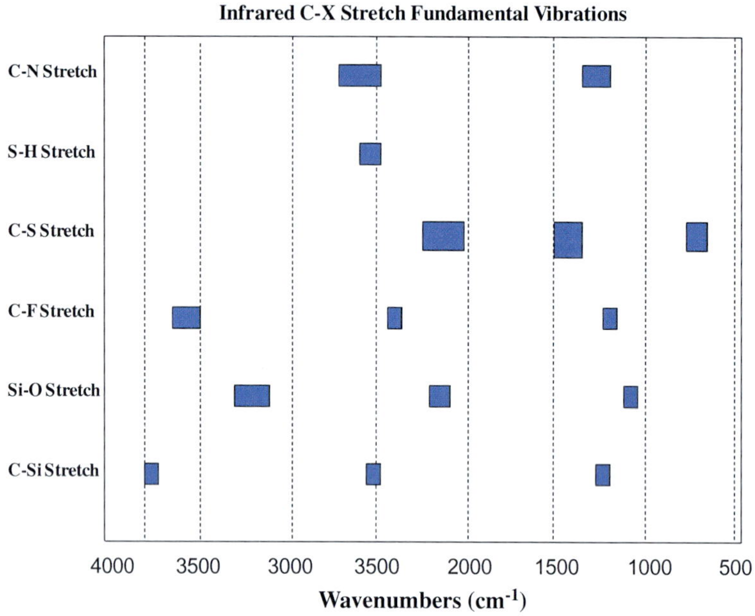

Fig. 7.17 IR Spectra-Structure Band Positions Chart — MIR Fundamental C-X absorptions indicated relative to wavenumbers (as 4000 to 500 cm^{-1}).

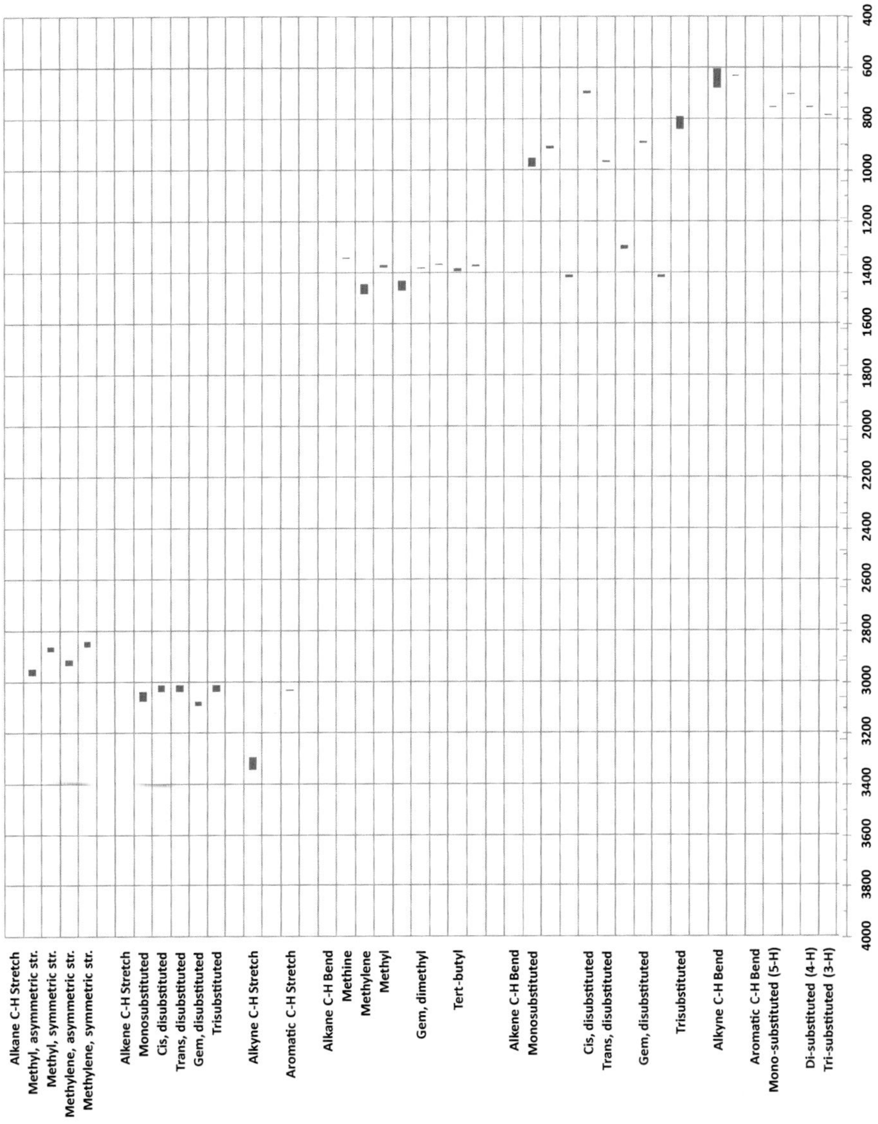

Fig. 7.18 MIR Spectra-Structure Band Positicns Chart — MIR Fundamental absorptions indicated relative to wavenumbers (as 4000 to 400 cm^{-1}).

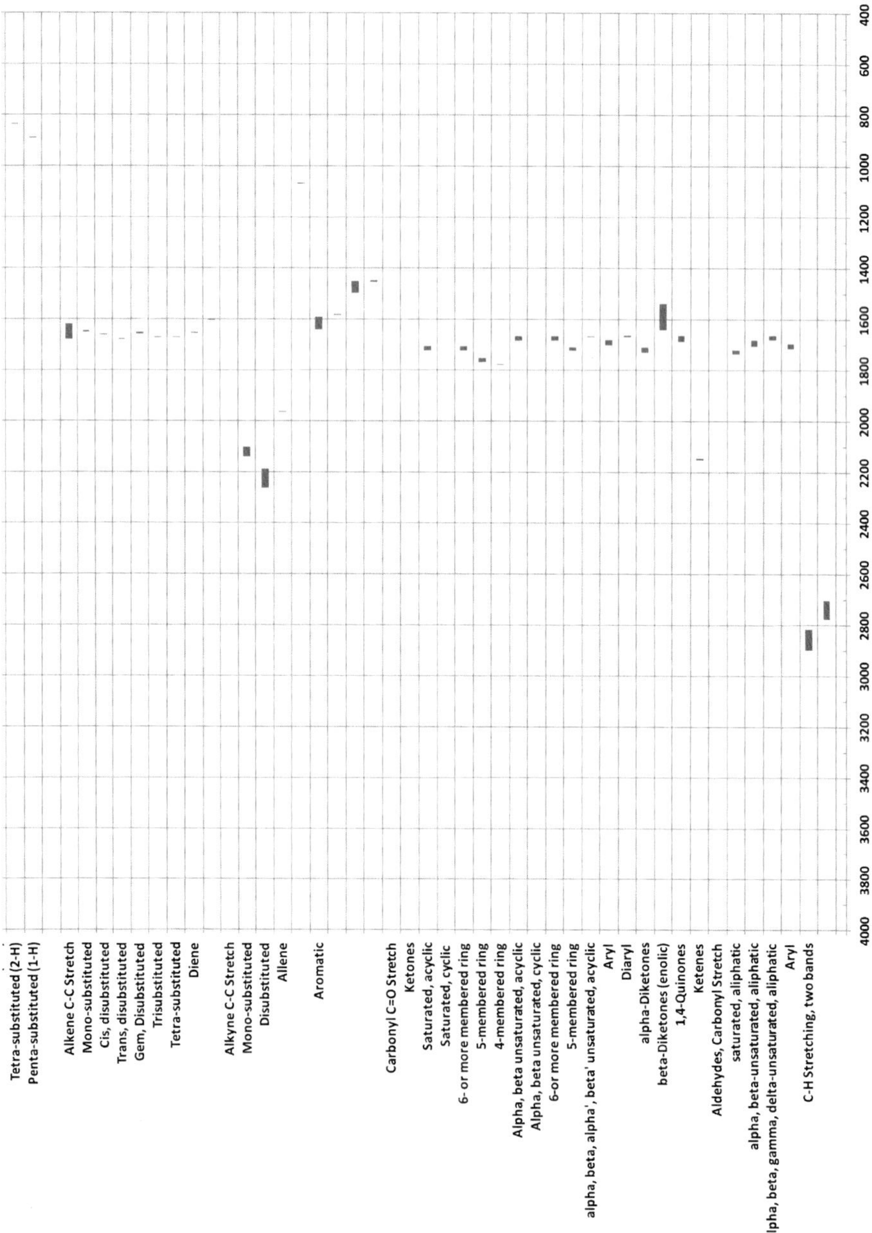

Fig. 7.19 MIR Spectra-Structure Band Positions Chart — MIR Fundamental absorptions indicated relative to wavenumbers (as 4000 to 400 cm^{-1}).

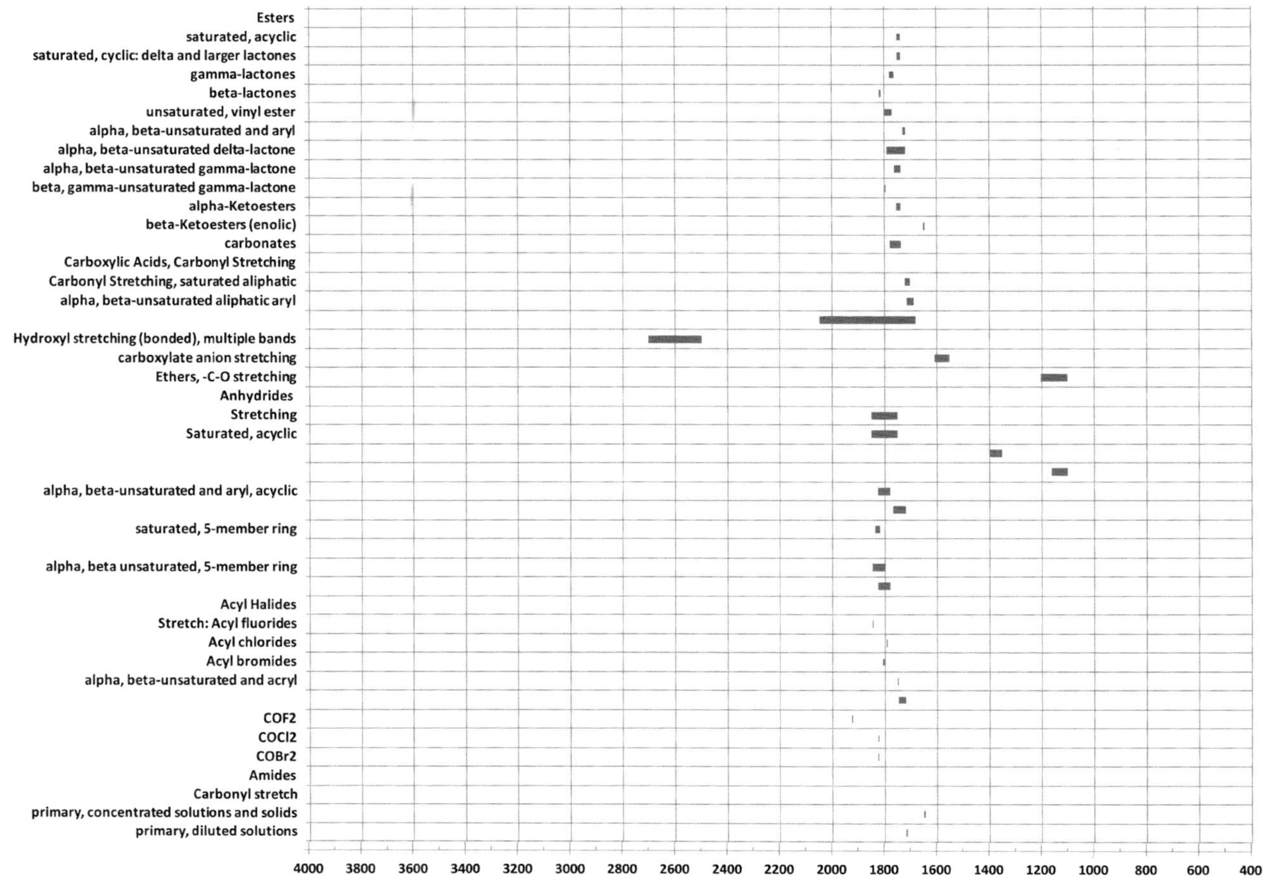

Fig. 7.20 MIR Spectra-Structure Band Positions Chart — MIR Fundamental absorptions indicated relative to wavenumbers (as 4000 to 400 cm^{-1}).

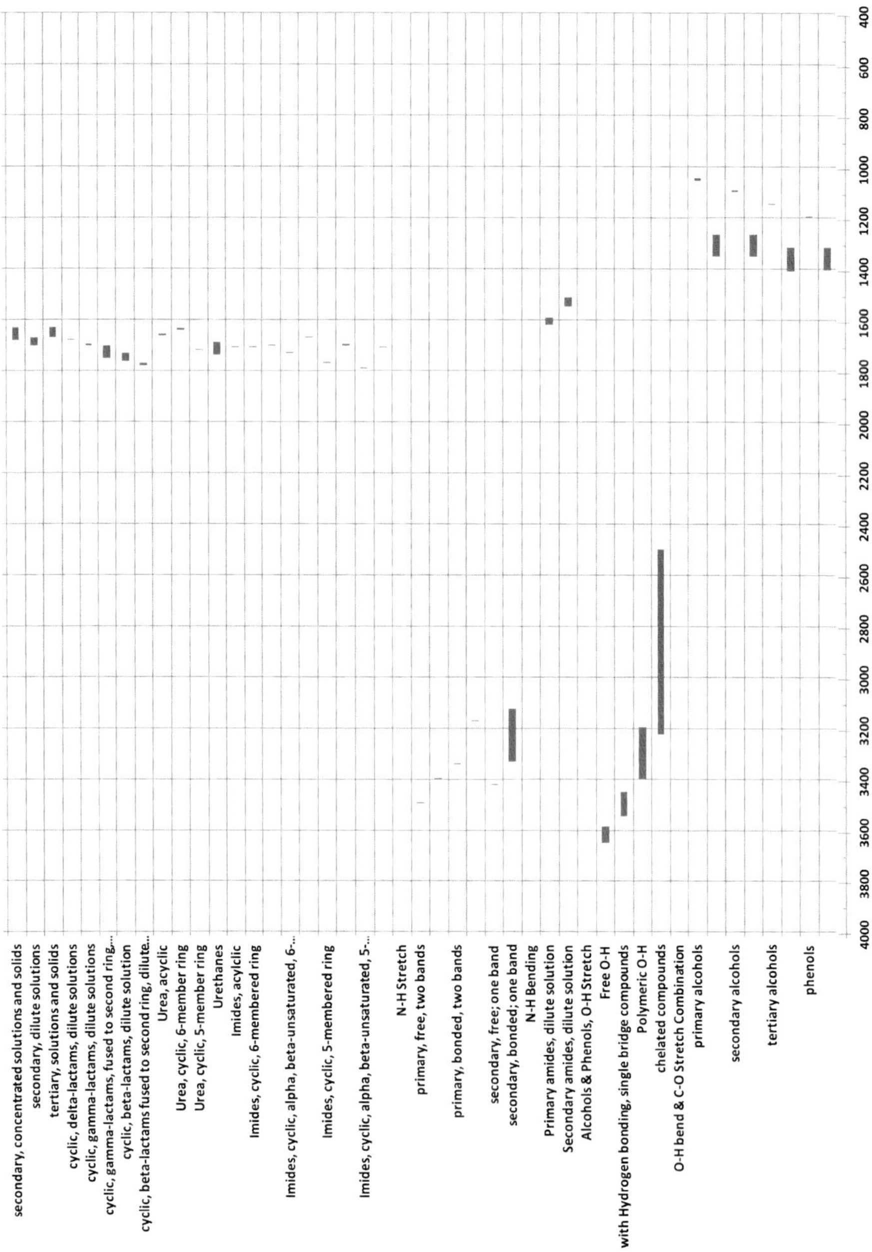

Fig. 7.21 MIR Spectra-Structure Band Positions Chart — MIR Fundamental absorptions indicated relative to wavenumbers (as 4000 to 400 cm^{-1}).

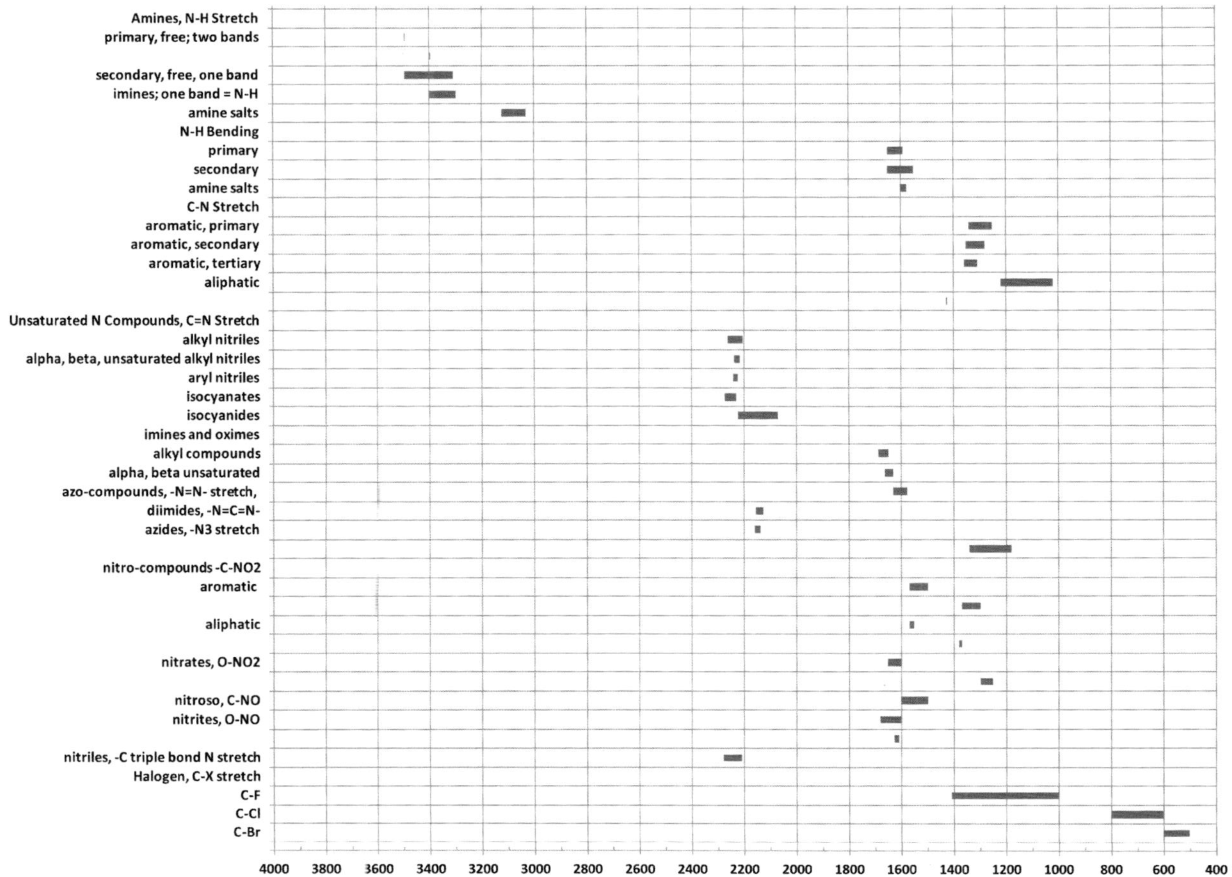

Fig. 7.22 MIR Spectra-Structure Band Positions Chart — MIR Fundamental absorptions indicated relative to wavenumbers (as 4000 to 400 cm^{-1}).

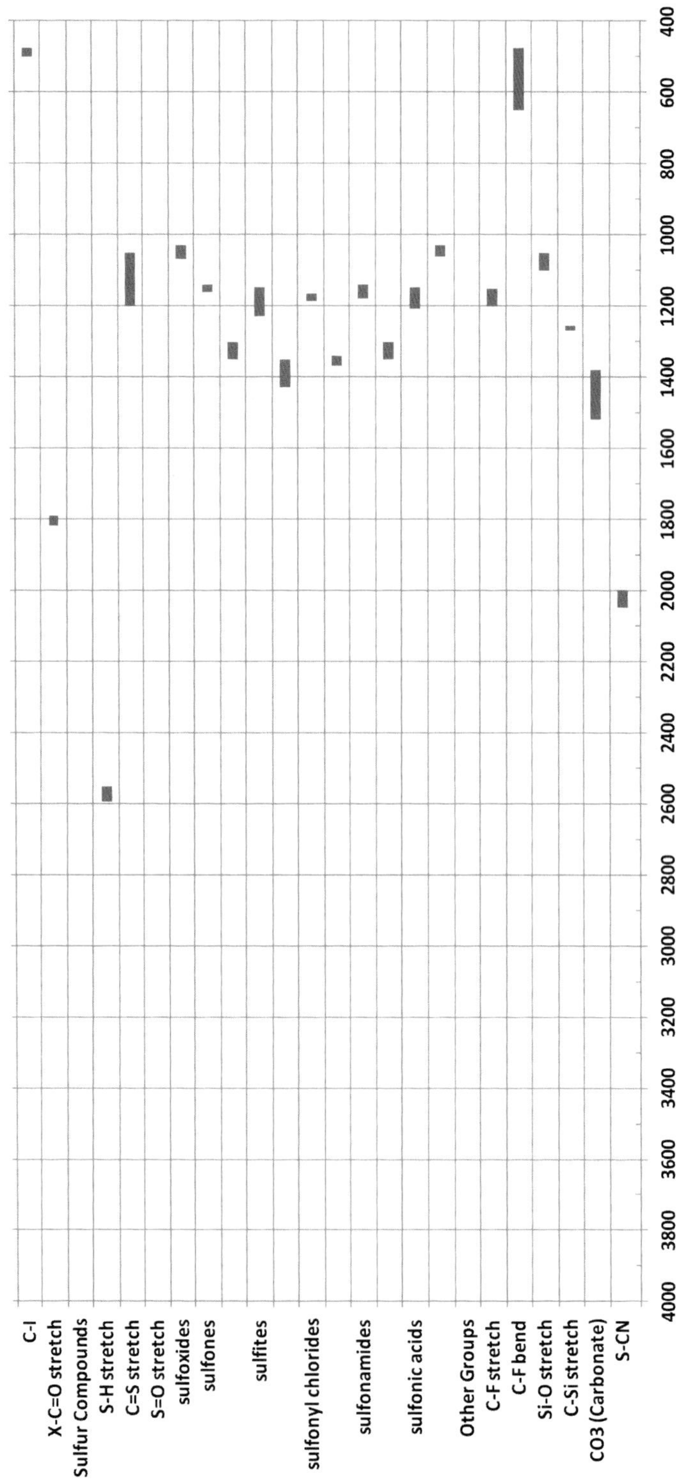

Fig. 7.23 MIR Spectra-Structure Band Positions Chart — MIR Fundamental absorptions indicated relative to wavenumbers (as 4000 to 400 cm^{-1}).

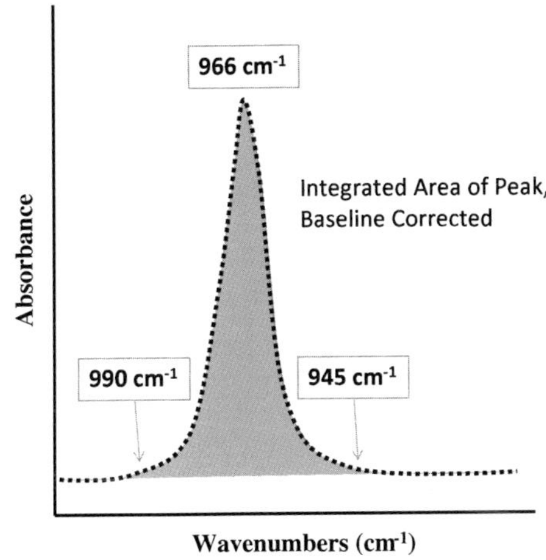

Fig. 7.24 Integrated peak area for 966 cm^{-1} baseline corrected band from 990 cm^{-1} to 945 cm^{-1} indicates presence of trans fat molecular structures. (From AOCS Cd 14d-99 and AOAC 2000.1 Methods.)[243,244].

7.1.2 *History and Applications of Infrared Detectors*

There have been several excellent comprehensive reviews of the history of infrared detectors, measurements and theory.[1–3] Such papers review in detail this history from the many times cited discovery of the existence of thermal radiation in the invisible beyond the red by W. Herschel in 1800, through the discovery of the thermoelectric effects using an antimony–copper pair by T. J. Seebeck in 1822. Reviews describe the voluminous developments that have occurred resulting in the development and production of basic infrared theoretical physics up to quantum well infrared photoconductors (QWIP) for hybrid second generation infrared systems in the 1990s. Recent production of third generation infrared systems and modern aspects of infrared spectrophotometry are described in multiple websites and publications. The references by A. Rogalski are highly recommended to the infrared historian and physicist for a list of references (the articles have been posted online in .PDF file format).

Beginning in the mid-twentieth century, the review articles describe the work by M.J.E. Golay to develop a sensitive infrared radiation detector in 1947; thermopile based infrared imaging cameras in 1954; mass produced PbS, PbTe, and InSb detectors for infrared guided missiles in 1955. In 1957 the HgCdTe detectors were developed by W.D. Lawson, S. Nelson, and A.S. Young. 1970 brought the discovery of CCD (charge-couple device) technology by W.S. Boyle and G.E. Smith. The first hybrid infrared focal plane array was designed and produced in 1975. 1977 and 1980, respectively brought about the broken gap type II InAs/GaSb detectors by G.A. Sai–Halasz, R. Tsu, and L. Esaki; and the hybrid HgCdTe(InSb)/Si(readout) Focal Plane Arrays. The 1980s and 1990s brought Schottky diode FPAs (platinum silicide), and quantum well infrared photoconductor (QWIP) detectors; while 1995 brought IR cameras with uncooled FPAs (focal plane arrays).

At the start of the mid-twentieth century the infrared region between 0.75 and 3.0 microns (μm) was referred to by physicists as the 'photoelectric infrared'.[4] This region was extended between the upper level of the visible eye sensitivity (i.e., 0.75 microns or 750 nm) to the upper end of detection for photoelectric detectors (i.e., 3 microns or 3000 nm). Energy above 3.0 microns, representing the infrared region, requires a different type of detector, namely as thermal or infrared detectors in the form of photoconductive or pyroelectric types. The major photoconductive type infrared detectors are Mercury cadmium telluride (MCT), and Mercury zinc telluride (MZT). The main pyroelectric types are Triglycine sulfate (TGS and DTGS), and Lithium tantalate ($LiTaO_3$).

The near infrared spectrum was referred to early on as the 'prism infrared' as dispersive instruments using low-OH quartz or salt prisms were used for this spectral region.[4] This term was initially applied to what is now more appropriately termed the 'mid infrared (MIR)' or most commonly the 'infrared' region. The same prisms did not transmit infrared energy and attenuated infrared frequencies quite readily. Prism materials are available to 25 μm, such as potassium bromide, but not practical as compared to interferometer based infrared spectrophotometers. The far infrared or Terahertz has traditionally been considered from the range of 25 μm to 1 mm in wavelength. See Table 7.2 for a summary of the names and spectral region definitions for the infrared spectral regions from the mid-Twentieth Century up until today.

For current terminology the defined molecular spectroscopy regions are as follows. In wavenumbers the near infrared is 14493 to 3333 cm^{-1}; the infrared is 4000 to 400 cm^{-1}; and the Far Infrared or Terahertz is from 400 to 10 cm^{-1}. These correspond to wavelength regions of 780 to 2500 nm (NIR), 2500 to 25000 nm (MIR), and 25000 to 10^6 nm (FIR). Microwaves or radar begins after the FIR range at approximately 1 mm in wavelength.

The modern infrared spectrophotometer, as described in Chapter 3 and other sections of this set of volumes, consists in either a Globar or Nernst lamp source combined with a Michelson interferometer including modernization and ruggedization improvements. The Globar or Nernst lamp sources have existed for many years and are generally specified as summarized in Table 7.3.

Table 7.2 Historical names and spectral regions pertaining to infrared spectroscopy.

Mid-Twentieth century name	Modern name	Common older definition	Common modern definition
Photoelectric Infrared	Near Infrared	0.75 to 3.0 μm	780 to 2500 nm
Prism Infrared	Mid Infrared or Infrared	3.0 to 25.0 μm	4000 to 400 cm^{-1}
Far Infrared	Terahertz	25.0 μm to 1.0 mm	300 GHz to 3 THz (1 mm to 100 μm)

Table 7.3 Characteristics of typical globar (Silicon Carbide), Nernst lamp, and modern emitter infrared radiation sources.

Parameter	Globar[4,5]	Nernst lamp[4,5]	Modern emitter[4,5]
Diameter of element	5 to 10 mm	3 to 5 mm	4 to7 mm
Length of element	20 to 50 mm	10 to 12 mm	4 to18 mm
Operating Temperature	1,000 to 1,650°C (1273.15 to 1923.15 K)	1600–2000 K	1000–1050 K
Useful Radiation Range	2 to 39 μm	1 to 25 μm	1 to 25 μm
Peak Wavelength	2.1 μm at 1400 K	1.3 μm at 2300 K	2.8 μm at 1025 K
Power, watts	150 to 250	22	140
Lamp Current (Amps)	3 to 5	0.2 to 0.5	12.0
Lamp Voltage	50 to 60	100	12.0

7.2 Applications of MIR (Mid Infrared or Infrared Spectrophotometry)

7.2.1 *Introduction*

A list of applications for infrared spectroscopy may include something like the following (in alphabetical order):[6] *Agriculture, Animal Sciences, Biotechnology, Cosmetics, Earth Sciences/Atmospheric Science/Mineralogy, Environmental Monitoring, Fine Chemicals/Chemical Production, Food and Beverages, Forensic Science, Gas Phase Analysis, Instrument Development Physics, Medical/Clinical Chemistry, Military Related Research, Moisture only, Petroleum /Natural Gas/Fuel Research, Pharmaceutical Production, Plant Sciences and Pulp and Paper, Polymer Science, Semiconductors and Materials Science, Surface Analysis, and Textiles.*

Methods or applications of measurement for the various applications would include (in alphabetical order): *Airborne, At-line, Automatic/Automated, Hyperspectral Imaging, Imaging/Image Analysis, In situ, In-line, Laboratory,*

Near-line, Non-invasive/Nondestructive, On-line, Open Path, Portable/Hand-held, Process Monitoring/Production Control, Quality Control/Quality Assurance, Quality Monitoring, Rapid, Real Time, Remote, Research.[6,7]

Infrared spectroscopic imaging includes: airborne imaging, hyperspectral imaging, remote monitoring, microspectroscopic imaging, and basic applications of standard spectroscopic techniques. Imaging techniques are used to produce a spectral image including the information for spatial dimensions combined with the precise location of chemical phenomena on surfaces. It has broad applications in biomedical, geological, agricultural, and industrial fields. There is a section on Infrared Imaging with more details in Chapter 6 of this volume.

7.2.2 *General Imaging*

A descriptive article surveying the field of fast FTIR imaging including theory and applications is presented by Koenig and Snively.[8] This basic review describes instrumental aspects, imaging methods, data processing, and applications with 20 references. An *Applied Spectroscopy* focal point article describing infrared spectroscopic imaging in terms of focal-plane arrays, remote sensing and astronomy, terrestrial and medical applications, is described in detail with 35 references by Colarusso *et al.*[9]

Crowley and Zimbelman, of the U.S. Geological Survey, Reston, VA[10] completed work entitled, "Mapping hydrothermally altered rocks on Mount Rainier, Washington, with Airborne Visible/Infrared Imaging Spectrometer (AVIRIS) data." Makipaa, Tanttu, and Virtanen[11] of Tampere University of Technology, Pori, Finland published, "IR-based method for copper electrolysis short-circuit detection." Ning *et al.*[12] of Oregon State Univ., Corvallis, OR published, "Five novel applications of imaging visible and short near infrared spectrophotometry and fluorometry in the plant sciences. Part I: Photographic and Histological Applications." Rowlands and Neville[13] of CAL Corporation, Ottawa, ON published, "Calcite and dolomite discrimination using airborne SWIR imaging spectrometer data." Otten[14] of Kestrel Corporation, Albuquerque, NM, "Measured performance of an airborne Fourier transform hyperspectral imager." Silk and Schildkraut[15] of Block Engineering, Marlborough, MA completed work entitled, "Imaging Fourier transform spectroscopy for remote chemical sensing." Rowan *et al.*[16] of the U.S. Geological Survey, Reston, VA completed, "Analysis of airborne visible-infrared imaging spectrometer (AVIRIS) data of the Iron Hill, Colorado, carbonatite-alkalic igneous complex."

7.2.3 *Mapping*

Mapping technologies provide a detail of data analysis where individual spatial images are combined across a broad 2- or 3-dimensional landscape to produce an actual map of the spectrally resolved data across a spatially resolved surface. Images are a part of the map, but images alone do not comprise a map until they are directly related to the spatial dimension. Kruse[17] of the Analytical Imaging and Geophysics LLC, Boulder, CO, USA published, "Geologic mapping using combined analysis of airborne visible/infrared imaging spectrometer (AVIRIS) and SIR-C/X-SAR data." Watanabe *et al.*

7.2.4 *Animal Sciences*

Photoacoustic infrared spectroscopy has been demonstrated to be useful in the analysis of surface-treated wool[18] by Carter and colleagues.

7.2.5 *Biotechnology*

DRIFTS and artificial neural networks have been reportedly combined for the rapid identification of bacterial species by Goodacre *et al.*[19] NIR spectroscopy has been shown by Macaloney *et al.* to be valuable in control and fault analysis during high cell density E. coli fermentation.[20] A general discussion by Dosi *et al.* of the use of on-line near infrared analysis for control of fermentation processes is given.[21] FTIR is used to by Szalontai *et al.* to investigate fatty acyl chains in model and biological membrane structures.[22] The air-water interface has been studied using polarization modulated Fourier transform infrared spectroscopy by Cornut *et al.*[23] in order to understand the basic structure and orientation of lipids and amphipathic peptides. This work is undertaken to gain a more fundamental understanding of the structure of biological membrane. Raman spectroscopy has been used by Berger *et al.* to determine concentration of biological materials in aqueous solutions.[24] FTIR spectroscopy has been used by Henderson *et al.*[25] to identify Candida at the species level using various spectroscopic discrimination techniques.

Polarization modulation has been described by Fournier *et al.* for in situ measurements of monolayer films in the study of biological membranes at the air-water interface.[26] Lipid monolayers have been studied by Axelsen *et al.* in situ using internal reflectance FTIR.[27] FTIR spectroscopy has been used to study biological membranes on solid surfaces as well as at the air-water interface by Pezolet and co-workers.[28]

7.2.6 *Earth Sciences and Mineralogy*

A review by Aldstadt and Martin containing 64 references describes the use of a cone penetrometer for in situ analysis of contaminants in soil and groundwater. The review contains details of samplers for soil gas, groundwater and soil and methods for determination of organic contaminants. Methods covered for organic analyses include laser-induced fluorescence, Raman and infrared spectrometry and sensors. Inorganic contaminants are analyzed using laser-induced breakdown spectrometry, XRF and passive gamma-ray spectrometry.[29] Infrared analysis of coals by Matuszewska is described in one reference.[30] Airborne infrared spectroscopy of wildfires in the western United States is described by Worden and co-workers in a paper.[31]

An FTIR open-path air monitoring system was used to monitor Volcanic fumarole gases by Chaffin *et al.*[32] The spectrometer described used a 1 mm^2 Hg-Cd-telluride detector cooled with liquid N_2 and an infrared field source. The spectrometer and notebook computer were operated from a 12 V DC power pack allowing portability. The applications of active and passive monitoring techniques are discussed. Of the potential analytes in volcanic plume gases, SO_2 and HCl were the only gases detected the best.

FTIR spectrometers were used by Watson and co-workers[33] for environmental analyses. Instrumentation was carried on tethered and free-flying manned hot-air balloons. These instruments were configured to include: mid-and far- IR cameras, multispectral imaging spectrometers, radiometers, forward-looking IR cameras, as well as a variety of other sensors. The instrumentation was powered from a portable generator suspended from the gondola or deep cycle batteries placed in the gondola. The experiments described include the detection of chemical effluents by FTIR spectrometry and Atomic Emission Spectrometry (AES). Other experiments are described such as the detection of Hg contamination, stack testing and multi-site air quality monitoring; and the *in situ* vertical profiling of the atmospheric concentration of various pollutants.

The use of far-infrared spectroscopy for studies of terrestrial and planetary or extraterrestrial atmospheres is described by Griffin.[34] Forestry studies conducted in North Carolina by Zwicker report using the remote sensing capabilities of open-path FTIR spectrometry to survey field data.[35]

FTIR spectroscopy has been used to characterize interfacial water species in situ at hydrophobic and hydrophilic surfaces in mineralogy by Yalamanchili *et al.*[36] Remote FTIR has been mounted on balloon-borne platforms for remote sensing of various analytes in the atmosphere as described by Camy-Peyret *et al.*[37] Atmospheric free radicals have been studied in situ by Blake using a mid infrared magnetic rotation spectrometer.[38] Infrared microspectroscopy has been shown by Gao *et al.* to be useful in rapid identification of gemstones.[39] Multispectral instrumentation has been used to record the burning of biomass in West Africa by De A. Franca *et al.*[40]

7.2.7 *Environmental Sciences*

Open path FTIR spectrometry is reviewed for field use in remote sensing of airborne gas and vapor contaminants by Levine and Russwurm.[41] The review includes 27 references covering open path FTIR applications and describing where additional work is required for further development of the technique. The advantages of open path FTIR spectroscopy are described by Marshall *et al.*[42] These benefits include (1) its versatility; (2) remote, long-path measurement; (3) in situ applications; and (4) near real time measurement. The technology is described in terms of the physical limitations of the instrumentation, and the limitations in current data processing techniques. FTIR is combined with column liquid chromatography for the identification of herbicide contamination in samples of 50-100 ml river water by Somsen *et al.*[43] Enough analyte was obtained for good-quality spectra down to water concentrations of 1-2 micrograms per liter. Of six library search programs tested for solute identification, the best results (39 out of 45 correctly identified) were given by the Peak-Search-Intensity-Ignored algorithm. FTIR emission spectroscopy has been employed by Wang *et al.* for the study of Freon-12 in an alcohol plus air flame.[44]

Infrared spectroscopy has been reported by Haschberger for use as a remote method for monitoring trace gases from aircraft emissions.[45] Environmental monitoring applications using fiber optic sensing and FTIR spectroscopy is described by Druy *et al.*[46] Ground based FTIR spectroscopy was used for remote sensing of pollution emission sources for application in environmental compliance monitoring by Schaefer.[47] Marshall *et al.* describe the effects of resolution on the performance of classical least squares (CLS) spectral interpretation when applied to volatile organic compounds (VOCs) of interest in remote sensing using open-air long-path Fourier transform infrared (FTIR) spectrometry.[48] A 1995 patent[49] describes the method and apparatus for remote infrared sensing and analysis of motor vehicle exhaust gases.

A real-time background correction algorithm is described for passive FTIR spectroscopy of chemical plumes for environmental monitoring applications by Polak *et al.*[50] Various trace gases are measured in situ using Airborne infrared diode laser spectrometry by Toci and co-workers.[51] Soil contamination is rapidly determined using reflectance Fourier transform infrared spectroscopy by Adams and

Bennett.[52] Theriault and Bissonnette report on the use of ground based FTIR for monitoring weather cloud parameters.[53] Remote FTIR is used to remotely monitor motor vehicle exhaust gases by Davies *et al.*[54]

7.2.8 *Fine Chemicals and Chemical Production*

The potential difference measured in situ using FTIR spectroscopy is presented by Barbour and Scherson[55] for the reduction of nitrite on a solid polycrystalline Pt electrode in aqueous 0.1M-$HClO_4$. The intensity of the N_2O asymmetric stretch band at 2231 cm^{-1} is monitored during a voltammetric sweep. The FTIR results correlated with dynamic differential electrochemical MS observations previously published Nishimura *et al.* (Electrochim. Acta, 1991, 36, 877).

May *et al.* report the use of an FTIR spectrophotometer equipped with an 8 m gas cell having solid nickel mirrors. The system is used for the determination of H_2O in anhydrous HCl samples. An on-line determination of impurities in gaseous Cl_2 using a temporary system is described. In addition, the on-line monitoring of impurities in anhydrous HCl as it passed through a carbon bed using a permanently-installed system are described. Conventional methods for the analysis of chlorine and hydrogen chloride are also reviewed.[56] Bae *et al.* report the use of two different in situ FTIR spectroscopic techniques to study the vibrational properties of the model redox-active species 2,5-dihydroxybenzyl mercaptan adsorbed on Au in aqueous 0.1M-$HClO_4$. The different spectroscopic methods included (1) potential difference ATR FTIR using a 4 nm thick gold layer sputtered onto a Au-patterned ZnSe internal reflection element as the electrode, and (2) potential difference FTIR external reflection absorption spectroscopy on a solid Au electrode. The spectra obtained using the two methods were nearly identical.[57]

The combination of chemometrics with FTIR spectrometry was used by Lindblom *et al.*[58] to monitor diphenylamine stabilizer concentrations in the presence of nitrocellulose, the main component of rocket propellants. The shelf-life of these propellants is normally determined by HPLC analysis of the diphenylamine stabilizer and its derivatives. A more rapid method of analysis was desirable, thus the feasibility of using a partial least squares calibration model of the IR spectra as calibrated using concentrations determined by HPLC was investigated. The combination of chemometrics with FTIR spectrometry allowed the monitoring of stabilizer concentrations to 0.2% diphenylamine. However this spectroscopic method was not as accurate as HPLC. Standard error using the FTIR method as compared to the HPLC method was 10% relative, with a correlation of 0.94 for diphenylamine and its first three derivatives.

FTIR was used by Bandekar and co-workers[59] to quantitatively measure sulfur oxide in white liquor. Liquid samples of approximately 1.5 ml were placed into Circle (Cylindrical Internal Reflectance Cell for Liquid Evaluation) cells comprising an open-boat arrangement with a zinc selenide crystal at 45 degrees. The infrared beam was configured for 12 internal reflections. The method was suitable for quantitative analysis calibration ranges to 70, 45 and 42 g/l were obtained for sulfate, thiosulfate and sulfite species, respectively. Corresponding minimum detection limits were 0.01, 0.02 and 0.02 g/l. The method should be useful for in situ analysis but is limited to samples which are water-soluble and have IR-active bands in the mid IR region.

A FTIR spectroscopic method was developed by Ge *et al.*[60] using an attenuated total reflectance (ATR) sample boat with a zinc selenide (ZnSe), or zinc sulfide (ZnS) crystal. Methanol/acetonitrile (1:1) or air were used as the background spectra for the two types of ATR crystals, respectively. The method was used to monitor the epoxidation step in the synthesis of cis-1-amino-2-indanol when indene is oxidized to indene oxide. For the ZnSe ATR a portion of the reaction mixture was analyzed by FTIR spectroscopy without sample preparation, and for the ZnS ATR the sample was centrifuged to separate the organic and aqueous layers with FTIR spectrometry being performed on the organic layer. Results were calibrated and cross validated by HPLC with details described. Spectral data were collected from 1500 to 900 cm^{-1} and processed using multiple linear regression (MLR) and partial least squares (PLS).

An FTIR method is described, with resultant data analysis techniques presented, for the prediction of organic vapor concentrations by Ruyken *et al.*[61] The method described includes techniques for identifying interferents present within the vapor mixtures. Basic principles of multivariate calibration using principal component regression (PCR) are outlined, and the effect of interfering species on the predicted concentration. The infrared spectra of organic vapors at 60°C were recorded at 4 cm^{-1} resolution. A simple two-component calibration model was constructed for hexane and ethanol, with additions of dichloroethane, acetone, methyl acetate and toluene as interferents. The prediction errors depended on the concentration of the interferent and the overlap of each interferents spectrum onto the analyte regression vector. A residual library search method was used to identify each interferent.

The FTIR spectrum of a ternary aqueous solution of chlorate, perchlorate and dichromate was measured from 900–1250 cm^{-1} by Kargosha *et al.*[62] After subtraction of the water spectrum and standard baseline correction, the spectrum was resolved using multicomponent analysis (i.e., partial least squares regression and multiple linear regression). The infrared bands at 973, 1110 and 950 cm^{-1} were used to quantify chlorate, perchlorate and dichromate, respectively. Calibrations were linear from 0.2–63.8, from 0–73.47 and from 0–149 g/l of chlorate, perchlorate and dichromate, respectively. The corresponding detection limits for each of the analytes were 1.064, 0.64 and 0.95 g/l. The relative standard deviation as 1 sigma (RSD) for 8 samples were from 1.19–2.7% relative. The method was applied to electrolytic solutions of chlorate and perchlorate.

In situ FTIR spectroscopy was used to study the structures of powdered humic acid salts by Woelki and Salzer.[63] Samples were prepared by placing the salt samples with KBr in a DRIFTS (diffuse reflectance) spectrometer accessory. The sample reaction chamber was flushed for 15 min with dry nitrogen (N_2) and the sample was heated under stopped and continuous flow N_2 at 10 and 65 C/min, respectively. The spectra were obtained at 4 cm^{-1} resolution by co-adding 50 scans at slow heating and 200 scans at fast heating. The structural changes caused by heating were recorded with results presented.

Remote high-resolution FTIR was used by Wang *et al.*[64] for spectral characterization of infrared flare material combustion. The flare materials were pressed into a 30 g pellet (20 × 15 mm) and positioned 15 m from a remote sensing (open path) FTIR instrument with a combustion flame. The emission spectra were recorded from 4700-740 cm^{-1} with a resolution of 0.24 cm^{-1}. Peaks corresponding to HCl, HF, H_2O, CO_2, CO and SiF_4 were obtained allowing qualitative and quantitative determination of these combustion products. The flame temperature was estimated and the spectral radiation distribution was reported.

Rapid IR- spectrophotometric determination of total nitrogen in silicon dioxide - silicon nitride (SiO_2-Si_3N_4) mixtures is presented by Wisniewski and co-workers.[65] Samples of 1–1.5 mg were ground with 200 mg KBr and pressed into a tablet. The IR spectra were recorded from 400–2000 cm^{-1}. The N content was quantified from the transmittance at 575 and 1450 cm^{-1} (background) using a formula described in the method paper. The calibrations were linear over the 20–35% nitrogen levels. The relative standard deviation (RSD) of 1 sigma were 7.5–44.5% relative. These results were compared to those obtained using a volumetric method where samples are dissolved in acid and the ammonia (NH_3) is released using distillation. The analysis time was 0.5 h. The method was applicable to mixtures containing lithium and magnesium salts and is suitable for evaluating the yield of Si_3N_4 synthesis from SiO_2 and NH_3.

In situ temperature-programmed diffuse-reflectance infrared Fourier-transform spectroscopy (TPDRIFTS) is used to monitor chemical changes in the surface structure of vanadium oxide/titanium oxide catalysts during reactions by Centeno *et al.*[66] A general article is presented by Brimmer describing the development and applications of near infrared spectroscopy in the chemical and pharmaceutical industries.[67] Fiber-optic-based Fourier transform infrared spectroscopy (FTIR) was used by Salim *et al.*[68] for in situ measurements to monitor the concentration of various reactants in a vertical OMVPE reactor. In situ attenuated total reflection (ATR) infrared spectroscopy was used to evaluate and study the various aspects of laminar flow in a channel-type electrochemical cell by Tolmachev *et al.*[69]

In situ IR reflectance spectroscopy was used by Lamy *et al.*[70] to study and monitor the electrooxidation of methanol at Pt-Ru electrodes. In situ diffuse reflectance infrared spectroscopy (DRIFTS) was used by Benitez *et al.*[71] for the study of the reversibility of CdGeON sensors towards oxygen. In situ ATR FTIR spectroscopy was described in a doctoral dissertation by Dunuwila[72] for the measurement of crystallization phenomena for research and development of batch crystallization processes. In situ infrared spectroscopy was used to study the solid-state formation reaction of polyglycolide by Epple *et al.*[73] In situ near infrared spectroscopic investigation of the kinetics and mechanisms of reactions between phenyl glycidyl ether (PGE) and multifunctional aromatic amines is described by Xu *et al.*[74]

Remote mid infrared spectroscopy is used by Mijovic and Andjelic to monitor reactive processing for polymers.[75] Developments of in situ molecular spectroscopy in catalysis studies is given by Xin and Qin.[76] Cross dispersion infrared spectrometry (CDIRS) for remote chemical sensing is described by Stevens *et al.*[77] In situ infrared spectroscopy is used by Eder-Mirth to study the mechanisms of selective alkylation of aromatics over zeolites.[78] Rapid infrared sampling methods are described by Scamehorn for the organic chemistry laboratory using a diffuse reflectance (DRIFTS) accessory.[79] In situ FTIR and FT-Raman studies of reactions in the hydrothermal environment are described by Schoppelrei.[80] The use of near infrared monitoring of dimethyl terephthalate in a low pressure methanolysis process is presented by Agyare.[81] In situ infrared studies on dehydroxylation of MCM-41 are described by Li *et al.*[82]

Infrared absorption spectroscopy is used as a method for rapid fault detection in borophosphosilicate glass thin films by Zhang *et al.*[83] In situ FTIR studies of the acidity and catalytic properties of Pt/zeolites is described by Chien *et al.*[84] In situ FTIR spectroscopy is related to kinetic studies of methanol synthesis from CO/H_2 over $ZnAl_2O_4$ and Cu-$ZnAl_2O_4$ catalysts by Le Peltier *et al.*[85]

7.2.9 *Food and Beverages*

A presentation of the uses of ATR FTIR spectroscopy for process control in the sugar industry is given by Veronique and Gilles.[86] A patent is presented for real-time on-line analysis of organic and non-organic compounds for food, fertilizers and pharmaceutical products are given.[87]

FTIR combined with a liquid ATR cell heated to 50°C was used by Mossoba *et al.* to determine the total trans-content of neat hydrogenated edible oils. The sloping background of the infrared trans-band at 966 cm^{-1} was eliminated by ratioing the single-beam spectrum of the sample to the corresponding spectrum of the unhydrogenated material; the trans-band could then be integrated between 990 and 945 cm^{-1}. The limits of detection and determination were 0.2 and 1%, respectively.[88]

An FTIR method is reported by Hewavitharana and van Brakel for the direct determination of casein in milk. FTIR spectra were recorded for homogenized milk samples while maintaining a sample temperature of 40°C. Data acquisition parameters, such as resolution and the types of background spectra, were studied to achieve optimum conditions. Two chemometric methods, viz., partial least squares (PLS) and principal components regression (PCR), were used for data processing. The best results were obtained using a resolution of 4 cm^{-1} with PLS calibration over the spectral regions 3000–2800, 1600–1500 and 1300–1000 cm^{-1}. For comparison, the results obtained agreed closely with those obtained by the International Dairy Federation reference method.[89]

A review is presented by Lipp containing 125 references describing the developments of methods for the determination and quantification of triglycerides in milk. The referenced papers appear in the literature during the period 1990–1994. In particular, gas- (GC) and high-pressure liquid-chromatographic (HPLC) methods are highlighted. Analytical methods such as infrared (IR) spectroscopy, mass spectrometry (MS), 12C/13C isotopic analysis, differential scanning calorimetry (DSC), gravimetry and titrimetry are also discussed. An evaluation of the different approaches to the detection of adulteration of milk fat was also described.[90]

A rapid, automated Fourier-transform infrared spectroscopic method was developed by van de Voort *et al.* for the determination of cis and trans content of edible fats and oils. A sample handling system for use in routine quality control measurement of fats and oils is described. Prior to analysis samples were warmed to within 5°C of the operating temperature of the instrument using a microwave oven. The heated samples were aspirated into an IR measurement liquid cell. The spectra were measured at a resolution of 4 cm^{-1} for 128 scans per spectrum. The liquid cell was emptied and reloaded with the next sequential sample. Spectra were calibrated by partial least squares (PLS) analysis for the cis and trans content with results given.[91]

Two infrared spectroscopic methods for cheese analysis are compared by McQueen *et al.* The methods were NIR filter (i.e., named optothermal spectroscopy;

cf., Int. Lab., 1990, 20, 16) and ATR FTIR spectroscopy. In NIR filter spectroscopy spectra were recorded using optical transmission filters of 1740, 1935 and 2180 nm. The ATR spectra were recorded using a conventional FTIR spectrometer at a resolution of 8 cm^{-1}. Partial least squares (PLS) regression analysis was used to correlate the spectral data to the composition of fat, protein and moisture determined in 24 cheeses using reference analytical methods. With the NIR filter spectroscopy, correlation coefficients (r) were 0.95–0.97 with the corresponding values obtained using ATR FTIR were 0.97–0.99. The NIR technique was more rapid and somewhat easier to use than FTIR.[92]

Carbon dioxide in water samples was analyzed using an Analytical Development non-dispersive IR CO_2 analyzer as reported by Abdullah. Coupled to a ChemLab segmented-flow Auto-Analyzer 2 system. The sample stream was introduced to the analyzer using a flow rate of 0.4 ml/min. Once introduced the sample was mixed with deionized H_2O (0.8 ml/min) and 0.5 M-HCl (0.1 ml/min). The stream was sent into a gas stripper with N_2 as carrier gas flowing at 800 ml/min. The gas phase was passed over H_2SO_4 then through a magnesium perchlorate column before IR analysis, calibrations were linear from 0-3mM total CO_2. This technique was used to analyze sea, estuarine and fresh waters and was reported superior to alternative methods based on measurements of total alkalinity.[93]

FTIR spectroscopy was investigated by Bellon-Maurel *et al.* for on-line monitoring of the concentrations of glucose, maltose, maltotriose and maltodextrin in starch hydrolysis mixtures. Measurements were made using a conventional FTIR spectrometer with a zinc selenide (ZnSe) ATR flat crystal accessory. Calibrations were accomplished using a partial least squares (PLS) regression approach with a calibration set of 30 mixtures. Acceptable results were obtained for the determination of glucose and maltose; the standard errors of prediction for glucose, maltose, maltotriose and maltodextrin were 4.1, 3.4, 4.9 and 4.4 g/kg respectively compared to the required precision of 8, 10, 5 and 5 g/kg respectively. Both repeatability and reproducibility were satisfactory.[94] FTIR is described as a method for the rapid determination of casein in raw milk by Hewavitharana and van Brakel.[95]

7.2.10 *Forensic Science*

Forensic science continues to refine the ability to identify original automobile topcoats using infrared spectroscopy with in situ measurement techniques. With ever increasing data bases and pigment search libraries identification becomes easier as reported by Suzuki and Marshall.[96] Infrared and other analytical methods are extensively described to aid in the rapid and conclusive identification of I homologues in illicit drug preparations. The hydrochloride salts of 3,4-methylenedioxycathinone (I) homologues were analyzed by IR spectroscopy using a KBr matrix. $CHCl_3$ solutions of the I homologues (as free bases) were analyzed by GC-MS-IR spectrometry to obtain vapor-phase spectra of the compounds. MS, NMR and GC measurement parameters and results are also presented. Melting points of the compounds, and presumptive tests were also completed with details described by Dal Cason.[97]

7.2.11 *Gas Phase Analysis*

A patent describing the method and apparatus for automatic identification of anesthetic gas mixtures in respiratory gas mixtures is described. The system uses a diode-array detection system combined with variable transmission filters at 4–5 um and 7–10 um.[98] Infrared spectroscopy combined with principal components based digital filtering has been presented by Hayden and Noll as a means for identifying trace gases in multispectral and hyperspectral imagery.[99] Near infrared, rather than infrared, spectroscopy has been utilized in a special application by Helland for on-line analysis of hydrocarbon gases. A method is described for maintaining calibration models over time.[100] A portable FTIR analyzer system is described by Ahonen *et al.* for gas analysis in industrial hygiene situations.[101] FTIR is described by workers at Argonne National Laboratory, Argonne, IL as a method for continuous monitoring of gas emissions by Mao *et al.*[102]

7.2.12 *Instrument Design Studies*

Shaffer and Small at the Center for Intelligent Chemical Instrumentation, Department of Chemistry, Clippinger Laboratories, Ohio University, Athens, have described the optimization of algorithms for discriminant analysis of remote FT -IR spectra. Five optimization methods: simplex optimization, simulated annealing, generalized simulated annealing, genetic algorithms, and simplex-genetic algorithm hybrid were compared for their ability to position the discriminant. Simplex optimization consistently out-performed the other discriminant methods.[103] The method and practice for describing the performance of an airborne Fourier transform hyperspectral imager is described by Otten *et al.*[104] Remote FTIR spectroscopy is described by Mijovic *et al.* as a means of in situ real-time monitoring of chemical processes.[105]

Rutter *et al.* at Lockheed Martin IR Imaging Systems, Lexington, MA have derived a multispectral MCT detection system for remote infrared sensing of atmospheric constituents.[106] A basic description of the uses of near infrared spectroscopy for process and production control is described by Hansen.[107]

Foulgoc *et al.* present work on the use of TeX based fiber optics for remote infrared spectroscopy.[108] A discussion on the selection of spectral bands to interpret hyperspectral remotely sensed images taken using hyperspectral methods is described by Valdez *et al.*[109]

An infrared reflectance device is described by Vohra *et al.* which is intended for eventual incorporation in a conventional cone penetrometer. Using a nichrome-wire source, an off-axis paraboloidal mirror focused the incident beam to a circular spot size of 5 mm at the sample surface. The diffuse reflectance from the sample was collimated using a second off-axis parabolic mirror and then focused by a third onto the entrance face of a 12 m long clad As/S-based chalcogenide fiber (core diameter 0.25 mm). The fiber conducted the reflected light to an FTIR spectrometer with a MCT detector. The operating wavenumber range for the system is 6000–2225 cm^{-1}. A calibration graph based on the C-H stretching absorption at 3100-2850 cm^{-1} was reported to be linear up to 1000 ppm of trichloroethylene when measured using dry sand; with the detection limit reported as 250 ppm. The H-S combination band at 3200 cm^{-1} was used as a reference.[110]

Genetic algorithms (GA) were used to optimize piecewise linear discriminant analysis (PLDA) as employed by Shaffer and Small in the automated classification of FTIR spectrometry remote sensing data.[111]

A transmission cell based on fiber optics and a stainless-steel union cross was developed by Tilotta *et al.* for use in on-line SFE-FTIR spectrometry. The sample was placed in the sample cell, pressurized with CO_2 and heated to the extraction temperature. After extraction (~30 min.) the valve between the SFE cell and the IR cell was opened and the IR spectra were recorded. The IR cell was cleaned by flushing with liquid CO_2. Example spectra of fresh orange peel, (+)-limonene, gasoline-contaminated soil and potato chips are presented. The method should be suitable for the determination of organic compounds in solid samples. A schematic diagram of the on-line SFE-FTIR system is given.[112]

Wilks has described small portable filter based IR analyzers which are dedicated to specific process analysis applications. They weigh less than 5 lbs. and consume less than 20 Watts of power drawn from a 12 Volt D.C. source. Applications described include the determination of fat in hamburger meat, oil and grease in water and soil samples. These measurements are completed using the C-H stretch band of fats and oils found at 3.4 microns.[113]

An overview is presented by Deeley *et al.* discussing ways in which manufacturers are facilitating the validation process of FTIR spectrometry. Three areas are examined: the monitoring of instrument performance; establishment of the validity of data manipulation procedures; and the generation of well-controlled operating procedures.[114] FTIR-ATR spectroscopy using 50 internal reflections at a Ge/I ATR interface prior to detection from 4400-700 cm^{-1} is described by Kulesza *et al.* for study of the spectroelectrochemistry of Prussian blue in the solid state.[115]

A two-fiber probe suitable for fluorescence, phosphorescence, Raman, mid IR and NIR (reflectance) spectroscopic measurements is described by Lin *et al.* The efficiency of this probe for fluorescence measurement across a window was demonstrated by polishing the fiber ends at predetermined angles and using different probe geometries. Best performance was obtained by maximized overlap of the light cones of the two fibers at the interface of the window and sample.[116]

A method is described in detail by Treffman and Morrison for hand grinding and polishing of KBr lenses designed for coupling an IR beam from a FTIR spectrometer into a hot filament chemical vapor deposition reactor for in situ measurements.[117]

A mid IR probe was fabricated by MacLaurin *et al.* from two single-strand 1.5 m × 750 micron core chalcogenide fibers, a zinc selenide (ZnSe) internal reflection element and an adjustable optical interface for the spectrometer. An integral internal water cooling system maintained the fibers at 35°C when the probe was immersed in a sample at 200°C. Its performance was illustrated by the in situ monitoring of the acid-catalyzed esterification reaction of but-2-enoic acid and butan-2-ol in toluene at 110°C for 8 h. The results were comparable with those obtained using an overhead ATR flow cell arrangement with an extractive sampling loop. Quantitative calibration of the spectroscopic data was achieved using GC reference data. Relative errors of less than 5% were obtained.[118]

Several sample preparation procedures for measuring infrared spectra obtained for water-containing or water-soluble materials and a soluble solid are compared by Dirksen and Gagnon. Traditional IR accessories are used with a disposable IR card (3M, St. Paul, MN, USA) which comprises a cardboard sample holder (5 × 10 cm) that fits into and aligns with most spectrometer optical systems. The use and performance of this technique is described.[119]

The choice of whether to use an FTIR instrument or a filter photometer depends on the application and the end-user requirements, however, there are some instances in which FTIR performs better and is more cost-effective than filter IR. This concept is presented by Coates.[120]

Wilks advocated the use of infrared filter photometers where possible when two or three wavelengths are needed for process analysis. Filter photometers are claimed to offer such advantages as: (1) very high signal levels; (2) better signal-to-noise ratios than with FTIR; (3) no moving parts; (4) easily temperature controlled; (5) insensitive to vibration; and (6) long-term unattended operation.[121] The merits of using an interferometer-based instrument as compared to grating-based instrument depends the type of sample and the end-user requirements. If an opaque sample has to be penetrated, then NIR spectrometry should be used. However, if surface measurements or clear liquid analysis (short pathlength) or gas analysis is required, then mid IR or FTIR spectrometry should be used. The only way to ensure proper calibration transfer between instruments is to have them built to the same performance specifications. Proper calibration of instruments should be provided by the manufacturer as described by Ciurczak.[122]

Remote detection of aluminum hydroxide was achieved by combining IR transmitting fiber optics with a FTIR spectrophotometer. Plastic-coated chalcogenide fibers 1 to 4 meters in length and having 15–25 cm of their length decoated were used for measurement. Five absorption bands were detected for aluminum hydroxide in the spectral range of 3350–3650 cm^{-1}. The method may be suitable for measuring industrial corrosion in aluminum alloys as reported by Namkung *et al.*[123]

The use of partial least squares (PLS) for the automatic classification of infrared spectra was presented by Luinge *et al.* The method was applied to IR spectra of pesticides divided into a training and test set of 175 and 184 spectra, respectively. Spectra were measured over the region of 4000–625 cm^{-1} and a data point resolution of 1 cm^{-1}. The PLS results were compared with those obtained using an error-back-propagation neural network method, and a combined principal-component analysis (PCA) neural network method described earlier (cf. Anal. Chim. Acta, 1993, 283, 508). The methods yield roughly equivalent performance but the training times are reduced from hours for neural networks to minutes for PCA-neural networks and PLS methods, respectively. The PLS method as employed was also useful for detecting spectral outliers.[124]

The use of FTIR-ATR spectrometry for on-line analysis is presented by Wilson *et al.* Method development is described and several possible applications are presented.[125] A special computational method for prefiltering infrared spectral database searches, using the flashcard algorithm (J. Chem. Inf. Comput. Sci., 1994, 34, 984) is described by Klawun and Wilkins. A back-propagation neural network was trained to recognize 35 different functional groups from a set of 609 matrix isolation FTIR spectra, each with 752 data points. The method was also modified for application to a larger training set of 2651 spectra, with use of 40 functional groups. This set required 38.5 min of training on a Cray C90 supercomputer.[126]

The application FTIR spectroscopy and discriminant analysis for quality control of industrial materials and products is discussed by Hartshorn. A compositional analysis testing system (CATS) is used to form training sets for discriminant analysis, in order to assess its use in quality control using infrared spectra with a discussion of results.[127] The development of an application using acousto-optic tunable-filter (AOTF) near infrared spectrometry is presented by Huehne *et al.* The system is demonstrated for the determination of water in acetone.[128]

A simple fiber-optic interface for on-line supercritical-fluid extraction-Fourier-transform infrared spectrometry as previously described by Heglund *et al.* (cf. Am. Lab. 1996, 28(6): 36R-36T). The interface consisted of a stainless-steel union cross through which two pieces of commercially available chalcogenide infrared transmitting fiber-optic cables were inserted along one of the axes of the cross. One end of this tubing was connected to the SFE extraction cell while the other end was connected to a vent-off valve. The IR spectra were recorded via the fiber-optic cables. The system was demonstrated for the analysis of caffeine in coffee and petroleum hydrocarbons in soil.[129]

Infrared spectrometry is presented by Driver *et al.* for use in on-line process monitoring. Spectra are presented for several sampling accessory designs including: (1) a side-stream flow-through cell, (2) a variable-path immersible probe, and (3) a high pressure gas cell. The parameters affecting measurement accuracy and process sampling discussed.[130] A device used to focus energy from an infrared spectrometer (FTIR) source into a fiber optic cable and to refocus the energy returning via a second optical fiber to the spectrometer detection system is described by Lowry *et al.*[131]

An in situ DRIFTS (diffuse reflectance infrared Fourier transform spectroscopy) cell was used by Dossi *et al.* to provide infrared analytical evidence of the oxygen-induced formation of iron carbonyl from carbon monoxide and the steel walls of the cell. Formation of $Fe(CO)_5$ can be detected by the use of zeolite as a blank sample. Intense analyte bands occur at 2034, 2014 and 1979 cm^{-1}.[132]

A set of algorithms and a computer program based on these algorithms for comparing IR spectra from different instruments and/or different data intervals is presented in Kavak and Esen. The various mathematical considerations for this application are described.[133]

Egevskaya from the Institute of Semiconductor Physics, Siberian Branch of Russian Academy of Sciences, Novosibirsk, Russia describe the construction of a portable "double cat's eye" FTIR spectrometer.[134] A patent describing an optical probe for in situ detection of hydrocarbon concentration is given.[135] Transient infrared spectroscopy is presented as a method for on-line analysis of solids and viscous liquids as described by Weeks *et al.*[136]

A patent describing a portable, battery-powered multiple filter infrared spectrometer consisting of an infrared source with a chopper, from which IR energy passes through compound parabolic concentrators (CPC's) and is directed by a beamsplitter to the sample. The beam interacts with the sample by reflectance and is returned through the CPC and beamsplitter to a stationary filter assembly and discrete detector array where the energy is converted to an A.C. (alternating current) signal. This current energizes an LED bar graph display for spectral identification. This method and apparatus is used to determine IR spectra of solids, liquids on a mirror, and gases in a container with a mirror to reflect light.[137]

A simple transmission infrared cell made of Swagelok components for in situ infrared spectroscopic studies of a heterogeneous catalysis process is presented by Komiyama and Obi.[138] A simple method to fabricate lenses for in situ Fourier transform infrared spectroscopy is described by Treffman and Morrison.[139] A general review of spaceborne, remote sensing FTIR instrumentation is given by Persky.[140]

Wilks has described portable infrared analyzers for repetitive analysis.[141] A description of the uses of FTIR for a long-term quality assurance program is presented by Childers *et al.*[142] A field portable Fourier transform infrared spectrometer system is described by Engel *et al.*[143] The vibration spectroscopy of surfaces using a non-linear optical process is described and an introduction to the use of IR-visible sum frequency generation spectroscopy (VSFG) is presented by Hirose.[144]

A discussion of the development of QA/QC performance standards for field use of open-path FTIR spectrometers is given by Cone *et al.*[145] A description of remote characterization of materials by vibrational spectrometry through optical fibers is given by Griffiths *et al.*[146] The spectral characterizations of the low and high resolution remote FTIR emission spectroscopy is presented by Wang *et al.*[147] An echelle grating spectrometer (EGS) for mid IR remote chemical detection is described by Stevens *et al.*[148] Tellurium halide (TeX) fibers for remote infrared spectroscopy is given by Zhang *et al.*[149] A general description of FTIR analysis and its uses is given by Wilson *et al.*[150]

7.2.13 *Medicine and Clinical Chemistry*

Fiber-optic evanescent-wave spectroscopy was combined with FTIR by Simhi *et al.* to measure the total protein, cholesterol, urea and uric acid in 2 mL of human serum. Multivariate calibration models derived from a partial least squares (PLS) algorithm were compared to a neural-network calibration. The method was reported as applicable for analysis of glucose and triglycerides in serum.[151] Lipids were determined directly in fecal homogenate by FTIR-ATR and partial least squares regression modeling by Franck *et al.* A zinc selenide ATR accessory was used for optical sampling of the slurries. The infrared spectral regions 1763–1730, 1680–1612, 1595–1568, 1564–1524, 1491–1448, 1429–1400 and 1390–1367 cm^{-1} were used for quantitative analysis. The infrared spectral results were within 2.5% of those obtained using the reference gravimetry method.[152]

A portable FTIR gas analyzer was evaluated by Ahonen *et al.* for industrial hygiene measurements made in different factories using organic solvent mixtures. The detection limit was approximately 1 mg per cubic meter. The results obtained were compared with those obtained by the generally used adsorption tube method. Continuous monitoring with this system allows exposure peaks to be identified and allows preventative measures to be taken.[153] Fourier-transform infrared spectrometry was used by Cole and Martin to determine gas-phase sidestream cigarette smoke components (i.e., the smoke generated during smolder from the lit end of a cigarette). Five gas-phase components, viz., NH_3, CO_2, CO, HCN and NO, were determined in the sidestream cigarette smoke from five different cigarette brands. Gas concentrations were calculated from absorbance measurements made every 30 s. A 20-fold decrease in sample analysis time was achieved with the described method.[154]

A general discussion of the uses of non-invasive infrared spectroscopic analysis in clinical chemistry is described by Stevens and Vadgama.[155] In situ characterization of beta-amyloid in Alzheimer's diseased tissue by synchrotron Fourier transform infrared microspectroscopy is described by Choo *et al.*[156]

A dissertation by Pastrana-Rios describes the in situ infrared spectroscopic test of the pulmonary surfactant system.[157] Laser diodes and a photoacoustic sensor head are presented by Spanner and Niessner for the noninvasive determination of blood constituents.[158] Fourier transform infrared microspectroscopy is used by LeVine and Wetzel for in situ chemical analyses of frozen tissue sections of the white matter in the rat brain.[159,160]

7.2.14 *Military, Explosives, and Propellants*

Remote high resolution FTIR was used to study the spectral characteristics of infrared flare material by Wang *et al.*[161]

7.2.15 *Moisture Measurement*

NIR is generally used for moisture analysis. Note that trace levels (ppm) of water in solvents may be quantified using mid infrared measurements.

7.2.16 *Petroleum, Natural Gas, and Fuels Research*

Automation of FTIR spectrometry for liquid samples is described including hardware and software by Stanek. Turnkey applications have been demonstrated for lubricating oil analysis, fuel, and edible oils and fats.[162] FTIR spectrometry and multivariate regression were reported by Andrarde *et al.* to be useful in the quality control of kerosene production. Results are presented of the progress from sequential univariate tests to multivariate approaches, leading to improved laboratory performance and reduced delay times for critical production information.[163] Hydrocarbon and CO_2 measurements were performed using GC, FID, FTIR spectrometry and the results compared by Stephens *et al.*[164]

An in situ infrared spectroscopic investigation of selective alkylation of toluene over basic zeolites by Palomares *et al.* is found.[165] In situ FTIR monitoring of a solar flux induced chemical process is presented by Markham *et al.*[166] A general description of the use of spectroscopic techniques used to study combustion-chamber chemistry is given by Masi.[167]

An in situ infrared absorption method for determination of high pressure phase diagrams of methane-heavy hydrocarbon mixtures is given.[168] Gas chromatographic and IR spectroscopic data combined with PLS (Partial Least Squares) regression analysis is presented for quality control of jet fuels by Seinsche and Luigart.[169]

The on-line determination of total olefins in gasoline by process gas chromatography and Fourier transform infrared spectroscopy is found in a paper by Bade *et al.*[170] In situ examination of coal macerals oxidation by infrared microspectroscopy is found in a paper by Landais.[171] A method for FTIR process monitoring of metal powder temperature and size distribution is detailed by Rosenthal *et al.*[172]

7.2.17 *Pharmaceutical Industry*

A clean method using flow-injection Fourier-transform infrared spectrometry for the simultaneous determination of propyphenazone and caffeine in powdered pharmaceutical tablets is given by Bouhsain *et al.*[173] The incorporation of HPLC–FTIR spectrometry into pharmaceutical research programs in such a way that it can compete with, e.g., NMR and MS, and can be automated is discussed by Pivonka and Kirkland.[174]

The coupling of chiral methods such as optical rotatory dispersion or circular dichroism with more conventional methods such as UV-visible or IR spectrometry was performed to obtain optical purity and concentration information by Erskine *et al.* Mathematical approaches such as partial least squares and principal-components regression methods were used to predict the enantiomeric purities of test mixtures.[175] In situ Fourier-transform infrared and calorimetric were used by Landau *et al.* to study the preparation of a pharmaceutical intermediate. A reaction effluent passed through an ATR flow cell and was monitored by FTIR spectrometry.

Periodic samples of the reaction effluent and the off-gas were analyzed by HPLC. The data obtained were used to give a mechanistic insight into the process and statistical tests on the data showed when the process deviated from the desired operating conditions.[176]

FTIR spectrometry was used by Severdia to detect traces of lubricating grease on the product contact surfaces of pharmaceutical tablet presses. Press surfaces were swabbed with Whatman No 1 filter paper soaked in n-hexane (HPLC solvent grade). The swab was allowed to dry in air and the lubricant residue was extracted by sonication of the swab in n-hexane. After removal of the filter paper, the extract was gently evaporated to dryness, the residue was dissolved in a known volume of $CDCl_3$ and the transmission IR spectrum of the solution was recorded (32 scans) at a spectral resolution of $4\,cm^{-1}$ against a solvent reference.[177]

In situ Fourier-transform infrared and calorimetric studies of the preparation of a pharmaceutical intermediate is detailed by Landau *et al.*[178]

7.2.18 *Plant Sciences, Paper and Pulp*

Rapid determination of effective alkali and dead-load concentrations in kraft liquors is presented by Leclerc and Hogikyan using attenuated total reflectance (ATR) infrared spectrometry.[179] Rapid characterization of dirt specks in kraft pulp is described by Leclerc and Ouchi using infrared spectrometry.[180]

7.2.19 *Polymer Science*

A new correction algorithm for FTIR on-line film thickness measurement is presented by Wang *et al.*[181] The structure of the conducting polymer poly(paraphenylene), synthesized under different electrochemical conditions, was studied in situ by Damlin *et al.* using external reflection FTIR spectrometry.[182] An FTIR microspectrometric system is described for the real-time on-polymer-bead monitoring of multistep combinatorial syntheses carried out in a flow cell, a diagram of which is presented by Pivonka *et al.*[183]

A FTIR spectrophotometer was used by Kirsch *et al.* to analyze the monomer and co-monomer in the reactor feed system of a polyolefin process. The system was used in closed-loop control of the co-monomer feed flow to attempt to control product density and type changes.[184] Formation of toxic gases during pyrolysis of polyacrylonitrile and nylons is analyzed by Nielsen *et al.* using an on-line optoacoustic FTIR gas analyzer with a fixed-bed pyrolysis reactor that consisted of a vertical quartz tube (1 cm i.d.) heated by a Lindberg furnace. The major toxic products of polyacrylonitrile were HCN and NH_3; nylons produced NH_3, but only minor amounts of HCN. Both types of polymer produced CO.[185]

Infrared microspectroscopy is combined with a microtome for assessing potential toxins in polymer packaging by Jickells. This instrumental method allows much more rapid analysis than is possible with dissolution techniques and can be used to study the penetration and migration of food and toxin materials within plastics.[186] Specular-reflectance infrared spectroscopy was used by Gaarenstroom *et al.* for rapid identification of automotive plastics in dismantling operations.[187]

External beam mid infrared spectroscopy is presented as a method for non-destructive rapid identification of plastics by Mucci.[188] Mid infrared reflectance spectroscopy is proposed by Zachmann and Turner as a method for rapid characterization of black polymeric material.[189] In situ Fourier transform infrared analysis of poly(ester-urethanes) at low temperature is described by Rehman *et al.* using a newly constructed liquid nitrogen cooled sample stage.[190] Evanescent wave FTIR spectroscopy is used by Linossier *et al.* for in situ study of the degradation of polymer-substrate systems.[191]

Step-scan photoacoustic FTIR spectroscopic studies were used by Niu and Urban, to study surfactant exudation and film formation in 50%/59% Sty/n-BA latex films.[192] Infrared and resonance Raman spectroscopic studies on the photopolymerization process of the Langmuir-Blodgett films of a diacetylene monocarboxylic acid are described by Saito *et al.*[193] Time-dependent polarization modulation infrared spectroscopy was used for an in situ study of photoinduced orientation in azopolymers by Buffeteau and Pezolet.[194]

On-line Fourier transform infrared spectroscopy is described by Kirsch *et al.* as a means of improved process control in polyolefin polymerization.[195] Spectroscopic attenuated total reflectance and photoacoustic Fourier transform IR approaches are described by Niu *et al.* for use in monitoring latex film formation at surfaces and interfaces.[196] Infrared spectrometric classification techniques are used by Van Every and Elder for validation in polyolefin formulation.[197] In situ external reflection Fourier transform infrared spectroscopy is presented by Damlin *et al.* as a method to study the structure of the conducting polymer poly(para-phenylene).[198] In situ, quantitative FTIR studies of polymers irradiated with synchrotron radiation are described by Wollersheim and Hormes.[199] In situ infrared spectroscopy is used by Yu *et al.* for studying the electric field-induced dipole reorientation in oriented nylon 11.[200]

A dissertation by Khettry describing the use of infrared fiber optics for the in-line monitoring of molten polymeric processes is presented.[201] In situ FTIR spectroscopy, synchrotron SAXS, and rheology are used to study the structure development during the reactive processing of model flexible polyurethane foam systems by Elwell *et al.*[202] Fourier transform infrared spectroscopy is described by Ge *et al.* as a means for quantitative monitoring of an epoxidation process.[203]

FTIR external reflection is used by Lutz *et al.* for a low-cost unit for rapid analysis of carbon-filled rubbers.[204] Process FTIR spectroscopy is used for real-time analysis of PC/ABS blend composition by Fidler.[205] IR-RAS is used for real-time in situ observation of vapor deposition polymerization of N-methylolacrylamide by Tamada *et al.*[206] A description of the value offered by molecular spectroscopy of polymers is offered by Siesler.[207]

7.2.20 *Semiconductors, Electronics*

An FTIR in-line sensor is used for modeling, simulation and control of single wafer process by Liu *et al.*[208] A doctoral dissertation describing the in situ FTIR studies of semiconductor/electrolyte interface (cadmium telluride, silicon) is given by Fan.[209] In situ FT-IR-ATR spectroscopy is used by Ping and Nauer to investigate the redox behavior of poly (thienylpyrrole) thin film electrodes in non-aqueous solutions.[210] In situ Fourier transform infrared spectroscopy and electrochemical analysis is used for the characterization of urea adlayers at Pt(100) electrodes by Climent *et al.*[211] Fourier transform infrared spectroscopy is used by Farquharson *et al.* for real-time process control of molecular beam epitaxial growth of mercury cadmium telluride (MCT) films.[212]

A quantitative in situ analysis method was developed by Sperline *et al.* to determine the adsorption and rinsing-removal of surfactants from Si surfaces. The method used was FT-IR-ATR spectroscopy at the silicon/aqueous solution interface in the 1550–1100 cm^{-1} spectral region. The method is useful for the wet cleaning of Si in semiconductor manufacturing. The surfactants tested were: dodecyl trimethyl ammonium bromide (I) and Triton X-100 (II). A zinc selenide (ZnSe) internal reflection element (IRE) was used after sputtering with a thin film of Si and Al_2O_3 as a buffer layer.[213]

A Ph.D. thesis by Papapanayiotou describes in situ infrared spectroscopic studies of electrodeposition additive adsorption on copper.[214] Fourier transform infrared spectroscopy used for process monitoring and control of integrated circuit manufacturing is given by Liu *et al.*[215]

7.2.21 *Surface Analysis*

In situ Fourier transform infrared-attenuated total reflection (ATR) spectroscopy is used by Fan and Ng for monitoring of the polyaniline synthesis mechanism on a p-type silicon electrode surface.[216] Spatially resolved infrared spectroscopy is used by Qin and Wolf for in situ study of spatial surface coverage during CO oxidation on supported catalysts.[217]

A patent describing the method and apparatus for analyzing minute foreign substances, and process for manufacturing semiconductor or LCD elements using spectroscopy is described.[218] The application of Fourier transform infrared spectroscopy to the optimization of chemically-amplified photoresist processing is described in a Ph.D. dissertation by Howes.[219] The surface corrosion of natural aluminum is studied via the use of optical fiber based FTIR evanescent wave absorption spectroscopy by Namkung.[220] In situ Fourier transform infrared-diffuse reflection spectroscopy of direct methanol fuel cell anodes and cathodes is described by Pan *et al.*[221] In situ high-resolution infrared spectroscopy of a photopolymerized C60 film is presented by Onoe and Takeuchi.[222]

Infrared (IR) spectroscopy and atomic force microscopy (AFM) are used for in situ and ex situ structural analysis of phospholipid-supported planar bilayers by Stephens and Dluhy.[223] In situ infrared spectroscopy is proposed by Pan *et al.* as a real time diagnostic for plasma polymer film deposition.[224] Infrared reflection spectroscopy is described by Mielczarski *et al.* as a method for qualitative and quantitative evaluation of heterogeneous adsorbed monolayers on semiconductor electrode surfaces.[225] In situ IR-ATR spectroscopy is used by Ozanam *et al.* to study the hydrogenated silicon surface in organic electrolytes.[226] The optics and cell design for in situ far-infrared spectroscopy of electrode surfaces is described by Russell *et al.*[227]

A dissertation describing the in situ Fourier transform infrared spectroscopy of chemistry and growth in chemical vapor deposition is given by Salim.[228] The in situ FTIR-diffuse reflection spectroscopy of the anode surface in a direct methanol/oxygen fuel cell is given by Fan *et al.*[229] In situ UV-visible and infrared spectroscopic techniques are described by Beden for studying corrosion products and corrosion inhibitors on materials surfaces.[230] Internal and external reflection modes are compared by Bae *et al.* for in situ Fourier transform infrared spectroscopy of molecular adsorbates at electrode-electrolyte interfaces.[231]

An in situ infrared spectroscopic study of the mechanism of CO-electrooxidation on Pt-Ru alloys is presented by Olligs *et al.*[232] An in situ IR spectroscopic study of the reaction of dimethylaluminum hydride with photochemically deposited amorphous silicon is presented by Wadayama *et al.*[233]

7.2.22 *Textiles*

A patent describing the use of a hand-held infrared spectrometer for identifying recyclable carpet materials is given.[234]

7.2.23 *Recent Advances in Infrared Spectroscopy*

Hyperspectral imaging systems are described as a scientific tool for food quality assessment, and recent advances up to 2012 are discussed.[235] Attenuated total reflection Fourier transform infrared (ATR-FTIR) spectroscopy can be applied as an imaging method for biomedical studies, including different tissue types, such as aorta, skin, and living tissues.[236] A review paper summarizing the unique optical properties of chalcogenide glasses for mid infrared sensing, integrated optics and ultrahigh-bandwidth signal processing is given.[237] A study to identify the differences in molecular structure of PANI and aniline oligomers and to relate them to the nanotube formation as supramolecular morphology was conducted with infrared band assignments given.[238]

7.3 IR Spectra-Structure Correlation Charts

The following figures demonstrate the spectra-structure correlation features of MIR (Mid Infrared) spectra for model hydrocarbon compounds. The figures are described as follows:

Fig. 7.1. MIR Spectral Chart: MIR Fundamental Group Frequency and Fingerprint Frequency locations, showing a Transmittance Spectrum of Cellulose Acetate Butyrate with Typical NIR Functional Groups and Band Locations Labeled (4000 to 500 cm^{-1}).

Fig. 7.2. MIR Spectral Chart: MIR Fundamental Group Frequency and Fingerprint Frequency locations, showing an Absorbance Spectrum of Cellulose Acetate Butyrate with Typical NIR Functional Groups and Band Locations Labeled (4000 to 500 cm^{-1}).

Fig. 7.3. MIR Spectral Chart: MIR Fundamental Group Frequency and Fingerprint Frequency locations, showing Transmittance Band Shapes and Intensities with Typical Functional Group Bands are Labeled (4000 to 500 cm^{-1}).

Fig. 7.4. MIR Spectral Chart: MIR Fundamental Group Frequency and Fingerprint Frequency locations, showing Transmittance C-H Band Shapes and Intensities for n-Alkanes with Typical Functional Group Bands are Labeled (4000 to 500 cm^{-1}).

Fig. 7.5. MIR Spectral Chart: MIR Fundamental Group Frequency and Fingerprint Frequency locations, showing Transmittance Band Shapes and Intensities for n-Alkanes C-H Stretching for Methyl (CH_3) and Methylene (CH_2) Functional Group Bands are Labeled (4000 to 500 cm^{-1}).

Fig. 7.6. MIR Spectral Chart: MIR Fundamental Group Frequency and Fingerprint Frequency locations, showing Transmittance Band Shapes and Intensities for Aromatic compounds with aliphatic attachments. Note Aromatic C-H, Aliphatic C-H, and Alkene (C=C) Functional Group Bands are Labeled (4000 to 500 cm^{-1}).

Fig. 7.7. MIR Spectral Chart: MIR Fundamental Group Frequency and Fingerprint Frequency locations, showing Transmittance Band Shapes and Intensities for O-H Stretching, Aliphatic C-H Stretching, C-H Bending, and C-O Stretching Functional Group Bands are Labeled (4000 to 500 cm^{-1}).

Fig. 7.8. MIR Spectral Chart: MIR Fundamental Group Frequency and Fingerprint Frequency locations, showing Transmittance Band Shape and Intensity for Dimethyl Ketone Carbonyl (>C=O) Stretching Functional Group Band is Labeled near 1715 cm^{-1} (4000 to 500 cm^{-1}).

Fig. 7.9. MIR Spectral Chart: MIR Fundamental Group Frequency and Fingerprint Frequency locations, showing Transmittance Band Shapes and Intensities for 2-Propanol O-H Stretching, Aliphatic C-H Stretching, and C-O Stretching Functional Group Bands are Labeled (4000 to 500 cm^{-1}); O-H band near 3340-3350 cm^{-1}.

Fig. 7.10. MIR Spectral Chart: MIR Fundamental Group Frequency and Fingerprint Frequency locations, showing Transmittance Band Shape and Intensity for 2-Propanol C-O Stretching (4000 to 500 cm^{-1}); Sharp C-O band near 1100–1170 cm^{-1}.

Fig. 7.11. MIR Spectral Chart: MIR Fundamental Group Frequency and Fingerprint Frequency locations, showing Transmittance Band Shapes and Intensities for Methyl Acetate (Ester) Carbonyl (>C=O) Stretching, and C-O Stretching Functional Group Bands are Labeled (4000 to 500 cm^{-1}); C=O band near 1735–1720 cm^{-1}, C-O Bands near 1250 and 1050 cm^{-1}.

Fig. 7.12. MIR Spectral Chart: MIR Fundamental Group Frequency and Fingerprint Frequency locations, showing Transmittance Band Shapes and Intensities for Ethanoic Acid (Carboxylic Acid) for O-H Stretching, Aliphatic C-H Stretching, Carbonyl (>C=O) Stretching, and C-O Stretching Functional Group Bands are Labeled (4000 to 500 cm^{-1}); O-H Stretching near 3050 cm^{-1}, C-H Stretching near 3200 to 3000 cm^{-1}, C=O Stretching near 1715 cm^{-1} and C-O stretching near 1280 cm^{-1}.

Fig. 7.13. MIR Spectral Chart: MIR Fundamental Group Frequency and Fingerprint Frequency locations, showing Transmittance Band Shapes and Intensities for 1-Aminobutane (Primary Amine) for N–H Stretching, and Aliphatic C-H Stretching Functional Group Bands are Labeled (4000 to 500 cm^{-1}); N–H bands near 3400 to 3100 cm^{-1}, C-H bands between 3000 and 2800 cm^{-1}.

Fig. 7.14. MIR Spectral Chart: MIR Fundamental Group Frequency and Fingerprint Frequency locations, showing Transmittance Band Shapes and Intensities for N-Methyl Acetamide (Secondary Amide) for N-H Stretching, and Aliphatic C-H Stretching, Carbonyl (>C=O) Stretching as Amide I Band; and NH, C-N, and N-H comprising the typical Amide II Band. These Functional Group Bands are Labeled (4000 to 500 cm^{-1}).

Fig. 7.15. IR Spectra-Structure Band Positions Chart — MIR Fundamental C-H absorptions indicated relative to wavenumbers (as 3300 to 2800 cm^{-1}).

Fig. 7.16. IR Spectra-Structure Band Positions Chart — Fundamental C=O, N-H, and O-H absorptions indicated relative to wavenumbers (as 3500 to 1000 cm^{-1}).

Fig. 7.17. IR Spectra-Structure Band Positions Chart — MIR Fundamental C-X absorptions indicated relative to wavenumbers (as 4000 to 500 cm^{-1}).

Fig. 7.18. MIR Spectra-Structure Band Positions Chart — MIR Fundamental absorptions indicated relative to wavenumbers (as 4000 to 400 cm^{-1}).

Fig. 7.19. MIR Spectra-Structure Band Positions Chart — MIR Fundamental absorptions indicated relative to wavenumbers (as 4000 to 400 cm^{-1}).

Fig. 7.20. MIR Spectra-Structure Band Positions Chart — MIR Fundamental absorptions indicated relative to wavenumbers (as 4000 to 400 cm^{-1}).

Fig. 7.21. MIR Spectra-Structure Band Positions Chart — MIR Fundamental absorptions indicated relative to wavenumbers (as 4000 to 400 cm^{-1}).

Fig. 7.22. MIR Spectra-Structure Band Positions Chart — MIR Fundamental absorptions indicated relative to wavenumbers (as 4000 to 400 cm^{-1}).

Fig. 7.23. MIR Spectra-Structure Band Positions Chart — MIR Fundamental absorptions indicated relative to wavenumbers (as 4000 to 400 cm^{-1}).

Fig. 7.24. Integrated peak area for 966 cm^{-1} baseline corrected band from 990 cm^{-1} to 945 cm^{-1} indicates presence of trans fat molecular structures. (From AOCS Cd 14d-99 and AOAC 2000.1 Methods.)

Table 7.4 Functional groupings and calculated lccations for infrared spectroscopy in alphabetical order by functional group name. (Adapted with permission from J. Workman, *Handbook of Organic Compounds: NIR, IR, Raman, and UV-Vis spectra featuring polymers and surfactants, Volume 1,* Academic Press-Elsevier publishers, 2001.).

Functional group	Molecular structure	Relative intensity	Fundamental band	1st overtone (2ν)	2nd overtone (3ν)
1. C-H Stretch					
(a) Alkane, C_nH_{2n+2}	CH_3-$(CH_2)_N$-CH_3, normal paraffins	m-s	2959-2849 cm^{-1}	5917-5698 cm^{-1} 1690-1755 nm	8873-8547 cm^{-1} 1127-1170 nm
methyl	-C-H_3	m-s			
asymmetric str.	"	"	2973-2950 cm^{-1}	5945-5900 cm^{-1} 1682-1695 nm	8913-8850 cm^{-1} 1122-1130 nm
symmetric str.	"	"	2882-2862 cm^{-1}	5964-5724 cm^{-1} 1735-1747 nm	8643-8584 cm^{-1} 1157-1165 nm
methylene	-C-H_2	m-s			
asymmetric str.	"	"	2936-2915 nm	5872-5831 cm^{-1} 1703-1715 nm	8811-8749 cm^{-1} 1135-1143 nm
symmetric str.	"	"	2864-2843 nm	5727-5685 cm^{-1} 1746-1759 nm	8591-8532 cm^{-1} 1164-1172 nm
(b) Alkene, C_nH_{2n}	>C=C<, olefinic group				
monosubstituted	H H >C=C< R H	m	3040-3012 cm^{-1}	6079-6024 cm^{-1} 1645-1660 nm	9116-9033 cm^{-1} 1097-1107 nm
		m	3096-3077 cm^{-1}	6192-6154 cm^{-1} 1615-1625 nm	9288-9231 cm^{-1}

(*Continued*)

Table 7.4 (*Continued*)

Functional group	Molecular structure	Relative intensity	Fundamental band	1st overtone (2ν)	2nd overtone (3ν)
cis, disubstituted	H H >C=C< R R	m	3040-3012 cm^{-1}	6079-6024 cm^{-1} 1645-1660 nm	9116-9033 cm^{-1} 1097-1107 nm
trans, disubstituted	H R >C=C< R H	m	3040-3012 cm^{-1}	6079-6024 cm^{-1} 1645-1660 nm	9116-9033 cm^{-1} 1097-1107 nm
gem, disubstituted	R H >C=C< R H	m	3096-3077 cm^{-1}	6192-6154 cm^{-1} 1615-1625 nm	9285-9234 cm^{-1} 1077-1083 nm
trisubstituted	R R >C=C< R H	m	3040-3012 cm^{-1}	6079-6024 cm^{-1} 1645-1660 nm	9116-9033 cm^{-1} 1097-1107 nm
(c) alkyne, C_nH_{2n-2}	$-C{\equiv}C-H$	s	3345-3295 cm^{-1}	6690-6595 cm^{-1} ~1515 nm	10,050-9880 cm^{-1} ~1010 nm
(d) aromatic, C_nH_n	Ar(C)-H	v	3030 cm^{-1}	6061 cm^{-1} ~1650 nm	9900 cm^{-1} ~ 1100
2. C-H Bend					
(a) Alkane, C_nH_{2n+2}					
methine	R-CHR′R″	w	1340 cm^{-1}	~2681 cm^{-1} ~3730 nm	~4021 cm^{-1} ~2487 nm

(*Continued*)

Table 7.4 (*Continued*)

Functional group	Molecular structure	Relative intensity	Fundamental band	1st overtone (2ν)	2nd overtone (3ν)
methylene	R-CH_2R′	m	1483-1445 cm^{-1}	2966-2889 cm^{-1}	4450-4335 cm^{-1}
					2247-2307 nm
methyl	R-CH_3	m	1379-1370 cm^{-1}	2759-2740 cm^{-1}	4137-4110 cm^{-1}
		s		3625-3650 nm	2267-2333 nm
			1470-1429 cm^{-1}	2940-2858 cm^{-1}	4411-4286 cm^{-1}
					2417-2433 nm
gem, dimethyl	CH_3-CHR-CH_3	s	1385-1379 cm^{-1}	2770-2759 cm^{-1}	4155-4144 cm^{-1}
		s		3610-3625 nm	2407-2413 nm
			1370-1364 cm^{-1}	2740-2729 cm^{-1}	4110-4093 cm^{-1}
				3650-3665 nm	2433-2443 nm
tert-butyl	$(CH_3)_3$-CH	m	1395-1385 cm^{-1}	2789-2770 cm^{-1}	4184-4155 cm^{-1}
		s		3585-3610 nm	2390-2407 nm
			~1370 cm^{-1}	~2740 cm^{-1}	~4110 cm^{-1}
				~3650 nm	~2433 nm
b) Alkene,					
C_nH_{2n}		s	985-952 cm^{-1}	1905-1970 cm^{-1}	2985-2956 cm^{-1}
monosubstituted	>C=C<	s		5250-5075 nm	3350-3383 nm
	H H	s	915-905 cm^{-1}	1830-1810 cm^{-1}	2745-2715 cm^{-1}
	>C=C<			5465-5525 nm	3643-3683 nm
	R H		1420-1410 cm^{-1}	1841-2821 cm^{-1}	4261-4232 cm^{-1}
				3520-3545 nm	2347-2363 nm

(*Continued*)

Table 7.4 (*Continued*)

Functional group	Molecular structure	Relative intensity	Fundamental band	1st overtone (2ν)	2nd overtone (3ν)
cis, disubstituted	H H >C=C< R R	s	$\sim$690 cm^{-1}	$\sim$1379 cm^{-1}	$\sim$2069 cm^{-1}
				$\sim$7250 nm	$\sim$4833 nm
trans, disubstituted	H R >C=C< R H	s	970-960 cm^{-1}	1940-1919 cm^{-1}	2910-2879 cm^{-1}
		m		5155-521 nm	3437-3473 nm
			1309-1295 cm^{-1}	2618-2591 cm^{-1}	3926-3887 cm^{-1}
				3820-3860 nm	2547-2573 nm
gem, disubstituted	R H >C=C< R H	s	895-885 cm^{-1}	1791-1770 cm^{-1}	2686-2655 cm^{-1}
				5585-5650 nm	3723-3767 nm
		s	1420-1410 cm^{-1}	2841-2821 cm^{-1}	4261-4232 cm^{-1}
				3520-3545 nm	2347-2363 nm
trisubstituted	R R >C=C< R H	s	840-790 cm^{-1}	1681-1580 cm^{-1}	2521-2370 cm^{-1}
				5950-6330 nm	3967-4220 nm
(c) Alkyne, C_nH_{2n-2}	$-C\equiv C-H$	s	$\sim$680-605 cm^{-1}	1360-1210 cm^{-1}	2040-1815 cm^{-1}
				$\sim$7950 nm	$\sim$5300 nm
			$\sim$629 cm^{-1}	$\sim$1258 cm^{-1}	$\sim$1887 cm^{-1}

(*Continued*)

Table 7.4 (*Continued*)

Functional group	Molecular structure	Relative intensity	Fundamental band	1st overtone (2ν)	2nd overtone (3ν)
(d) Aromatic, C_nH_n	**Ar(C)-H**				
mono-substituted (5 hydrogens)	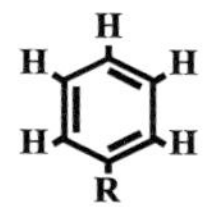	v, s	~752 cm^{-1} ~699 cm^{-1}	~1504 cm^{-1} ~6650 nm ~1399 cm^{-1} ~7150 nm	~2256 cm^{-1} ~4433 nm ~2098 cm^{-1} ~4767 nm
di-substituted (4 hydrogens)	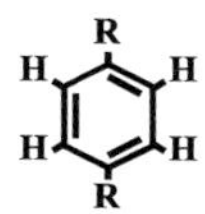	v, s	~752 cm^{-1}	~1504 cm^{-1} ~6650 nm	~2256 cm^{-1} ~4433 nm
tri-substituted (3 hydrogens)	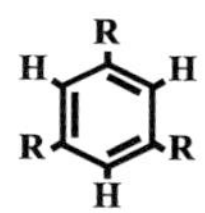	v, s	781 cm^{-1}	1563 cm^{-1} ~6400 nm	2344 cm^{-1} ~4267 nm
tetra-substituted (2 hydrogens)	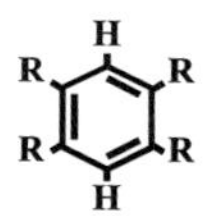	v, m	~833 cm^{-1}	~1667 cm^{-1} ~6000 nm	~2500 cm^{-1} ~4000 nm
penta-substituted (1 hydrogen)	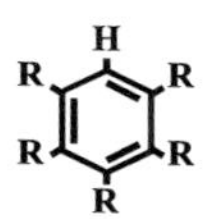	v, w	~885 cm^{-1}	~1770 cm^{-1} ~5650 nm	~2655 cm^{-1} ~3767 nm
3. C-C Stretch					
(a) Alkene, C_nH_{2n}	>C=C<	v	1681-1621 cm^{-1}	3361-3241 cm^{-1} 2975-3085 nm	5043-4861 cm^{-1} 1983-2057 nm

(*Continued*)

Table 7.4 (*Continued*)

Functional group	Molecular structure	Relative intensity	Fundamental band	1st overtone (2ν)	2nd overtone (3ν)
mono-substituted	As shown above	m	$\sim$1645 cm^{-1}	$\sim$3289 cm^{-1}	$\sim$4933 cm^{-1}
				$\sim$3040 nm	$\sim$2027 nm
cis, disubstituted	"	m	$\sim$1658 cm^{-1}	$\sim$3317 cm^{-1}	$\sim$4975 cm^{-1}
				$\sim$3015 nm	
trans, disubstituted	"	m	$\sim$1675 cm^{-1}	$\sim$3350 cm^{-1}	$\sim$5025 cm^{-1}
				$\sim$2985 nm	$\sim$1990 nm
gem, disubstituted	"	m	$\sim$1653 cm^{-1}	$\sim$3306 cm^{-1}	$\sim$6024 cm^{-1}
				$\sim$3025 nm	$\sim$1660 nm
trisubstituted	"	m	$\sim$1669 cm^{-1}	$\sim$3339 cm^{-1}	$\sim$5008 cm^{-1}
				$\sim$2995 nm	$\sim$1997 nm
tetra-substituted	"	w	$\sim$1669 cm^{-1}	$\sim$3339 cm^{-1}	$\sim$5008 cm^{-1}
				$\sim$2995 nm	$\sim$1997 nm
diene	>C=C<. . .>C=C<	w	$\sim$1650 cm^{-1}	$\sim$3300 cm^{-1}	$\sim$4950 cm^{-1}
		w		$\sim$3030 nm	$\sim$2020 nm
			$\sim$1600 cm^{-1}	$\sim$3200 cm^{-1}	$\sim$4801 cm^{-1}
				$\sim$3125 nm	$\sim$2083 nm
(b) Alkyne, C_nH_{2n-2}					
mono-substituted		m	2141-2101 cm^{-1}	4283-4202 cm^{-1}	6423-6301 cm^{-1}
				2335-2380 nm	1557-1587 nm
disubstituted		v, w	2262-2188 cm^{-1}	4525-4376 cm^{-1}	6789-6566 cm^{-1}
				2210-2285 nm	1473-1523 nm
Allene	>C=C=C<	m	$\sim$1961 cm^{-1}	$\sim$3922 cm^{-1}	$\sim$5444 cm^{-1}
		m		$\sim$2550 nm	$\sim$1837 nm
			$\sim$1064 cm^{-1}	$\sim$2128 cm^{-1}	$\sim$3192 cm^{-1}
				$\sim$4700 nm	$\sim$3133 nm

(*Continued*)

Table 7.4 (*Continued*)

Functional group	Molecular structure	Relative intensity	Fundamental band	1st overtone (2ν)	2nd overtone (3ν)
Aromatic containing		v	1640-1590 cm^{-1}	3280-3180 cm^{-1}	4920-4770 cm^{-1}
		v		~3125 nm	~2083 nm
		m	~1580 cm^{-1}	~3160 cm^{-1} ~3165 nm	~4739 cm^{-1} ~2110 nm
		m	1499-1450 cm^{-1}	~2999 cm^{-1} ~3335 nm	4498 cm^{-1} ~2223 nm
			~1449 cm^{-1}	~2899 cm^{-1} ~3450 nm	~4348 cm^{-1} ~2300 nm
4. Carbonyl >C=O					
(a) Ketones	R >C=O R'				
saturated, acyclic	alkane (R) >C=O alkane(R')	s	1724-1704 cm^{-1}	3448-3407 cm^{-1} 2900-2935 nm	5173-5110 cm^{-1} 1933-1957 nm
saturated, cyclic 6- or more membered ring	(or) AR >C=O R'	s	1724-1704 cm^{-1}	3448-3407 cm^{-1} 2900-2935 nm	5173-5110 cm^{-1} 1933-1957 nm
5-membered ring	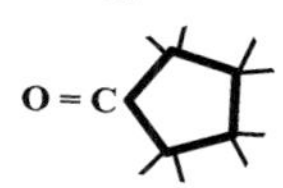	s	1751-1770 cm^{-1}	3503-3540 cm^{-1} 2855-2825 nm	5255-5216 cm^{-1} 1903-1917 nm

(*Continued*)

Table 7.4 (*Continued*)

Functional group	Molecular structure	Relative intensity	Fundamental band	1st overtone (2ν)	2nd overtone (3ν)
4-membered ring	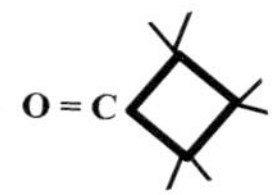	s	~1776 cm^{-1}	~3552 cm^{-1} ~2815 nm	~5328 cm^{-1} ~1877 nm
alpha, beta unsaturated, acyclic		s	1684-1664 cm^{-1}	3367-3328 cm^{-1} 2970-3005 nm	5051-4993 cm^{-1} 1980-2003 nm
alpha, beta unsaturated, cyclic					
6- or more membered ring		s	1684-1664 cm^{-1}	3367-3328 cm^{-1} 2970-3005 nm	5051-4993 cm^{-1} 1980-2003 nm
5-member ring		s	1724-1709 cm^{-1}	3448-3419 cm^{-1} 2900-2925 nm	5173-5128 cm^{-1} 1933-1950 nm
alpha, beta, alpha', beta', unsaturated, acyclic		s	1669-1664 cm^{-1}	3339-3328 cm^{-1} 2995-3005 nm	5008-4993 cm^{-1} 1997-2003 nm
Aryl		s	1701-1681 cm^{-1}	3401-3361 cm^{-1} 2940-2975 nm	5102-5043 cm^{-1} 1960-1983 nm
Diaryl		s	1669-1661 cm^{-1}	3339-3322 cm^{-1} 2995-3010 nm	5008-4983 cm^{-1} 1997-2007 nm
alpha-Diketones		s	1730-1709 cm^{-1}	3460-3419 cm^{-1} 2890-2925 nm	5189-5128 cm^{-1} 1927-1950 nm
beta-Diketones (enolic)		s	1639-1538 cm^{-1}	3279-3077 cm^{-1} 3050-3250 nm	4919-4615 cm^{-1} 2033-2167 nm

(*Continued*)

Table 7.4 (*Continued*)

Functional group	Molecular structure	Relative intensity	Fundamental band	1st overtone (2ν)	2nd overtone (3ν)
1,4-Quinones			1689-1661 cm^{-1}	3378-3322 cm^{-1} 2960-3010 nm	5068-4983 cm^{-1} 1973-2007 nm
Ketenes			~2151 cm^{-1}	~4301 cm^{-1} ~2325 nm	~6667 cm^{-1} ~1500 nm
(b) Aldehydes, Carbonyl stretch	>C=O				
saturated, aliphatic		s	1739-1721 cm^{-1}	3478-3442 cm^{-1} 2875-2905 nm	5216-5163 cm^{-1} 1917-193 nm
alpha, beta-unsaturated, aliphatic		s	1704-1681 cm^{-1}	3407-3361 cm^{-1} 2935-2975 nm	5110-5043 cm^{-1} 1957-1983 nm
alpha, beta, gamma, delta-unsaturated, aliphatic		s	1681-1661 cm^{-1}	3361-3322 cm^{-1} 2975-3010 nm	5043-4983 cm^{-1} 1983-2007 nm
Aryl		s	1715-1695 cm^{-1}	3431-3390 cm^{-1} 2915-2950 nm	5147-5084 cm^{-1} 1943-1967 nm
C-H Stretching, two bands		w	2899-2817 cm^{-1}	5797-5634 cm^{-1}	8696-8453 cm^{-1}
		w		1725-1775 nm	1150-1183 nm
			2778-2703 cm^{-1}	5556-5405 cm^{-1} 1800-1850 nm	8333-8110 cm^{-1} 1200-1233 nm
(c) ESTERS	R-C=O-OR'				
saturated, acyclic		s	1751-1736 cm^{-1}	3503-3472 cm^{-1} 2855-2880 nm	5255-5208 cm^{-1} 1903-1920 nm
saturated, cyclic: delta and larger lactones		s	1751-1736 cm^{-1}	3503-3472 cm^{-1} 2855-2880 nm	5255-5208 cm^{-1} 1903-1920 nm

(*Continued*)

Table 7.4 (*Continued*)

Functional group	Molecular structure	Relative intensity	Fundamental band	1st overtone (2ν)	2nd overtone (3ν)
gamma-lactones		s	1779-1761 cm^{-1}	3559-3521 cm^{-1} 2810-2810 nm	5339-5283 cm^{-1} 1873-1893 nm
beta-lactones		s	~1818 cm^{-1}	~3636 cm^{-1} ~2750 nm	~5456 cm^{-1} ~1833 nm
unsaturated, vinyl ester		s	1799-1770 cm^{-1}	3597-3540 cm^{-1} 2780-2825 nm	5397-5311 cm^{-1} 1853-1883 nm
alpha, beta-unsaturated and aryl		s	1730-1718 cm^{-1}	3460-3436 cm^{-1} 2890-2910 nm	5189-5155 cm^{-1} 1927-1940 nm
alpha, beta-unsaturated delta-lactone		s	1792-1718 cm^{-1}	3584-3436 cm^{-1} 2790-2910 nm	5189-5155 cm^{-1} 1927-1940 nm
alpha, beta unsaturated gamma-lactone		s	1761-1739 cm^{-1}	3521-3478 cm^{-1} 2840-2875 nm	5283-5216 cm^{-1} 1893-1917 nm
beta, gamma-unsaturated gamma-lactone		s	~1799 cm^{-1}	~3597 cm^{-1} ~2780 nm	~5397 cm^{-1} ~1853 nm
alpha-Ketoesters		s	1754-1739 cm^{-1}	3509-3478 cm^{-1} 2850-2875 nm	5263-5216 cm^{-1} 1900-1917 nm
beta-Ketoesters (enolic)		s	~1650 cm^{-1}	~3300 cm^{-1} ~3030 nm	~4950 cm^{-1} ~2020 nm
carbonates		s	1779-1739 cm^{-1}	3559-3478 cm^{-1} 2810-2875 nm	5339-5216 cm^{-1} 1873-1917 nm

(*Continued*)

Table 7.4 (*Continued*)

Functional group	Molecular structure	Relative intensity	Fundamental band	1st overtone (2ν)	2nd overtone (3ν)
5. CARBOXYLIC ACIDS -O-C=O–, AND ETHERS R-O-R					
carbonyl stretching:		s	1724-1701 cm^{-1}	3448-3401 cm^{-1}	5173-5102 cm^{-1}
saturated				2900-2940 nm	1933-1960 nm
aliphatic					
alpha,		s	1715-1689 cm^{-1}	3431-3378 cm^{-1}	5147-5068 cm^{-1}
beta-unsaturated				2915-2960 nm	1960-1983 nm
aliphatic aryl			2049-1681 cm^{-1}	4098-3361 cm^{-1}	5102-5043 cm^{-1}
				2440-2975 nm	1943-1973 nm
hydroxyl stretching (bonded), several bands		w	2703-2500 cm^{-1}	5405-5000 cm^{-1}	8110-7502 cm^{-1}
				1850-2000 nm	1233-1333 nm
carboxylate anion stretching		s	1610-1550 cm^{-1}	3221-3101 cm^{-1}	4831-4651 cm^{-1}
				3105-3225 nm	2070-2150 nm
Ethers, -C–O stretch	R'-O;R"	s	1205-1100 cm^{-1}	2410-2200 cm^{-1}	3615-3300 cm^{-1}
6. ANHYDRIDES, R'-C=O-O-C=O-R'					
stretching		s	1850-1750 cm^{-1}	3700-3500 cm^{-1}	5550-5250 cm^{-1}
				3575-3845 nm	2383-2563 nm
saturated, acyclic		s	1850-1750 cm^{-1}	3700-3500 cm^{-1}	5550-5250 cm^{-1}
		s		2705-2780 nm	1803-1853 nm
		s	1400-1350 cm^{-1}	2800-2700 cm^{-1}	4200-4050 cm^{-1}
				2795-2875 nm	1863-1917 nm
			1160-1100 cm^{-1}	2320-2200 cm^{-1}	3480-3300 cm^{-1}

(*Continued*)

Table 7.4 (*Continued*)

Functional group	Molecular structure	Relative intensity	Fundamental band	1st overtone (2ν)	2nd overtone (3ν)
alpha,		s	1828-1779 cm^{-1}	3656-3559 cm^{-1}	5485-5339 cm^{-1}
beta-unsaturated		s		2735-2810 nm	1823-1873 nm
and aryl, acyclic			1770-1721 cm^{-1}	3540-3442 cm^{-1}	5311-5163 cm^{-1}
				2825-2905 nm	1883-1937 nm
saturated,		s	1869-1821 cm^{-1}	3738-3643 cm^{-1}	5609-5464 cm^{-1}
5-member ring		s		2675-2745 nm	1783-1830 nm
			1767-1751 cm^{-1}	3534-3503 cm^{-1}	5299-5255 cm^{-1}
				2830-2855 nm	1887-1903 nm
alpha, beta		s	1848-1799 cm^{-1}	3697-3597 cm^{-1}	5546-5397 cm^{-1}
unsaturated,		s		2705-2780 nm	1803-1853 nm
5-member ring			1828-1779 cm^{-1}	3656-3559 cm^{-1}	5485-5068 cm^{-1}
				2735-2810 nm	1823-1973 nm
7. ACYL HALIDES					
Stretch: Acyl		s	~1848 cm^{-1}	~3697 cm^{-1}	~5546 cm^{-1}
fluorides				~2705 nm	~1803 nm
Acyl chlorides		s	~1795 cm^{-1}	~3591 cm^{-1}	~5385 cm^{-1}
				~2785 nm	~1857 nm
Acyl bromides		s	~1808 cm^{-1}	~3617 cm^{-1}	~5426 cm^{-1}
				~2765 nm	~1843 nm
alpha,		s	1751-1748 cm^{-1}	3503-3497 cm^{-1}	5348-5244 cm^{-1}
beta-unsaturated		m		2855-2860 nm	1870-1907 nm
and aryl			1748-1718 cm^{-1}	3497-3436 cm^{-1}	5244-5155 cm^{-1}
				2860-2910 nm	1907-1940 nm

(*Continued*)

Table 7.4 (*Continued*)

Functional group	Molecular structure	Relative intensity	Fundamental band	1st overtone (2ν)	2nd overtone (3ν)
COF_2		s	~1927 cm^{-1}	~3854 cm^{-1} ~2595 nm	~5780 cm^{-1} ~1730 nm
$COCl_2$		s	~1828 cm^{-1}	~3656 cm^{-1} ~2735 nm	~5485 cm^{-1} ~1823 nm
$COBr_2$		s	~1828 cm^{-1}	~3656 cm^{-1} ~2735 nm	~5485 cm^{-1} ~1823 nm
8. AMIDES					
Carbonyl stretch					
primary, concentrated solutions and solids		s	~1650 cm^{-1}	~3300 cm^{-1} ~3030 nm	~4950 cm^{-1} ~2020 nm
primary, dilute solutions		s	~1718 cm^{-1}	~3436 cm^{-1} ~2910 nm	~5068 cm^{-1} ~1973 nm
secondary, concentrated solutions and solids		s	1681-1629 cm^{-1}	3361-3257 cm^{-1} 2975-3070 nm	5043-4885 cm^{-1} 1983-2047 nm
secondary, dilute solutions		s	1701-1669 cm^{-1}	3401-3339 cm^{-1} 2940-2995 nm	5102-5008 cm^{-1} 1960-1997 nm
tertiary, solutions and solids		s	1669-1629 cm^{-1}	3339-3257 cm^{-1} 2995-3070 nm	5008-4885 cm^{-1} 1997-2047 nm

(*Continued*)

Table 7.4 (*Continued*)

Functional group	Molecular structure	Relative intensity	Fundamental band	1st overtone (2ν)	2nd overtone (3ν)
cyclic, delta-lactams, dilute solutions		s	~1681 cm^{-1}	~3361 cm^{-1} ~2975 nm	~5043 cm^{-1} ~1983 nm
cyclic, gamma-lactams, dilute solutions		s	~1701 cm^{-1}	~3401 cm^{-1} ~2940 nm	~5102 cm^{-1} ~1960 nm
cyclic, gamma-lactams fused to second ring, dilute solution		s	1751-1701 cm^{-1}	3503-3401 cm^{-1} 2855-2940 nm	5255-5102 cm^{-1} 1903-1960 nm
cyclic, beta-lactams, dilute solution		s	1761-1730 cm^{-1}	3521-3460 cm^{-1} 2840-2890 nm	5283-5189 cm^{-1} 1893-1927 nm
cyclic, beta-lactams fused to second ring, dilute solutions		s	1779-1770 cm^{-1}	2559-3540 cm^{-1} 2810-2825 nm	5339-5311 cm^{-1} 1873-1883 nm
Urea, acyclic		s	~1661 cm^{-1}	~3322 cm^{-1}	~4983 cm^{-1} ~2007 nm
Urea, cyclic, 6-membered ring		s	~1639 cm^{-1}	~3279 cm^{-1} ~3050 nm	~4919 cm^{-1} ~2033 nm
Urea, cyclic 5-member ring		s	~1721 cm^{-1}	~3442 cm^{-1} ~2905 nm	~5163 cm^{-1} ~1937 nm
Urethanes		s	1739-1689 cm^{-1}	3478-3378 cm^{-1} 2875-2960 nm	5216-5068 cm^{-1} 1917-1973 nm
Imides, acyclic		s	~1709 cm^{-1}	~3419 cm^{-1} ~2925 nm	~5128 cm^{-1} ~1950 nm

(*Continued*)

Table 7.4 (*Continued*)

Functional group	Molecular structure	Relative intensity	Fundamental band	1st overtone (2ν)	2nd overtone (3ν)
Imides, cyclic,		s	~1709 cm^{-1}	~3419 cm^{-1}	~5128 cm^{-1}
6-membered ring		s		~2925 nm	~1950 nm
		s	~1701 cm^{-1}	~3401 cm^{-1}	~5102 cm^{-1}
				~2940 nm	~1960 nm
Imides, cyclic,		s	~1730 cm^{-1}	~3460 cm^{-1}	~5189 cm^{-1}
alpha,		s		~2890 nm	~1927 nm
beta-unsaturated,			~1669 cm^{-1}	~3339 cm^{-1}	~5008 cm^{-1}
6-membered ring				~2995 nm	~1997 nm
Imides, cyclic,		s	~1770 cm^{-1}	~3540 cm^{-1}	~5311 cm^{-1}
5-membered ring					
		s		~2825 nm	~1883 nm
			~1701 cm^{-1}	~3401 cm^{-1}	~5102 cm^{-1}
				~2940 nm	~1960 nm
Imides, cyclic,		s	~1792 cm^{-1}	~3584 cm^{-1}	~5368 cm^{-1}
alpha,		s		~2790 nm	~1863 nm
beta-unsaturated,			~1709 cm^{-1}	~3419 cm^{-1}	~5128 cm^{-1}
				~2925 nm	~1950 nm
N-H stretch					
primary, free, two		m	~3497 cm^{-1}	~6993 cm^{-1}	~10493 cm^{-1}
bands		m		~1430 nm	~953 nm
			~3401 cm^{-1}	~6803 cm^{-1}	~10204 cm^{-1}
				~1470 nm	~980 nm
primary, bonded,		m	~3344 cm^{-1}	~6689 cm^{-1}	~10030 cm^{-1}
two bands				~1495 nm	~997 nm
		m	~3175 cm^{-1}	~6349 cm^{-1}	~9524 cm^{-1}
				~1575 nm	~1050 nm

(*Continued*)

Table 7.4 (*Continued*)

Functional group	Molecular structure	Relative intensity	Fundamental band	1st overtone (2ν)	2nd overtone (3ν)
secondary, free; one band		m	~3425 cm^{-1}	~6849 cm^{-1} ~1460 nm	~10256 cm^{-1} ~975 nm
secondary, bonded; one band		m	3333-3125 cm^{-1}	6667-6250 cm^{-1} 1500-1600 nm	10,000-9372 cm^{-1} 1000-1067 nm
N-H Bending					
Primary amides, dilute solution		s	1621-1590 cm^{-1}	3241-3180 cm^{-1} 3085-3145 nm	4861-4769 cm^{-1} 2057-2097 nm
secondary amides, dilute solution		s	1548-1511 cm^{-1}	3101-3021 cm^{-1} 3225-3310 nm	4651-4531 cm^{-1} 2150-2207 nm
9. ALCOHOLS & PHENOLICS, O-H stretch					
Free O-H		v, sh	3650-3584 cm^{-1}	7299-7168 cm^{-1} 1370-1395 nm	10953-10753 cm^{-1} 913-930 nm
with hydrogen bonding, single bridge compounds	O-H....H-	v, sh	3546-3448 cm^{-1}	7092-6897 cm^{-1} 1410-1450 nm	10638-10341 cm^{-1} 940-967 nm
Polymeric O-H.		s,b	3401-3195 cm^{-1}	6803-6390 cm^{-1} 1470-1565 nm	10204-9588 cm^{-1} 980-1043 nm
chelated compounds		w,b	3226-2500 cm^{-1}	6452-5000 cm^{-1} 1550-2000 nm	9671-7496 cm^{-1} 1034-1334 nm
O-H bend & C-O stretch Combination					
primary alcohols		s s	~1053 cm^{-1}	~2105 cm^{-1} ~4750 nm	~3158 cm^{-1} ~3167 nm
			1351-1266 cm^{-1}	2703-2532 cm^{-1} 3700-3950 nm	4054-3798 cm^{-1} 2467-2633 nm
secondary alcohols		s s	~1099 cm^{-1}	~2198 cm^{-1} ~4550 nm	~3297 cm^{-1} ~3033 nm
			1351-1266 cm^{-1}	2703-2532 cm^{-1} 3700-3950 nm	4054-3798 cm^{-1} 2467-2633 nm

(*Continued*)

Table 7.4 (*Continued*)

Functional group	Molecular structure	Relative intensity	Fundamental band	1st overtone (2ν)	2nd overtone (3ν)
tertiary alcohols		s	~1149 cm^{-1}	~2299 cm^{-1}	~3448 cm^{-1}
		s		~4350 nm	~2900 nm
			1408-1316 cm^{-1}	2817-2632 cm^{-1}	4225-3948 cm^{-1}
				3550-3300 nm	2367-2533 nm
phenols		s	~1198 cm^{-1}	~2395 cm^{-1}	~3614 cm^{-1}
		s		~4175 nm	~2767 nm
			1405-1316 cm^{-1}	2810-2632 cm^{-1}	4225-3948 cm^{-1}
				3559-3300 nm	2367-2533 nm
10. AMINES, N-H stretch					
primary, free; two bands		m	~3497 cm^{-1}	~6993 cm^{-1}	~10493 cm^{-1}
				~1430 nm	~953 nm
		m	~3401 cm^{-1}	~6803 cm^{-1}	~10204 cm^{-1}
				~1470 nm	~980 nm
secondary, free; one band		m	3497-3311 cm^{-1}	6993-6623 cm^{-1}	10493-9930 cm^{-1}
				1430-1510 nm	953-1007 n
Imines; one band	=N-H	m	3401-3300 cm^{-1}	6803-6601 cm^{-1}	10204-9901 cm^{-1}
				1470-1515 nm	980-1010 nm
amine salts		m	3125-3030 cm^{-1}	6250-6061 cm^{-1}	9372-9091 cm^{-1}
				1600-1650 nm	1067-1100 nm
N-H bend					
primary		s-m	1650-1590 cm^{-1}	3300-3180 cm^{-1}	4950-4769 cm^{-1}
				3030-3145 nm	2020-2097 nm
secondary		w	1650-1550 cm^{-1}	3300-3101 cm^{-1}	4950-4651 cm^{-1}
				3030-3225 nm	2020-2150 nm
amine salts		s	1600-1575 cm^{-1}	3200-3150 cm^{-1}	4801-4724 cm^{-1}
				125-3175 nm	2083-2117 nm

(*Continued*)

Table 7.4 (*Continued*)

Functional group	Molecular structure	Relative intensity	Fundamental band	1st overtone (2ν)	2nd overtone (3ν)
		s	~1499 cm^{-1}	~2999 cm^{-1} ~3335 nm	~4498 cm^{-1} ~2223 nm
C-N stretch					
aromatic, primary		s	1340-1250 cm^{-1}	2681-2500 cm^{-1} 3730-4000 nm	4021-3750 cm^{-1} 2487-2667 nm
aromatic, secondary		s	1350-1280 cm^{-1}	2699-2561 cm^{-1} 3705-3905 nm	4049-3842 cm^{-1} 2470-2603 nm
aromatic, tertiary		s	1359-1309 cm^{-1}	2717-2618 cm^{-1} 3680-3820 nm	4077-3926 cm^{-1} 2453-2547 nm
aliphatic		w	1220-1020 cm^{-1}	2439-2041 cm^{-1} 4100-4900 nm	3741-3061 cm^{-1} 2673-3267 nm
		w	~1427 cm^{-1}	~2853 cm^{-1} ~3505 nm	~4225 cm^{-1} ~2367 nm
11. UNSATURATED N COMPOUNDS, C=N stretch					
alkyl nitriles		m	2262-2205 cm^{-1}	4525-4410 cm^{-1} 2210-2230 nm	6789-6615 cm^{-1} 1473-1487 nm
alpha, beta, unsaturated alkyl nitriles		m	2237-2217 cm^{-1}	4474-4435 cm^{-1} 2235-2255 nm	6711-6653 cm^{-1} 1490-1503 nm
aryl nitriles		m	2242-2222 cm^{-1}	4484-4444 cm^{-1} 2230-2250 nm	6725-6667 cm^{-1} 1487-1500 nm
isocyanates		m	2273-2232 cm^{-1}	4545-4464 cm^{-1} 2200-2240 nm	6817-6725 cm^{-1} 1467-1487 nm
isocyanides		m	2222-2070 cm^{-1}	4444-4141 cm^{-1} 2250-2415 nm	6667-6211 cm^{-1} 1500-1610 nm

(*Continued*)

Table 7.4 (*Continued*)

Functional group	Molecular structure	Relative intensity	Fundamental band	1st overtone (2ν)	2nd overtone (3ν)
imines & oximes					
alkyl compounds		v	1689-1649 cm^{-1}	3378-3297 cm^{-1}	5068-4919 cm^{-1}
				2960-3033 nm	1973-2033 nm
alpha, beta		v	1661-1629 cm^{-1}	3322-3257 cm^{-1}	4831-4885 cm^{-1}
unsaturated				3010-3070 nm	2070-2047 nm
-N=N-stretch, azo-		v	1629-1575 cm^{-1}	3257-3150 cm^{-1}	4885-4724 cm^{-1}
compounds				3070-3175 nm	2047-2117 nm
-N=C=N-,		s	2155-2128 cm^{-1}	4310-4255 cm^{-1}	6464-6423 cm^{-1}
diimides				2320-2350 nm	1547-1557 nm
-N3 stretch, azides		s	2160-2137 cm^{-1}	4320-4274 cm^{-1}	6481-6382 cm^{-1}
		w		2315-2340 nm	1543-1567 nm
			1340-1179 cm^{-1}	2681-2358 cm^{-1}	4021-3537 cm^{-1}
				3730-4240 nm	2487-2827 nm
C-NO2, nitro-					
compounds					
aromatic		s	1570-1499 cm^{-1}	3140-2999 cm^{-1}	4577-4498 cm^{-1}
		s		3185-3335 nm	2185-2223 nm
			1370-1299 cm^{-1}	2740-2597 cm^{-1}	4110-3896 cm^{-1}
				3650-3850 nm	2433-2567 nm
aliphatic		s	1570-1550 cm^{-1}	3140-3101 cm^{-1}	4710-4651 cm^{-1}
		s		3185-3225 nm	2123-2150 nm
			1379-1370 cm^{-1}	2759-2740 cm^{-1}	4137-4110 cm^{-1}
				3625-3650 nm	2417-2433 nm
O-NO2, nitrates		s	1650-1600 cm^{-1}	3300-3200 cm^{-1}	4950-4801 cm^{-1}
		s		3030-3125 nm	2020-2083 nm
			1299-1250 cm^{-1}	2597-2500 cm^{-1}	3896-3750 cm^{-1}
				3850-4000 nm	2567-2667 nm

(*Continued*)

Table 7.4 (*Continued*)

Functional group	Molecular structure	Relative intensity	Fundamental band	1st overtone (2ν)	2nd overtone (3ν)
C-NO, nitroso		s	1600-1499 cm^{-1}	3200-2999 cm^{-1}	4801-4498 cm^{-1}
				3125-3335 nm	2083-2223 nm
O-NO, nitrites		s	1681-1600 cm^{-1}	3361-3200 cm^{-1}	5043-4801 cm^{-1}
		s		2975-3125 nm	1983-2083 nm
			1626-1610 cm^{-1}	3252-3221 cm^{-1}	4878-4830 cm^{-1}
				3075-3105 nm	2050-2070 nm
d. Nitriles, -C≡N stretch		s,w	2280-2210 cm^{-1}	4560-4420 cm^{-1}	6840-6630 cm^{-1}
12. HALOGEN GROUPS, C-X stretch					
C-F		s	1408-1000 cm^{-1}	2817-2000 cm^{-1}	4227-3000 cm^{-1}
				3550-5000 nm	2366-3333 nm
C-Cl		s	800-602 cm^{-1}	1600-1250 cm^{-1}	2400-1807 cm^{-1}
				6250-8300 nm	4167-5533 nm
C-Br		s	602-500 cm^{-1}	1205-1000 cm^{-1}	1807-1500 cm^{-1}
				8300-10000 nm	5533-6667 nm
C-I		s	~500 cm^{-1}	~1000 cm^{-1}	~1500 cm^{-1}
				~10000 nm	~6667 nm
X-C=O stretch		v	1820-1790 cm^{-1}	3640-3580 cm^{-1}	5460-5370 cm^{-1}
C-F		s	1408-1000 cm^{-1}	2817-2000 cm^{-1}	4227-3000 cm^{-1}
				3550-5000 nm	2366-3333 nm
13. SULFUR COMPOUNDS					
S-H stretch		w	2597-2551 cm^{-1}	5195-5102 cm^{-1}	7794-7651 cm^{-1}
				1925-1960 nm	1283-1307 nm
C=S stretch		s	1200-1050 cm^{-1}	2401-2101 cm^{-1}	3601-3152 cm^{-1}
				4165-4760 nm	2777-3173 nm
S=O stretch					
sulfoxides		s	1070-1030 cm^{-1}	2139-2060 cm^{-1}	3150-3089 cm^{-1}
				4675-4855 nm	3175-3237 nm
sulfones		s	1160-1140 cm^{-1}	2320-2281 cm^{-1}	3481-3421 cm^{-1}
		s		4310-4385 nm	2873-2923 nm
			1350-1300 cm^{-1}	2699-2601 cm^{-1}	4049-3902 cm^{-1}
				3705-3845 nm	2470-2563 nm

(*Continued*)

Table 7.4 (*Continued*)

Functional group	Molecular structure	Relative intensity	Fundamental band	1st overtone (2ν)	2nd overtone (3ν)
sulfites		s	1230-1149 cm^{-1}	2460-2299 cm^{-1}	3690-3448 cm^{-1}
		s		4065-4350 nm	2710-2900 nm
			1429-1350 cm^{-1}	2857-2699 cm^{-1}	4286-4049 cm^{-1}
				3500-3705 nm	2333-2470 nm
sulfonyl chlorides		s	1185-1164 cm^{-1}	2370-2328 cm^{-1}	3568-3493 cm^{-1}
		s		4220-4295 nm	2813-2863 nm
			1370-1340 cm^{-1}	2740-2681 cm^{-1}	4110-4021 cm^{-1}
				3650-3730 nm	2433-2487 nm
sulfonamides		s	1179-1140 cm^{-1}	2358-2281 cm^{-1}	3537-3421 cm^{-1}
		s		4240-4385 nm	2827-2923 nm
			1350-1300 cm^{-1}	2699-2601 cm^{-1}	4049-3902 cm^{-1}
				3705-3845 nm	2470-2563 nm
sulfonic acids		s	1209-1149 cm^{-1}	2418-2299 cm^{-1}	3627-3448 cm^{-1}
		s		4135-4350 nm	2757-2900 nm
		s	1060-1030 cm^{-1}	2121-2060 cm^{-1}	3182-3089 cm^{-1}
				4715-4855 nm	3143-3237 nm
			~649 cm^{-1}	~1299 cm^{-1}	~1948 cm^{-1}
				~7700 nm	~5133 nm
14. C-F, SI-O, C-SI, CO_3, S-CN GROUPS					
C-F stretch	C-F	s	~1200 cm^{-1}	~2400-2300 cm^{-1}	~3600 cm^{-1}
C-F bend		v, b	650-475 cm^{-1}	1300-950 cm^{-1}	1950-1425 cm^{-1}
Si-O stretch		s	1100-1050 cm^{-1}	2200-2100 cm^{-1}	3300-3150 cm^{-1}
C-Si stretch		s	1270-1255 cm^{-1}	2540-2510 cm^{-1}	3810-3765 cm^{-1}
			(~1261 cm^{-1})	(~2522 cm^{-1})	
CO_3 (Carbonate)		s, b	1520-1380 cm^{-1}	3040-2760 cm^{-1}	4560-4140 cm^{-1}
S-CN		s	2050-2000 cm^{-1}	4100-4000 cm^{-1}	6150-6000 cm^{-1}

Note: The following abbreviations are used for Relative Intensities of infrared bands: w = *weak*, m = medium, m-s = medium-strong, s = strong, υ = variable.

7.4 Published IR Standard Reference Methods and Practices

The following information is provided to indicate the standard reference methods, practices, and guidelines published which relate directly to NIR spectroscopy.

7.4.1 *United States Pharmacopeia (USP) Methods*[239–240]

General methods are described for identification of compendial chemical substances. The test procedures that follow are applicable to substances that absorb infrared or ultraviolet radiation. Identification using different sample preparation and sample presentation methods is described, to include: mixing the sample with potassium bromide (KBr) (Method 197K); finely ground and dispersed in mineral oil (197M); and suspended neat between transmittance plates such as sodium chloride or potassium bromide (197F). The sample may be dissolved in a suitable solvent and placed into a 0.1 mm or other suitable pathlength transmittance cell for spectroscopic measurements (197S). Other methods used for infrared identification include: Attenuated Total Reflectance (197A); thin sample onto a transmittance plate (197E); and ground material mixed with infrared transparent salts for diffuse reflectance (DRIFTS) as (197D). The measurements are made between 3800 and 650 cm^{-1} or 2.6 to 15 microns (μm).[239] The presence of polymorphs may indicate significant differences in materials and so the user is referred to reference.[240] The ultraviolet method is described in Volume 1 of this text set and is referred to as monograph (197U).[240]

7.4.2 *International Organization for Standardization (ISO) Standards*[241,242]

For the International Standard Organization (ISO) there are 34 methods listed for 2014 as "infrared methods". These are listed below with both ISO designation numbers and titles of the respective methods.

- **ISO 4935:1989**

Steel and iron — Determination of sulfur content — Infrared absorption method after combustion in an induction furnace

- **ISO 6781:1983**

Thermal insulation — Qualitative detection of thermal irregularities in building envelopes — Infrared method

- **ISO 9889:1994**

Determination of carbon content in uranium dioxide powder and sintered pellets — Resistance furnace combustion — Titrimetric/coulometric/infrared absorption method

- **ISO 15106-2:2003**

Plastics — Film and sheeting — Determination of water vapour transmission rate — Part 2: Infrared detection sensor method

- **ISO 11873:2005**

Hardmetals — Determination of sulfur and carbon contents in cobalt metal powders — Infrared detection method

- **ISO 4689-3:2004**

Iron ores — Determination of sulfur content — Part 3: Combustion/infrared method

- **ISO 9686:2006**

Direct reduced iron — Determination of carbon and/or sulfur — High-frequency combustion method with infrared measurement

- **ISO 17053:2005**

Steel and iron — Determination of oxygen — Infrared method after fusion under inert gas

- **ISO 21258:2010**

Stationary source emissions — Determination of the mass concentration of dinitrogen monoxide (N_2O) — Reference method — Non-dispersive infrared method

- **ISO 22963:2008**

Titanium and titanium alloys — Determination of oxygen — Infrared method after fusion under inert gas

- **ISO 17328:2014**

Optics and photonics — Optical materials and components — Test method for refractive index of infrared optical materials

- **ISO 7829:1986**

Photography — Black-and-white aerial camera films — Determination of ISO speed and average gradient

- **ISO 10291:1994**

Glass in building — Determination of steady-state U values (thermal transmittance) of multiple glazing — Guarded hot plate method

- **ISO 16794:2003**

Nuclear energy — Determination of carbon compounds and fluorides in uranium hexafluoride infrared spectrometry

- **ISO 22754:2008**

Pulp and paper — Determination of the effective residual ink concentration (ERIC number) by infrared reflectance measurement

- **ISO 19579:2006**

Solid mineral fuels — Determination of sulfur by IR spectrometry

- **ISO 21614:2008**

Determination of carbon content of UO_2, $(U, Gd)O_2$ and $(U, Pu)\ O_2$ powders and sintered pellets — Combustion in a high-frequency induction furnace — Infrared absorption spectrometry

- **ISO 1126:2006**

Rubber compounding ingredients — Carbon black — Determination of loss on heating

- **ISO/TS 23556:2007**

Performance test method for diesel engine soot-removal devices in lubricating oils — Initial filtration efficiency

- **ISO/TS 10867:2010**

Nanotechnologies — Characterization of single-wall carbon nanotubes using near infrared photoluminescence spectroscopy

- **ISO 15063:2011**

Plastics — Polyols for use in the production of polyurethanes — Determination of hydroxyl number by NIR spectroscopy

- **ISO 4650:2012**

Rubber — Identification — Infrared spectrometric methods

- **ISO 13199:2012**

Stationary source emissions — Determination of total volatile organic compounds (TVOCs) in waste gases from non-combustion processes — Non-dispersive infrared analyser equipped with catalytic converter

- **ISO 15064:2010**

Plastics — Aromatic isocyanates for use in the production of polyurethanes — Determination of the isomer ratio in toluenediisocyanate

- **ISO 9712:2012**

Non-destructive testing — Qualification and certification of NDT personnel

- **ISO 16565:2013**

Rubber — Determination of 5-ethylidenenorbornene (ENB) or dicyclopentadiene (DCPD) in ethylene-propylene-diene (EPDM) terpolymers

- **ISO 16378:2013**

Space systems — Measurements of thermo-optical properties of thermal control materials

- **ISO 17184:2014**

Soil quality — Determination of carbon and nitrogen by near infrared spectrometry (NIRS)

- **ISO 9869-1:2014**

Thermal insulation — Building elements — In situ measurement of thermal resistance and thermal transmittance — Part 1: Heat flow meter method

- **ISO 9556:1989**

Steel and iron — Determination of total carbon content — Infrared absorption method after combustion in an induction furnace

- **ISO 13902:1997**

Steel and iron — Determination of high sulfur content — Infrared absorption method after combustion in an induction furnace

- **ISO 15350:2000**

Steel and iron — Determination of total carbon and sulfur content — Infrared absorption method after combustion in an induction furnace (routine method)

- **ISO 15349-2:1999**

Unalloyed steel — Determination of low carbon content — Part 2: Infrared absorption method after combustion in an induction furnace (with preheating)

- **ISO 4224:2000**

Ambient air — Determination of carbon monoxide — Non-dispersive infrared spectrometric method

7.4.3 *Association of Official Analytical Chemists (AOAC) International Methods*[243,244]

The Association of Official Analytical Chemists (AOAC) International has several methods for using infrared spectroscopy for analysis. These include the measurement of trans fat in oils, and fat, lactose, protein, and solids in milk. The following text describe these methods.

Trans Fat Measurement Infrared Methods

A more accurate alternative method for determination of isolated trans isomers was developed. It is recommended to replace the previous single-bounce attenuated total reflectance (SB-ATR) Fourier transform infrared (FTIR) of the AOAC for the American Oil Chemists' Society (AOCS) group. The new method is based on transmission measurements using spectral reconstitution (SR).[243]

AOCS Cd 14d-99 and AOAC 2000.1, Trans Fat ATR-FTIR Official Methods

The introduction of ATR (also known as internal reflection) instead of transmission cells, resulted in an alternative mid IR procedure that was developed, validated in a multi-laboratory collaborative study, and adopted as official method Cd 14d-99 by AOCS and official method 2000.10 by AOAC International. In comparative studies, better precision such as lower reproducibility relative standard deviation, RSD(R), values were obtained by using the ATR-FTIR method relative to older transmission FTIR methods. The ATR method requires a 5 minute measurement at the 966 cm^{-1} *trans* fat absorption using horizontal baseline correction (Figure 7.24).

AOAC International Method 972.16, Fat, lactose, protein and solids in milk

Mid infrared (MIR) milk analysis using the classical fat B, fat A, protein, and lactose measurement wavelengths combined with separate reference wavelengths is a method that provides rapid determination of milk composition (AOAC, 2000; method 972.16; 33.2.31; IDF, 2000). Analysis of milk by MIR is based on the principle that different functional groups absorb MIR energy at different wavelengths.

7.4.4 *ASTM International Standards*[245,246]

For the American Society for Testing and Materials ASTM) International there are 182 results listed for 2014 for "infrared spectroscopy". The most cited Infrared Practices include the following, with designation numbers, titles, and brief descriptions.

ASTM E131-10, Standard Terminology Relating to Molecular Spectroscopy

Description of terms used in standards documents related to Molecular Spectroscopy; all units are in SI units.

ASTM E204-98(2007), Standard Practices for Identification of Material by Infrared Absorption Spectroscopy, Using the ASTM Coded Band and Chemical Classification Index (Withdrawn 2014)

This qualitative method was withdrawn and replaced with more modern practices and techniques.

ASTM E168-06, Standard Practices for General Techniques of Infrared Quantitative Analysis

These practices cover the techniques most often used in infrared quantitative analysis. Practices associated with the collection and analysis of data on a computer are included as well as practices that do not use a computer.

ASTM E932-89(2013), Standard Practice for Describing and Measuring Performance of Dispersive Infrared Spectrometers

This practice is intended for all infrared spectroscopists who are using dispersive instruments for qualitative or quantitative areas of analysis. The purpose of this practice is to set forth performance guidelines for testing instruments used in developing an analytical method. These guidelines can be used to compare an instrument in a specific application with the instrument(s) used in developing the method.

ASTM E1252-98(2013)e1, Standard Practice for General Techniques for Obtaining Infrared Spectra for Qualitative Analysis

This practice covers the spectral range from 4000 to 50 cm^{-1} and includes techniques that are useful for qualitative analysis of liquid-, solid-, and vapor-phase samples by infrared spectrometric techniques for which the amount of sample available for analysis is not a limiting factor. These techniques are often also useful for recording spectra at frequencies higher than 4000 cm^{-1}, in the near infrared region.

ASTM E1421-99(2009), Standard Practice for Describing and Measuring Performance of Fourier Transform Mid Infrared (FT-MIR) Spectrometers: Level Zero and Level One Tests

This practice describes two levels of tests to measure the performance of laboratory Fourier transform mid infrared (FT-MIR) spectrometers equipped with a standard sample holder used for transmission measurements. This practice is not directly applicable to Fourier transform infrared (FTIR) spectrometers equipped with various specialized sampling accessories such as flow cells or reflectance optics, nor to Fourier transform near infrared (FT-NIR) spectrometers, nor to FTIR spectrometers run in step scan mode. If the specialized sampling accessory can be removed and replaced with a standard transmission sample holder, then this practice can be used.

ASTM E1655-05(2012), Standard Practices for Infrared Multivariate Quantitative Analysis

These practices cover a guide for the multivariate calibration of infrared spectrometers used in determining the physical or chemical characteristics of materials. These practices are applicable to analyses conducted in the near infrared (NIR) spectral region (roughly 780 to 2500 nm) through the mid infrared (MIR) spectral region (roughly 4000 to 400 cm^{-1}). While the practices described herein deal specifically with mid- and near infrared analysis, much of the mathematical and procedural detail contained herein is also applicable for multivariate quantitative analysis done using other forms of spectroscopy. The user is cautioned that typical and best practices for multivariate quantitative analysis using other forms of spectroscopy may differ from practices described herein for mid- and near infrared spectroscopies.

ASTM E1790-04(2010), Standard Practice for Near Infrared Qualitative Analysis

This practice covers the use of near infrared (NIR) spectroscopy for the qualitative analysis of liquids and solids. The practice is written under the assumption that most NIR qualitative analyses will be performed with instruments designed specifically for this region and equipped with computerized data handling algorithms.

ASTM E1866-97(2013), Standard Guide for Establishing Spectrophotometer Performance Tests

This guide covers basic procedures that can be used to develop spectrophotometer performance tests. The guide is intended to be applicable to spectrophotometers operating in the ultraviolet, visible, near infrared and mid infrared regions. This guide is not intended as a replacement for specific practices such as Practices E275, E925, E932, E958, E1421, or E1683 that exist for measuring performance of specific types of spectrophotometers. These tests are designed to be used as rapid, routine checks of spectrophotometer performance. They are designed to uncover malfunctions or other changes in instrument operation, but do not specifically diagnose or quantitatively assess the malfunction or change. The tests are not intended for the comparison of spectrophotometers of different manufacture.

ASTM E2224-10, Standard Guide for Forensic Analysis of Fibers by Infrared Spectroscopy

Infrared (IR) spectrophotometery is a valuable method of fiber polymer identification and comparison in forensic examinations. The use of IR microscopes coupled with Fourier transform infrared (FTIR) spectrometers has greatly simplified the IR analysis of single fibers, thus making the technique feasible for routine use in the forensic laboratory.

ASTM E2310-04(2009), Standard Guide for Use of Spectral Searching by Curve Matching Algorithms with Data Recorded Using Mid Infrared Spectroscopy

This guide presents the use of spectral searching by curve matching search algorithms for data recorded using mid infrared spectroscopy. The methods described may be applicable to the use of search algorithms for other types of spectroscopic data, but each type of data search should be assessed separately. The purpose of this evaluation is the classification and, where possible, identification of the unknown. Spectral searching is intended as a screening method to assist the analyst, and is not an absolute identification technique, and hence, not intended to replace an expert in infrared spectroscopy and should not be used without suitable training. The Euclidean distance algorithm and the first derivative Euclidean distance algorithm are described and their use discussed. The theory and common assumptions made when using search algorithms are also discussed, along with guidelines for the use and interpretation of the search results.

ASTM E2937-13, Standard Guide for Using Infrared Spectroscopy in Forensic Paint Examinations

This guide applies to the forensic IR analysis of paints and coatings and is intended to supplement information presented in the Forensic Paint Analysis and Comparison Guidelines (**1**)[2] written by Scientific Working Group on Materials Analysis (SWGMAT). This guideline is limited to the discussion of Fourier Transform Infrared (FTIR) instruments and provides information on FTIR instrument setup, performance assessment, sample preparation, analysis and data interpretation. It is intended to provide an understanding of the requirements, benefits, limitations and proper use of IR accessories and sampling methods available for use by forensic paint examiners. The following accessory techniques will be discussed: FTIR microspectroscopy (transmission and reflectance), diamond cell and attenuated total reflectance.

ASTM D6122-13, Standard Practice for Validation of the Performance of Multivariate Online, At-Line, and Laboratory Infrared Spectrophotometer Based Analyzer Systems

This practice covers requirements for the validation of measurements made by laboratory or process (online or at-line) near- or mid infrared analyzers, or both, used in the calculation of physical, chemical, or quality parameters (that is, properties) of liquid petroleum products. The properties are calculated from spectroscopic data using multivariate modeling methods. The requirements include verification of adequate instrument performance; verification of the applicability of the calibration model to the spectrum of the sample under test; and verification of equivalence between the result calculated from the infrared measurements and the result produced by the primary test method used for the development of the calibration model.

ASTM D7418-12, Standard Practice for Set-Up and Operation of Fourier Transform Infrared (FTIR) Spectrometers for In-Service Oil Condition Monitoring

This practice covers the instrument set-up and operation parameters for using FTIR spectrometers for in-service oil condition monitoring for both direct trend analysis and differential trend analysis approaches. This practice describes how to acquire the FTIR spectrum of an in-service oil sample using a standard transmission cell and establishes maximum allowable spectral noise levels.

ASTM D7806-12, Standard Test Method for Determination of the Fatty Acid Methyl Ester (FAME) Content of a Blend of Biodiesel and Petroleum-Based Diesel Fuel Oil Using Mid Infrared Spectroscopy

This test method covers the determination of the content of biodiesel (fatty acid methyl esters — FAME) in diesel fuel oils. It is applicable to concentrations from 1 to 30 volume percent. This procedure is applicable only to FAME. This test method is not appropriate for the determination of the concentration of biodiesel that is in the form of fatty acid ethyl esters (FAEE).

References

1. A. Rogalski, Infrared Detectors at the Beginning of the Next Millenium, Opto–Electron. Rev. **9**(2) (2001) 173–187.
2. A. Rogalski, History of infrared detectors, *Opto–Electron. Rev.* **20**(3) (2012) 279–308.
3. E. S. Barr, Historical survey of the early development of the infrared spectral region, *Amer. J. Phys.* **28** (1960) 42–54.
4. G. R. Harrison, R. C. Lord, and J. R. Loofbourow, *Practical Spectroscopy*, 1 edn. (Prentice-Hall, New York, 1948).

5. Newport Corporation, 1791 Deere Avenue, Irvine, CA 92606, United States (2014).
6. J. Workman, Review of Process and Non-invasive Near Infrared and Infrared Spectroscopy, *Applied Spectroscopy Reviews*, **34**(1 & 2) (1999) 1–89.
7. J. B. Callis, D. L. Illman and B. R. Kowalski, Process Analytical Chemistry, *Anal. Chem.* **59** (1987) 624A.
8. J. L. Koenig and C. M. Snively, Fast FTIR Imaging: Theory and Applications, *Spectroscopy* **13**(11) (1998) 22–28.
9. P. Colarusso, L. H. Kidder, I. W. Levin, J. C. Fraser, J. F. Arens, and E. N. Lewis, Infrared Spectroscopic Imaging: From Planetary to Cellular Systems, *Appl. Spectrosc.* **52**(3) (1998) 106A–120A.
10. J. K. Crowley and D. R. Zimbelman, Mapping hydrothermally altered rocks on Mount Rainier, Washington, with Airborne Visible/Infrared Imaging Spectrometer (AVIRIS) data, *Geology* **25**(6) (1997) 559–562.
11. E. Makipaa, J. T. Tanttu, and H. Virtanen, IR-based method for copper electrolysis short-circuit detection, Proc. SPIE-Int. Soc. Opt. Eng. Vol. 3056: Thermosense XIX: An International Conference on Thermal Sensing and Imaging Diagnostic Applications (1997) 100–104.
12. L. Ning, A. M. Chozinski, A. Azarenko, L. S. Daley, W. J. Bowyer, T. Buban, G. E. Edwards, J. B. Callis, and G. B. Strobel, Five novel applications of imaging visible and short near infrared spectrophotometry and fluorimetry in the plant sciences. Part I: Photographic and Histological Applications, *Spectroscopy* **11**(9) (1996) 44.
13. N. Rowlands and R. Neville, Calcite and dolomite discrimination using airborne SWIR imaging spectrometer data, Proc. SPIE-Int. Soc. Opt. Eng. Vol. 2819: Imaging Spectrometry II (1996) 36–44.
14. L. J. Otten, III, A. D. Meigs, R. G. Sellar, and J. B. Rafert, Measured performance of an airborne Fourier transform hyperspectral imager, Proc. SPIE-Int. Soc. Opt. Eng., Vol. 2819: Imaging Spectrometry II (1996) 291–299.
15. K. Silk and E. R. Schildkraut, Imaging Fourier transform spectroscopy for remote chemical sensing, Proc. SPIE-Int. Soc. Opt. Eng., Vol. 2763: Electro-Optical Technology for Remote Chemical Detection and Identification (1996) 169–177.
16. L. C. Rowan, T. L. Bowers, J. K. Crowley, C. Anton-Pacheco, P. Gumiel, and M. J. Kingston, Analysis of airborne visible-infrared imaging spectrometer (AVIRIS) data of the Iron Hill, Colorado, carbonatite-alkalic igneous complex, Econ. Geol. **90**(7) (1995) 1966–82.
17. F. A. Kruse, Geologic mapping using combined analysis of airborne visible/infrared imaging spectrometer (AVIRIS) and SIR-C/X-SAR data, Proc. SPIE-Int. Soc. Opt. Eng., Vol. 2819: Imaging Spectrometry II (1996) 24–35.
18. E. A. Carter, P. M. Fredericks, J. S. Church, Fourier transform infrared photoacoustic spectroscopy of surface-treated wool, *Text. Res. J.* **66**(12) (1996) 787–794.
19. R. Goodacre, E. M. Timmins, P. J. Rooney, J. J. Rowland, D. B. Kell, Rapid identification of Streptococcus and Enterococcus species using diffuse reflectance-absorbance Fourier transform infrared spectroscopy and artificial neural networks, *FEMS Microbiol. Lett.* **140**(2–3) (1996) 233–239.
20. G. Macaloney, I. Draper, J. Preston, K. B. Anderson, M. J. Rollins, B. G. Thompson, J. W. Hall, B. Mcneil, At-line control and fault analysis in an industrial high cell density Escherichia coli fermentation, using NIR spectroscopy, *Food Bioprod. Process.* **74**(C4) (1996) 212–220.
21. E. Dosi, G. Vaccari, A. L. Campi, G. Mantovani, A. Gonzalez-Vara y R, A. Trilli, Control of fermentations by means of on-line near infrared spectrometry, Near infrared Spectrosc.: Future

Waves, Proc. Int. Conf. Near infrared Spectrosc., 7th, A. M. C. Davies, and P. Williams (Eds) (1996) 249–254.
22. B. Szalontai, F. Joo, L. Vigh, In situ deuteration of fatty acyl chains in model and biological membranes as a new approach to investigate membrane structures. A Fourier transform infrared spectroscopic study, *6th Spectrosc. Biol. Mol., Eur. Conf.*, M. J. Claude, S. Turrell, J. P. Huvenne (Eds) (1995) 377–378.
23. I. Cornut, B. Desbat, J. M. Turlet, J. Dufourcq, In situ study by polarization modulated Fourier transform infrared spectroscopy of the structure and orientation of lipids and amphipathic peptides at the air-water interface, *Biophys. J.* **70**(1) (1996) 305–312.
24. A. J. Berger, Y. Wang, M. S. Feld, Rapid, noninvasive concentration measurements of aqueous biological analytes by near infrared Raman spectroscopy, *Appl. Opt.* **35**(1) (1996) 209–212.
25. D. O. Henderson, R. Mu, M. Gunasekaran, A rapid method for the identification of Candida at the species level by Fourier transform infrared spectroscopy, *Biomed. Lett.* **51**(204) (1995) 223–228.
26. P. Fournier, T. Buffeteau, A. M. Ritcey, M. A. Pezolet, Polarization modulation FTIR spectrophotometer for in situ measurements of monolayer films of model biological membranes at the air-water interface, *6th Spectrosc. Biol. Mol., Eur. Conf.*, M. J. Claude, S. Turrell, J. P. Huvenne (Eds) (1995) 373–374.
27. P. H. Axelsen, W. D. Braddock, H. L. Brockman, C. M. Jones, R. A. Dluhy, B. K. Kaufman, F. J. Puga II, Use of internal reflectance infrared spectroscopy for the in situ study of supported lipid monolayers, *Appl. Spectrosc.* **49**(4) (1995) 526–531.
28. M. Pezolet, J. Labrecque, M. Subirade, B. Desbat, Infrared spectroscopy of monolayer films of model biological membranes on solid substrates and in situ at the air-water interface, *Spectrosc. 6th Spectrosc. Biol. Mol., Eur. Conf.*, M. J. Claude, S. Turrell, J. P. Huvenne (Eds) (1995) 363–6.
29. J. H. Aldstadt, A. F. Martin, Analytical chemistry and the cone penetrometer: in situ chemical characterization of the subsurface. *Mikrochim. Acta* **127**(1–2) (1997) 1–18.
30. A. Matuszewska, The estimation of changes occurring in the coalification process on the ground of the results of infrared spectroscopy and ultimate analysis of the selected coals and macerals, *Pol. J. Appl. Chem.* **41**(1–2) (1997) 75–94.
31. H. Worden, R. Beer, C. P. Rinsland, Airborne infrared spectroscopy of 1994 western wildfires, *J. Geophys. Res.* **102**(D1) (1997) 1287–1299.
32. C. T. Chaffin, T. L. Marshall, W. G. Fateley, R. M. Hammaker, Infrared analysis of volcanic plumes: a case study in the application of open-path FTIR monitoring techniques, *Spectrosc. Eur.* **7**(3) (1995) 18, 20–24.
33. S. M. Watson, R. T. Kroutil, C. A. Traynor, E. S. Edgerton, J. J. Bowser, R. B. Zweidinger, R. N. Olson, R. P. Dalley, W. J. Bone, R. Price, Remote-sensing and in situ atmospheric chemistry studies with the use of a manned hot-air balloon platform, *Field Anal. Chem. Technol.* **1**(1) (1996) 13–22.
34. M. Griffin, Detectors for far-infrared spectroscopy of the Earth's atmosphere and beyond, *Spectrum* **34**(2) (1996) 22–23.
35. J. O. Zwicker, Intercomparison of data from open-path Fourier transform infrared spectrometers collected during the EPA/API cooperative remote sensing/dispersion field study at Duke Forest, North Carolina January 9 to 27, 1995, Proc. SPIE-Int. Soc. Opt. Eng., 2883 (Optical Remote Sensing for Environmental and Process Monitoring) (1996) 550–567.
36. M. R. Yalamanchili, A. A. Atia, J. Drelich, J. D. Miller, Characterization of interfacial water at hydrophilic and hydrophobic surfaces by in situ FTIR/internal reflection spectroscopy, 1st

Process. Hydrophobic Miner. Fine Coal, Proc. UBC-McGill Bi-Annu. Int. Symp. Fundam. Miner. Process., J. S. Laskowski, and G. W. Poling (Eds) (1995) 3–16.

37. C. Camy-Peyret, P. Jeseck, T. Hawat, G. Durry, S. Payan, G. Berube, L. Rochette, D. Huguenin, The LPMA balloon-borne FTIR spectrometer for remote sensing of atmospheric constituents, Eur. Space Agency, (Spec. Publ.) ESA SP, ESA SP-370 (12th ESA Symposium on European Rocket and Balloon programs and Related Research) (1995) 323–8.
38. T. A. Blake, C. Chackerian, J. R. Podolske, Prognosis for a mid infrared magnetic rotation spectrometer for the in situ detection of atmospheric free radicals, *Appl. Opt.* **35**(6) (1996) 973–985.
39. Y. Gao, J. Li, B. Zhang, The infrared microscope and rapid identification of gemstones, *J. Gemol.* **24**(6) (1995) 411–414.
40. J. R. De A. Franca, J.-M. Brustet, J. Fontan, Multispectral remote sensing of biomass burning in West Africa, *J. Atmos. Chem.* **22**(1 & 2) (1995) 81–110.
41. S. P. Levine, G. M. Russwurm, Fourier-transform infrared optical remote sensing for monitoring airborne gas and vapor contaminants in the field, *TrAC, Trends Anal. Chem.* **13**(7) (1994) 258–262.
42. T. L. Marshall, C. T. Chaffin, R. M. Hammaker, W. G. Fateley, An introduction to open-path FTIR. Atmospheric monitoring, *Environ. Sci. Technol.* **28**(5) (1994) 224A–232A.
43. G. W. Somsen, I. Jagt, C. Gooijer, N. H. Velthorst, U. A. T. Brinkman, T. Visser, Identification of herbicides in river water using on-line trace enrichment combined with column liquid chromatography -Fourier-transform infrared spectrometry, *J. Chromatogr. A.* **756** (1996) 145–157.
44. J. Wang, J. Kang, Z. Chem, M. Huang, J. Zhang, T. Wang, The study of Freon-12 in alcohol/air flame with remote sensing Fourier transform infrared emission spectroscopy, *J. Environ. Sci. Health, Part A: Environ. Sci. Eng. Toxic Hazard. Subst.* Control **A30**(10) (1995) 2111–22.
45. P. Haschberger, Remote measurement of trace gases from aircraft emissions using infrared spectroscopy, Mitt. — Dtsch. Forschungssanst. Luft-Raumfahrt, (94–06, Impact of Emissions from Aircraft and Spacecraft Upon the Atmosphere) (1994) 100–5.
46. M. A. Druy, P. J. Glatkowski, R. Bolduc, W. A. Stevenson, T. C. Thomas, Applications of remote fiber optic spectroscopy using infrared fibers and Fourier transform infrared (FTIR) spectroscopy to environmental monitoring, *Proc. SPIE-Int. Soc. Opt. Eng.* 2293 (Chemical, Biochemical, and Environmental Fiber Sensors VI) (1994) 23–31.
47. K. Schaefer, R. Haus, J. Heland, Remote sensing of emission sources by ground based Fourier-transform infrared spectrometry, Transp. Transform. Pollut. Troposphere, Proc. 3rd EUROTRAC Symp.P.M. Borrell (Ed) (1994) 944–8.
48. T. L. Marshall, C. T. Chaffin, V. D. Makepeace, R. M. Hoffman, R. Hammaker, W. G. Fateley, P. Saarinen, J. Kauppinen, Investigation of the effects of resolution on the performance of classical least squaress (CLS) spectral interpretation programs when applied to volatile organic compounds (VOCs) of interest in remote sensing using open-air long-path Fourier transform infrared (FTIR) spectrometry, *J. Mol. Struct.* **324**(1–2) (1994) 19–28.
49. M. D. Jack, D. R. Nelson, R. D. Stephens, G. A. Walter, C. B. Tacelli, J. A. Santana, Optical IR sensing apparatus for remote analysis of vehicle exhaust gases, Santa Barbara Research Center; General Motors Corp. Canada Pat Appl, CA 2131865 AA DATE: 11/03/95; CA 2131865 (09/12/94), US 119788 (10/09/93); 33 pp.

50. M. L. Polak, J. L. Hall, K. C. Herr, Passive Fourier-transform infrared spectroscopy of chemical plumes: an algorithm for quantitative interpretation and real-time background removal, *Appl. Opt.* **34**(24) (1995) 5406–12.
51. G. Toci, P. Mazzinghi, M. Vannini, Airborne infrared diode laser spectrometer for in situ measurement of stratospheric trace gas concentration, *AIP Conf. Proc.* 386 (Spectral Line Shapes. **9** (1997) 437–438.
52. J. W. Adams, H. H. Bennett, Using reflectance Fourier transform infrared spectroscopy (FTIR) for rapid analysis of contaminated soils, *Contam. Soils* **1** (1996) 241–252.
53. J.-M. Theriault, L. Bissonnette, G. Roy, Monitoring cloud parameters using a ground-based FTIR system, *Proc. SPIE-Int. Soc. Opt. Eng.* 3106 (Spectroscopic Atmospheric Monitoring Techniques) (1997) 7–16.
54. N. M. Davies, M. Hilton, A. H. Lettington, Vehicle exhaust gas monitoring by remote FTIR spectroscopy, *Mikrochim. Acta, Suppl. Progress in Fourier Transform Spectroscopy* **14** (1997) 551–553.
55. I. T. Bae, R. L. Barbour, D. A. Scherson, In situ Fourier-transform infrared spectroscopic studies of nitrite reduction on platinum electrodes in perchloric acid, *Anal. Chem.* **69** (1997) 249–252.
56. L. J. May, L. A. Lentz, B. A. Kirsch, R. O'Reilly, Long-path (8-metre) gas cell analysis of corrosive gases by Fourier-transform infrared spectroscopy, *Process Control Qual.* **7**(2) (1995) 101–111.
57. I. T. Bae, M. Sandifer, Y. W. Lee, D. A. Tryk, C. N. Sukenik, D. A. Scherson, In situ Fourier-transform infrared spectroscopy of molecular adsorbates at electrode/electrolyte interfaces: a comparison between internal and external reflection modes, *Anal. Chem.* **67**(24) (1995) 4508–4513.
58. T. Lindblom, A. A. Christy, F. O. Libnau, Quantitative determination of stabiliser in a single base propellant by chemometric analysis of Fourier transform infrared spectra, *Chemom. Intell. Lab. Syst.* **29**(2) (1995) 243–254.
59. J. Bandekar, R. Sethna, M. Kirschner, Quantitative determination of sulfur oxide species in white liquor by FTIR, *Appl. Spectrosc.* **49**(11) (1995) 1577–1582.
60. Z. Ge, R. Thompson, S. Cooper, D. Ellison, P. Tway, Quantitative monitoring of an epoxidation process by Fourier-transform infrared spectroscopy, *Process Control Qual*, **7**(1) (1995) 3–12.
61. M. M. A. Ruyken, J. A. Visser, A. K. Smilde, On-line detection and identification of interferences in multivariate predictions of organic gases using FTIR spectroscopy, *Anal. Chem.* **67**(13) (1995) 2170–2179.
62. K. Kargosha, S. H. Davarani, R. Moosavi, Quantitative analysis of binary aqueous sodium chlorate and sodium perchlorate solution in the presence of sodium dichromate using Fourier transform infrared spectrometry, *Analyst* **120**(7) (1995) 1945–1948.
63. G. Woelki, R. Salzer, Thermal investigations of the structure of two humic acid salts by in situ FTIR spectroscopy, *Fresenius' J. Anal. Chem.* **352**(5) (1995) 529–531.
64. T. S. Wang, C. J. Zhu, J. D. Wang, F. M. Xu, Z. O. Chen, Y. H. Luo, Study on spectral characterization of infrared flare material combustion with remote high-resolution Fourier-transform infrared spectrometry, *Anal. Chim. Acta* **306**(2–3) (1995) 249–258.
65. W. Wisniewski, E. Malinowska, S. Podsiadlo, M. Jarosz, Rapid IR- spectrophotometric determination of total nitrogen in silicon dioxide — silicon nitride mixtures, *Quim. Anal. (Barcelona)* **14**(1) (1995) 64–66.

66. M. A. Centeno, J. J. Benitez, P. Malet, I. Carrizosa, J. A. Odriozola, In situ temperature-programmed diffuse-reflectance infrared Fourier-transform spectroscopy (TPDRIFTS) of vanadium oxide/titanium oxide catalysts, *Appl. Spectrosc.* **51** (1997) 416–422.
67. P. J. Brimmer, NIR method development in the chemical/polymer industry: meeting the needs of the analysis and the analyst, In Leaping Ahead Near infrared Spectrosc., (Proc. 6th Int. Conf. Near infrared Spectrosc.), G. D. Batten (Ed.) (1995) 360–363.
68. S. Salim, C. A. Wang, R. D. Driver, K. F. Jensen, In situ concentration monitoring in a vertical OMVPE reactor by fiber-optics-based Fourier transform infrared spectroscopy, *J. Cryst. Growth* **169**(3) (1996) 443–449.
69. Y. V. Tolmachev, Z. Wang, D. A. Scherson, Theoretical aspects of laminar flow in a channel-type electrochemical cell as applied to in situ attenuated total reflection-infrared spectroscopy, *J. Electrochem. Soc.* **143**(10) (1996) 3160–3166.
70. C. Lamy, J.-M. Leger, F. Hahn, B. Beden, Electrooxidation of methanol at Pt-Ru electrodes: an in situ IR reflectance spectroscopic study, Laboratoire Chimie I, Universite Poitiers, 86022, Poitiers, Fr. Proc. — Electrochem. Soc. 96-8 (1996) 356–370.
71. J. J. Benitez, M. A. Centeno, C. L. dit Picard, O. Merdrignac, Y. Laurent, J. A. Odriozola, In situ diffuse reflectance infrared spectroscopy (DRIFTS) study of the reversibility of CdGeON sensors towards oxygen, *Inst. Ciencia Materiales Sevilla, Univ. Sevilla, Seville, Spain, 41071 Sens. Actuators*, B. **B31**(3) (1996) 197–202.
72. D. D. Dunuwila, An investigation of the feasibility of using in situ ATR ATIR spectroscopy in the measurement of crystallization phenomena for research and development of batch crystallization processes, Michigan State Univ., East Lansing, MI, USA, 139 pp. *Diss. Abstr. Int.*, **B57**(5) (1996) 3314.
73. M. Epple, H. Kirschnick, J. M. Thomas, An in situ IR-spectroscopic study of the solid-state formation reaction of polyglycolide. Thermal elimination of NaCl from sodium chloroacetate, Inst. Inorg. Appl. Chem., Univ. Hamburg, Hamburg, Germany, D-20146 *Therm. Anal.* **47**(2) (1996) 331–338.
74. L. Xu, J. H. Fu, J. R. Schlup, In Situ Near Infrared Spectroscopic Investigation of the Kinetics and Mechanisms of Reactions between Phenyl Glycidyl Ether (PGE) and Multifunctional Aromatic Amines, *Ind. Eng. Chem. Res.* **35**(3) (1996) 963–72.
75. J. Mijovic, S Andjelic, Monitoring of reactive processing by remote mid infrared spectroscopy, *Polymer* **37**(8) (1996) 1295–303.
76. Q. Xin, X. Gao, Developments of in situ molecular spectroscopy in catalysis studies, *Prog. Nat. Sci.* **5**(3) (1995) 279–90.
77. C. G. Stevens, P. J. Kuzmenko, W. E. Conaway, F. Magnotta, N. L. Thomas, J. Galkowski, I. T. Lewis, T. W. Alger, Cross dispersion infrared spectrometry (CDIRS) for remote chemical sensing, *Proc. SPIE-Int. Soc. Opt. Eng.* 2552(Pt. 1) (1995) 284–91.
78. G. Eder-Mirth, Using in situ IR spectroscopy to understand selective alkylation of aromatics over zeolites, Collect. Czech. Chem. **60**(3) (1995) 421–7.
79. R. Scamehorn, Rapid IR sampling methods for the organic chemistry laboratory using a diffuse reflectance accessory, *J. Chem. Educ.* **72**(6) (1995) 566.
80. J. W. Schoppelrei, In situ FTIR and FT-Raman studies of reactions in the hydrothermal environment, Doctoral Dissertation, Univ. of Delaware, Newark, DE; 175 pp.; *Diss. Abstr. Int., B.* **58**(5) (1997) 2441; UMI Order No. DA9733589.

81. Y. S. Agyare, Near infrared monitoring of dimethyl terephthalate in a low pressure methanolysis process, *Process Control Qual.* **9**(4) (1997) 189–195.
82. Q. Li, Y. Zhu, H. Han, Z. Li, In situ infrared studies on dehydroxylation of MCM-41, *Yanbian Daxue Xuebao (Bianjibu), Z. iran Kexueban.* **23**(1) (1997) 35–37.
83. S. Zhang, J. E. Franke, T. M. Niemczyk, D. M. Haaland, J. N. Cox, I. Banerjee, Testing of a rapid fault detection model for quality control: Borophosphosilicate glass thin films monitored by infrared absorption spectroscopy, *J. Vac. Sci. Technol.*, B. **15**(4) (1997) 955–960.
84. S. H. Chien, S.-Y. Yang, Y.-K. Tseng, H.-W. Chen, In situ FTIR studies of the acidity and catalytic properties of Pt/zeolites, *Bull. Inst. Chem.*, Acad. Sin. **44** (1997) 35–44.
85. F. Le Peltier, P. Chaumette, J. Saussey, M. M. Bettahar, J. C. Lavalley, In situ FTIR spectroscopy and kinetic study of methanol synthesis from CO/H_2 over $ZnAl_2O_4$ and Cu-$ZnAl_2O_4$ catalysts, *J. Mol. Catal. A: Chem.* **122**(2–3) (1997) 131–139.
86. B.-M. Veronique, T. Gilles, Process control in sugar industry by FTIR spectrometer coupled with an ATR accessory, *Dev. Food Sci.* **36** (1994) 11–18.
87. S. Soloman, Real-time on-line analysis of organic and non-organic compounds for food, fertilizers and pharmaceutical product, WO Patent No. 9740361 A1; Oct. 30. (1997) 19 pp.
88. M. M. Mossoba, M. P. Yurawecz, R. E. McDonald, Rapid determination of the total trans-content of neat hydrogenated oils by attenuated total reflection spectroscopy, *J. Am. Oil Chem. Soc.* **73** (1996) 1003–1009.
89. A. K. Hewavitharana, B. van Brakel, Fourier transform infrared spectrometric method for the rapid determination of casein in raw milk, *Analyst* **122** (1997) 701–704.
90. M. Lipp, Review of methods for the analysis of triglycerides in milk fat: application for studies of milk quality and adulteration, *Food Chem.* **54**(2) (1995) 213–221.
91. F. R. van de Voort, A. A. Ismail, J. Sedman, A rapid, automated method for the determination of cis and trans content of fats and oils by Fourier-transform infrared spectroscopy, *J. Am. Oil Chem. Soc.* **72**(8) (1995) 873–880.
92. D. H. McQueen, R. Wilson, A. Kinnunen, J. E. Paaske, Comparison of two infrared spectroscopic methods for cheese analysis, *Talanta* **42**(12) (1995) 2007–2015.
93. M. I. Abdullah, E. Eek, Automatic method for the determination of total carbon dioxide in natural waters, *Water Res.* **29**(5) (1995) 1231–1234.
94. V. Bellon-Maurel, C. Vallat, D. Goffinet, Quantitative analysis of individual sugars during starch hydrolysis by FT-IR-ATR Spectrometry. Part I: multivariate calibration study — repeatability and reproducibility, *Appl. Spectrosc.* **49**(5) (1995) 556–562.
95. A. K. Hewavitharana, B. van Brakel, Fourier transform infrared spectrometric method for the rapid determination of casein in raw milk, *Analyst* **122**(7) (1997) 701–704.
96. E. M. Suzuki, W. P. Marshall, Infrared spectra of U.S. automobile original topcoats (1974–1989): III. In situ identification of some organic pigments used in yellow, orange, red, and brown nonmetallic and brown metallic finishes-benzimidazolones, *J. Forensic Sci.* **42**(4) (1997) 619–648.
97. T. A. Dal Cason, The characterization of some 3,4-methylenedioxycathinone (MDCATH) homologues, *Forensic Sci. Int.* **87** (1997) 9–53.
98. R. J. Fischer, P. F. Crane, C. L. Yu, S. D. Walker, Apparatus for automatic identification of gas samples, Ohmeda Inc., US Patent No. 5,731,581, Issued 03/24/98, 17 pp.

99. A. Hayden, R. Noll, Remote trace gas quantification using thermal IR spectroscopy and digital filtering based on principal components of background scene clutter, *Proc. SPIE-Int. Soc. Opt. Eng.* 3071(Algorithms for Multispectral and Hyperspectral Imagery III) (1997) 158–168.
100. K. Helland, On-line analysis of hydrocarbon gases with emphasis on calibration model maintenance, In Leaping Ahead Near infrared Spectrosc., (Proc. 6th Int. Conf. Near Infrared Spectrosc. G. D. Batten (1995) 355–359.
101. I. Ahonen, H. Riipinen, A. Roos, Portable Fourier transform infrared spectrometer for use as a gas analyzer in industrial hygiene, *Analyst* **121**(9) (1996) 1253–1255.
102. Z. Mao, J. Demirgian, A. Mathew, R. Hyre, Use of Fourier transform infrared spectrometry as a continuous emission monitor, *Waste Manage.* **15**(8) (1995) 567–577.
103. R. E. Shaffer, G. W. Small, Comparison of optimization algorithms for piecewise linear discriminant analysis: application to Fourier transform infrared remote sensing measurements, *Anal. Chim. Acta* **331**(3) (1996) 157–175.
104. L. J. Otten III, A. D. Meigs, R. G. Sellar, J. B. Rafert, Measured performance of an airborne Fourier transform hyperspectral imager, *Proc. SPIE-Int. Soc. Opt. Eng.* 2819 (Imaging Spectrometry II) (1996) 291–299.
105. J. Mijovic, S. Andelic, S. Pejanovic, In situ real-time monitoring of processes by remote Fourier transform infrared spectroscopy, *J. Serb. Chem. Soc.* **61**(12) (1996) 1193–1202.
106. J. Rutter, D. Jungkman, J. Stobie, E. Krueger, J. Garnett, M. Reine, B. Denley, M. Jasmin, A. Sofia, Anthony, A multispectral hybrid HgCdTe FPA/Dewar assembly for remote sensing in the atmospheric infrared Sounder (AIRS) instrument, *Proc. SPIE-Int. Soc. Opt. Eng.* 2817 (Infrared Spaceborne Remote Sensing IV) (1996) 200–213.
107. W. G. Hansen, Production control by near infrared spectroscopy, Unichema Int., 2800 AA, Gouda, Neth. In Leaping Ahead Near infrared Spectrosc., (Proc. 6th Int. Conf. Near infrared Spectrosc.), G. D. Batten (Ed.) (1995) 376–381.
108. K. L. Foulgoc, L. L. Neindre, X. H. Zhang, J. Lucas, Tapered TeX glass optical fibers for remote IR spectroscopic analysis, *Proc. SPIE-Int. Soc. Opt. Eng.* 2836 (Chemical, Biochemical, and Environmental Fiber Sensors VIII) (1996) 26–36.
109. P. F. Valdez, G. W. Donohoe, M. R. Descour, S. Motomatsu, Selection of spectral bands for interpretation of hyperspectral remotely sensed images, *Proc. SPIE-Int. Soc. Opt. Eng,* 2819 (Imaging Spectrometry II) (1996) 195–203.
110. S. T. Vohra, F. Bucholtz, G. M. Nau, K. J. Ewing, I. D. Aggarwal, Remote detection of trichloroethylene in soil by a fiber-optic infrared reflectance probe, *Appl. Spectrosc.* **50** (1996) 985–990.
111. R. E. Shaffer, G. W. Small, Genetic algorithms for the optimization of piecewise linear discriminants, *Chemom. Intell. Lab. Syst.* **35** (1996) 87–104.
112. D. C. Tilotta, D. L. Heglund, S. B. Hawthorne, On-line SFE-FT-IR spectrometry with a fiber optic transmission cell, *Am. Lab.* **28**(6) (1996) 36R-36T.
113. P. Wilks, Portable infrared analysers for repetitive analysis, *Int. Lab. (International Laboratory)* **26** (1996) 11A–11B.
114. C. M. Deeley, J. Sellors, R. A. Spragg, Systems validation for analytical FTIR measurements, *Int. Labmate (International Labmate)* **21** (1996) 20–22.
115. P. J. Kulesza, M. A. Malik, A. Denca, J. Strojek, In situ FTIR/ATR spectroelectrochemistry of Prussian blue in the solid state, *Anal. Chem.* **68** (1996) 2442–2446.
116. J. Lin, S. J. Hart, J. E. Kenny, Improved two-fiber probe for *in situ* spectroscopic measurement, *Anal. Chem.* **68** (1996) 3098–3103.

117. L. M. Treffman, P. W. Morrison Jr., A simple method to fabricate lenses for *in situ* Fourier-transform infrared spectroscopy, *Rev. Sci. Instrum.* **67**(4) (1996) 1454–1457.
118. P. MacLaurin, N. C. Crabb, I. Wells, P. J. Worsfold, D. Coombs, Quantitative in situ monitoring of an elevated-temperature reaction using a water-cooled mid infrared fiber-optic probe, *Anal. Chem.* **68**(7) (1996) 1116–1123.
119. T. A. Dirksen, J. E. Gagnon, A sample-preparation technique for infrared analysis of aqueous samples, *Spectroscopy* **11**(2) (1996) 58–62.
120. J. P. Coates, Clarifying the issues. Further reflections on filter analysers in process IR spectroscopy, *Spectroscopy* **10**(9) (1995) 24–26.
121. P. A. Wilks, Filter analyzers vs. process FTIR spectrometers, continued, *Spectroscopy* **10**(9) (1995) 23.
122. E. W. Ciurczak, Questions and answers on process analysis. Tips for selecting and calibrating IR instruments, *Spectroscopy* **10**(9) (1995) 19–20.
123. J. S. Namkung, M. Hoke, R. S. Rogowski, S. Albin, FTIR optical fiber remote detection of aluminium hydroxide by evanescent wave absorption spectroscopy, *Appl. Spectrosc.* **49**(9) (1995) 1305–1310.
124. H. J. Luinge, J. H. van der Maas, T. Visser, Partial least squares regression as a multivariate tool for the interpretation of infrared spectra, *Chemom. Intell. Lab. Syst.* **28**(10) (1995) 129–138.
125. R. H. Wilson, J. Holland, J. Potter, Lining up for FTIR analysis. FTIR spectroscopy is no longer just a laboratory technique, it can now be used to provide 'on-line' quality and process control in industrial processes, *Chem. Br.* **30**(12) (1994) 993–996.
126. C. Klawun, C. L. Wilkins, Neural network assisted rapid screening of large infrared spectral databases, *Anal. Chem.* **67**(2) (1995) 374–378.
127. J. H. Hartshorn, Infrared discriminant analysis techniques for determining product quality, *Spectroscopy* **10**(3) (1995) 32–35.
128. M. Huehne, U. Eschenauer, H. W. Siesler, Performance and selected applications of an acousto-optic tunable-filter near infrared spectrometer, *Appl. Spectrosc.* **49**(2) (1995) 177–180.
129. D. L. Heglund, D. C. Tilotta, S. B. Hawthorne, D. J., Miller, Simple fiber-optic interface for on-line supercritical-fluid extraction-Fourier-transform infrared spectrometry, *Anal. Chem.* **66**(20) (1994) 3543–3551.
130. R. D. Driver, G. L. Dewey, D. A. Greenberg, J. D. Stark, "Sample interface in on-line process monitoring. Liquid- and gas-phase sampling of optically homogeneous processes, *Spectroscopy* **9**(4) (1994) 36–40.
131. S. Lowry, T. May, A. Bornstein, Y. Weissman, R. Harman, I. Tugenthaft, New accessory for characterizing optical fibers with an FTIR spectrometer, *Appl. Spectrosc.* 48(7) (1994) 852–856.
132. C. Dossi, S. Recchia, A. Fusi, Infrared analytical evidence of oxygen-induced formation of iron carbonyl from carbon monoxide and steel walls of an in situ DRIFT cell, *Ann. Chim. (Rome)* **84**(7–8) (1994) 347–352.
133. H. Kavak, R. Esen, Spectrum comparison of IR data taken from different spectrometers with various precision, *J. Chem. Inf. Comput. Sci.* **33**(4) (1993) 595–597.
134. T. B. Egevskaya, Portable 'double cat's eye' FTIR spectrometer, *Mikrochim. Acta, Suppl.* **14** (Progress in Fourier Transform Spectroscopy) (1997) 767–768.
135. R. W. Dibble, R. K. Mongia, Q.-v. Nguyen, Optical probe for in situ detection of hydrocarbon concentration, WO Patent No. 9725609, Issue Date:07/17/1997; Filing Date:01/09/1997; 21 pp.

136. S. J. Weeks, J. F. Mcclelland, S. L. Wright, R. W. Jones, Transient infrared spectroscopy for on-line analysis of solids and viscous liquids, *Mikrochim. Acta, Suppl*, **14** (Progress in Fourier Transform Spectroscopy) (1997) 801–802.
137. C. C. Alexay, W. L. Truett, C. D. Prozzo, B. O'Dwyer, Portable filter infrared spectrometer, Janos Technology Inc.; U.S. Patent No. 5,519,219; 05/21/96: 6 pp.
138. M. Komiyama, Y. Obi, Simple transmission infrared cell made of Swagelok for in situ infrared spectroscopic studies of heterogeneous catalysis, *Rev. Sci. Instrum.* **67**(4) (1996) 1590–2.
139. L. M. Treffman, P. W. Morrison, A simple method to fabricate lenses for in situ Fourier transform infrared spectroscopy, *Rev. Sci. Instrum.* **67**(4) (1996) 1454–7.
140. M. J. Persky, A review of spaceborne infrared Fourier transform spectrometers for remote sensing, *Rev. Sci. Instrum.* **66**(10) (1995) 4736–97.
141. P. Wilks, Portable infrared analyzers for repetitive analysis, *Am. Lab.* **28**(3) (1996) 52–53.
142. J. W. Childers, G. M. Russwurm, E. L. Thompson Jr., Quality assurance considerations in a long-term FTIR monitoring program, *Proc. SPIE-Int. Soc. Opt. Eng.* **2365** (Optical Sensing for Environmental and Process Monitoring) (1995) 389–95.
143. J. R. Engel, R. K. Dorval, D. L. Carlson, T. G. Quinn, Thomas G., Field portable Fourier transform infrared spectrometer system, *AT-ONSITE* **1**(1) (1995) 71–5.
144. C. Hirose, Vibration spectroscopy of surfaces using non-linear optical process: IR-visible sum frequency generation spectroscopy (VSFG), *Adv. Multi-Photon Processes Spectrosc.* **9** (1995) 145–97.
145. A. L. Cone, S. K. Farhat, L. A. Todd, Development of QA/QC performance standards for field use of open-path FTIR spectrometers, *Proc. SPIE-Int. Soc. Opt. Eng.* **2365** (Optical Sensing for Environmental and Process Monitoring) (1995) 334–8.
146. P. R. Griffiths, I. R. Lewis, N. C. Chaffin, N. W. Daniel Jr., J. D. Jegla, Remote characterization of materials by vibrational spectrometry through optical fibers, *J. Mol. Struct.* **347** (1995) 169–85.
147. J. Wang, J. Kag, T. Wang, The spectral characterizations of the low and high resolution remote FTIR emission spectroscopy, *Spectrosc. Lett.* **28**(6) (1995) 839–848.
148. C. G. Stevens, N. Thomas, P. Kuzmenko, T. Alger, An echelle grating spectrometer (EGS) for mid IR remote chemical detection, *Proc. SPIE-Int. Soc. Opt. Eng.* **2266** (Optical Spectroscopic Techniques and Instrumentation for Atmospheric and Space Research) (1994) 2–12.
149. X. H. Zhang, H. L. Ma, C. Blanchetiere, K. L. Foulgoc, J. Lucas, J. Heuze, P. Collardelle, P. Froissard, D. Picque, *et al.*, Tellurium halide (TeX) IR fibers for remote spectroscopy, *Proc. SPIE-Int. Soc. Opt. Eng.* **2131** (Biomendical Fiber Optic INstrumentation) (1994) 90–9.
150. R. H. Wilson, J. K. Holland, J. Potter, Lining up for FTIR analysis, *Chem. Br.* **30**(12) (1994) 993–6.
151. R. Simhi, Y. Gotshal, D. Bunimovich, B.-A. Sela, A. Katzir, Fiber-optic evanescent-wave spectroscopy for fast multicomponent analysis of human blood, *Sci. Total Environ*, **187** (1996) 3421–3425.
152. P. Franck, J.-L. Sallerin, H. Schroeder, M.-A. Gelot, P. Nabet, Rapid determination of faecal fat by Fourier-transform infrared analysis (FTIR) with partial least squaress regression and an attenuated total reflectance accessory, *Clin. Chem.* **42** (1996) 2015–2020.
153. I. Ahonen, H. Riipinen, A. Roos, Portable Fourier-transform infrared spectrometer for use as a gas analyser in industrial hygiene, *Analyst* **121** (1996) 1253–1255.
154. S. K. Cole, P. Martin, Determination of gas-phase sidestream cigarette smoke components using Fourier-transform infrared spectrometry, *Analyst* **121**(4) (1996) 495–500.

155. J. F. Stevens, P. Vadgama, Infrared analysis in clinical chemistry: its use in the laboratory and in non-invasive near patient testing, *Ann. Clin. Biochem.* **34**(3) (1997) 215–221.
156. L. P. Choo, D. L. Wetzel, W. C. Halliday, M. Jackson, S. M. LeVine, H. H. Mantsch, In situ characterization of beta-amyloid in Alzheimer's diseased tissue by synchrotron Fourier transform infrared microspectroscopy, *Biophys. J.* **71**(4) (1996) 1672–1679.
157. B. Pastrana-Rios, Lipid-lipid and lipid-protein interaction in the pulmonary surfactant system an in situ infrared spectroscopic test of the "squeeze-out" hypothesis (dipalmitoylphosphatidylcholine), Doctoral Dissertation, Rutgers, State Univ., Newark, NJ, 1996: 153 pp.; Diss. Abstr. Int., B 1996, 56(10), 5510; Univ. Microfilms Int. Order No. DA9603554.
158. G. Spanner, R. Niessner, Noninvasive determination of blood constituents using an array of modulated laser diodes and a photoacoustic sensor head, *Fresenius' J. Anal. Chem.* **355**(3–4) (1996) 327–328.
159. S. M. LeVine, D. L. Wetzel, In situ chemical analyses from frozen tissue sections by Fourier transform infrared microspectroscopy: Examination of white matter exposed to extravasated blood in the rat brain, *Am. J. Pathol.* **145**(5) (199) 1041–1047.
160. S. M. LeVine, D. L. Wetzel, In situ chemical analysis of brain tissue by Fourier transform infrared microspectroscopy, *NeuroProtocols* **5**(1) (1994) 72–79.
161. T. Wang, C. Zhu, J. Wang, F. Xu, Z. Chen, Y. Luo, Study on spectral characterization of infrared flare material combustion with remote high resolution Fourier transform infrared spectrometry, *Anal. Chim. Acta* **306**(2–3) (1995) 249–258.
162. Z. Stanek, Automating laboratory FTIR spectroscopy with a liquid analysis system, *Am. Lab.* **28**(8) (1996) 25–28.
163. J. M. Andrarde, S. Muniategui, P. Lopez-Maria, D. Prada, Use of multivariate techniques in quality control of kerosine production, *Fuel* **76**(1) (1997) 51–59.
164. R. D. Stephens, P. A. Mulawa, M. T. Giles, K. G. Kennedy, P. J. Groblicki, S. H. Cadle, K. T. Knapp, An experimental evaluation of remote sending-based hydrocarbon measurements: a comparison to FID measurements, *J. Chem. Inf. Comput. Sci.* **36**(1) (1996) 148–158.
165. A. E. Palomares, G. Eder-Mirth, J. A. Lercher, Selective alkylation of toluene over basic zeolites: an in situ infrared spectroscopic investigation, *J. Catal.* **168**(2) (1997) 442–449.
166. J. R. Markham, J. E. Cosgrove, C. M. Nelson, A. S. Bonanno, R. E. Schlief, M. A. Stoy, G. C. Glatzmaier, C. E. Bingham, A. A. Lewandowski, In situ FTIR monitoring of a solar flux induced chemical process, *J. Sol. Energy Eng.* **119**(3) (1997) 219–224.
167. C. G. Masi, Spectroscopic techniques peer into combustion-chamber chemistry, *R&D (Cahners Publishers)* **38**(13) (1996) 14–16.
168. Ph. Marteau, P. Tobaly, V. Ruffier-Meray, A. Barreau, In situ determination of high pressure phase diagrams of methane-heavy hydrocarbon mixtures using an infrared absorption method, *Fluid Phase Equilib.* **119**(1–2) (1996) 213–230.
169. K. Seinsche, F. Luigart, P. Bartl, Quality control of jet fuels. PLS (Partial least squares) regression analysis of gas chromatographic and IR spectroscopic data, *Erdoel, Erdgas, Kohle* **112**(6) (1996) 261–263.
170. R. K. Bade, K. L. Gallaher, S. B. Hunt, G. L. Combs, E. L. Kesselhuth, The on-line determination of total olefins in gasoline by process gas chromatography and Fourier transform infrared spectroscopy, *Adv. Instrum. Control* **51**(1) (1996) 33–42.
171. P. Landais, In situ examination of coal macerals oxidation by infrared microspectroscopy, Conf. Proc. — 7th Int. Conf. Coal Sci., K. H. Michaelian (Ed.) **2** (1993) 375–378.

172. P. A. Rosenthal, J. E. Cosgrove, J. R. Haigis, J. R. Markham, P. R. Solomon, S. Farquharson, P. W. Morrison Jr., S. D. Ridder, F. S. Biancaniello, FTIR process monitoring of metal powder temperature and size distribution, *Proc. SPIE-Int. Soc. Opt. Eng.* 2367(Optical Sensors for Environmental and Chemical Process Monitoring) (1995) 183–193.

173. Z. Bouhsain, S. Garrigues, M. de la Guardia, Clean method for the simultaneous determination of propyphenazone and caffeine in pharmaceuticals by flow-injection Fourier-transform infrared spectrometry, *Analyst* **122** (1997) 441–445.

174. D. E. Pivonka, K. M. Kirkland, Research strategy for the HPLC/FTIR analysis of drug metabolites, *Appl. Spectrosc.* **51** (1997) 866–873.

175. S. R. Erskine, B. M. Quencer, K. R. Beebe, Rapid optical purity determination using chiral spectroscopy, achiral spectroscopy, and multivariate analysis, *Appl. Spectrosc.* **49**(11) (1995) 1682–1691.

176. R. N. Landau, P. F. McKenzie, A. L. Forman, R. R. Dauer, M. Futran, A. D. Epstein, In situ Fourier-transform infrared and calorimetric studies of the preparation of a pharmaceutical intermediate, *Process Control Qual.* **7**(3–4) (1995) 133–142.

177. A. G. Severdia, Transmission FTIR spectroscopy applied to the determination of hydrocarbon residues on pharmaceutical process equipment, *Appl. Spectrosc.* **49**(4) (1995) 540–541.

178. R. N. Landau, P. F. McKenzie, A. L. Forman, R. R. Dauer, M. Futran, A. D. Epstein, In situ Fourier-transform infrared and calorimetric studies of the preparation of a pharmaceutical intermediate, *Process Control Qual.* **7**(3–4) (1995) 133–42.

179. D. F. Leclerc, R. M. Hogikyan, Rapid determination of effective alkali and dead-load concentrations in kraft liquors by attenuated total reflectance infrared spectrometry, *J. Pulp Pap. Sci.* **21**(7) (1995) J231–J237.

180. D. F. Leclerc, M. D. Ouchi, Rapid characterization of dirt specks in kraft pulp by infrared spectrometry, *J. Pulp Pap. Sci.* **22**(3) (1996) J112–J117.

181. D. Wang, Y. Yang, J. Zou, New correction method for FTIR on-line film thickness measurement, *Proc. SPIE-Int. Soc. Opt. Eng,* **2857** (Advanced Materials for Optics and Precision Structures) (1996) 12–20.

182. P. Damlin, C. Kvarnstroem, A. Ivaska, In situ external reflection Fourier-transform infrared spectroscopic study on the structure of the conducting polymer poly(paraphenylene), *Analyst* **121** (1996) 1881–1884.

183. D. E. Pivonka, K. Russell, T. Gero, Tools for combinatorial chemistry: in situ infrared analysis of solid-phase organic reactions, *Appl. Spectrosc.* **50** (1996) 1471–1478.

184. B. A. Kirsch, J. P. Chauvel, D. P. Denton, S. V. Lange, D. R. Lafevor, R. A. Bredeweg, W. K. Winnett, J. L. Dinaro, Improved process control in polyolefin polymerization by on-line Fourier transform infrared spectroscopy, *Process Control Qual.* **8** (1996) 75–83.

185. M. Nielsen, P. Jurasek, J. Hayashi, E. Furimsky, Formation of toxic gases during pyrolysis of polyacrylonitrile and nylons, *J. Anal. Appl. Pyrolysis* **35**(1) (1995) 43–51.

186. S. Jickells, Toxins tracked down in polymer packaging, *Lab. Equip. Dig.* **32**(6) (1994) 23.

187. S. W. Gaarenstroom, P. B. Coleman, C. M. Duranceau, Rapid identification of automotive plastics in dismantling operations. Evaluation of specular-reflectance infrared spectroscopy systems, *Soc. Automot. Eng. (Spec. Publ.)* SP, SP-1263(Design for Environmentally Safe Automotive Products and Processes) (1997) 47–54.

188. P. E. R. Mucci, Rapid identification of plastics using external beam mid infrared spectroscopy, *Spec. Publ. — R. Soc. Chem.* **199** (Chemical Aspects of Plastics Recycling) (1997) 53–70.

189. G. Zachmann, P. Turner, Rapid characterization of black polymeric material by mid infrared reflectance spectroscopy, *Spec. Publ. — R. Soc. Chem.* **199** (Chemical Aspects of Plastics Recycling) (1997) 71–76.
190. I. Rehman, C. Barnardo, R. Smith, Fourier transform infrared analysis of poly(ester-urethanes) at low temperature in situ by using a newly constructed liquid nitrogen cooled sample stage, *J. Mater. Sci.* **32**(10) (1997) 2617–2621.
191. I. Linossier, F. Gaillard, M. Romand, Use of evanescent waves FTIR spectroscopy for in situ study of the degradation of polymer-substrate systems, *Proc. Annu. Meet. Adhes. Soc.* **18** (1995) 86–88.
192. B.-J. Niu, M. W. Urban, Surfactant exudation and film formation in 50%/59% Sty/n-BA latex films; step-scan photoacoustic FTIR spectroscopic studies, *Polym. Mater. Sci. Eng.* **75** (1996) 41–42.
193. A. Saito, Y. Urai, K. Itoh, Infrared and Resonance Raman Spectroscopic Study on the Photopolymerization Process of the Langmuir-Blodgett Films of a Diacetylene Monocarboxylic Acid, 10,12-Pentacosadiynoic Acid, *Langmuir* **12**(16) (1996) 3938–3944.
194. T. Buffeteau, M. Pezolet, In situ study of photoinduced orientation in azopolymers by time-dependent polarization modulation infrared spectroscopy, *Appl. Spectrosc.* **50**(7) (1996) 948–955.
195. B. A. Kirsch, J. P. Chauvel, D. P. Denton, S. V. Lange, Improved process control in polyolefin polymerization by on-line Fourier transform infrared spectroscopy, *Process Control Qual.* **8**(2–3) (1996) 75–83.
196. B.-J. Niu, L. R. Martin, L. K. Tebelius, M. W. Urban, Latex film formation at surfaces and interfaces. Spectroscopic attenuated total reflectance and photoacoustic Fourier transform IR approaches, *ACS Symp. Ser.* **648** (Film Formation in Waterborne Coatings): (1996) 301–331.
197. K. W. Van Every, M. J. Elder, Polyolefin formulation validation using infrared spectrometric classification techniques, *J. Vinyl Addit. Technol.* **2**(3) (1996) 224–228.
198. P. Damlin, C. Kvarnstroem, A. Ivaska, In situ external reflection Fourier transform infrared spectroscopic study on the structure of the conducting polymer poly(para-phenylene), *Analyst* **121**(12) (1996) 1881–1884.
199. O. Wollersheim, J. Hormes, Quantitative in situ FTIR studies of polymers irradiated with synchrotron radiation, *Chem. Phys.* **204**(1) (1996) 129–134.
200. H. H. Yu, L. J. Fina, Electric field-induced dipole reorientation in oriented nylon 11 by in situ infrared spectroscopy, *J. Polym. Sci., Part B: Polym. Phys.* **34**(4) (1996) 781–788.
201. A. Khettry, In-line monitoring of molten polymeric processes (infrared spectroscopy, fiber optics), Doctoral Dissertation, Univ. of Tennessee, Knoxville, TN, USA; 1995 PAGES: 167 pp.; Diss. Abstr. Int., B 1995, 56(8), 4460 AVAIL: Univ. Microfilms Int., Order No. DA9540099.
202. J. Mijovic, S. Andjelic, J. M. Kenny, In situ real-time monitoring of epoxy/amine kinetics by remote near infrared spectroscopy, *Polym. Adv. Technol.* **7**(1) (1996) 1–16.
203. Z. Ge, R. Thompson, S. Cooper, D. Ellison, P. Tway, Quantitative monitoring of an epoxidation process by Fourier transform infrared spectroscopy, *Process Control Qual.* **7**(1) (1995) 3–12.
204. B. Lutz, H.-J. Luinge, J. van der Maas, R. van Agen, A low-cost external reflection unit for rapid analysis of carbon-filled rubbers, *Proc. SPIE-Int. Soc. Opt. Eng.* **2089** (9th International Conference on Fourier Transform Spectroscopy) (1993) 546–547.
205. R. A. Fidler, Real-time analysis of PC/ABS blend composition via process-FTIR spectroscopy, Compalloy '93, *Proc. 8th Int. Congr. Compat. React. Polym. Alloying* (1993) 119–131.

206. M. Tamada, H. Omichi, N. Okui, Real-time in situ observation of vapor deposition polymerization of N-methylolacrylamide with IR-RAS, *Thin Solid Films* **260**(2) (1995) 168–173.
207. H. W. Siesler, Vibrational spectroscopy of polymers. Analysis, physics, and process control, *Adv. Chem. Ser.* **236** (Structure-Property Relations in Polymers) (1993) 41–87.
208. S. Liu, P. Solomon, R. Carpio, B. Fowler, D. Simmons, J. Wang, R. Wise, G. Imper, N. B. Riley, *et al.*, Modeling, simulation and control of single wafer process in cluster tool base on FTIR-in-line sensor, *Mater. Res. Soc. Symp. Proc.* **389** (Modeling and Simulation of Thin-Film Processing) (1995) 269–274.
209. Q. Fan, In situ FTIR studies of semiconductor/electrolyte interface (cadmium telluride, silicon), Doctoral Dissertation, Cleveland State Univ., Cleveland, OH, USA 1995; 132 pp.; Diss. Abstr. Int., B 1995, 56(6), 3163; Univ. Microfilms Int., Order No. DA9535860.
210. Z. Ping, G. E. Nauer, In situ FTIR-ATR spectroscopic investigations on the redox behavior of poly(thienylpyrrole) thin film electrodes in non-aqueous solutions, *Synth. Met.* **84**(1–3) (1997) 843–844.
211. V. Climent, A. Rodes, J. M. Orts, J. M. Feliu, J. M. Perez, A. Aldaz, On the electrochemical and in situ Fourier transform infrared spectroscopy characterization of urea adlayers at Pt(100) electrodes, *Langmuir* **13**(8) (1997) 2380–2389.
212. S. Farquharson, K. Kinsella, J. R. Markham, P. R. Solomon, M. D. Carangelo, J. R. Haigis, N. M. Ravindra, F.-M. Tong, M. J. Bevan, G. H. Westphal, Real-time process control of molecular beam epitaxial growth of mercury cadmium telluride films by Fourier transform infrared spectroscopy, *Transient Therm. Process. Tech. Electron. Mater. Proc. Symp.* (1996) 133–136.
213. R. P. Sperline, J. S. Jeon, S. Raghavan, FTIR-ATR analysis of the silicon/aqueous solution interface using sputtered silicon thin films to access the 1550–1100 cm^{-1} spectral region, *Appl. Spectrosc.* **49**(8) (1995) 1178–1182.
214. D. Papapanayiotou, In situ infrared spectroscopic studies of electrodeposition additive adsorption on copper, Doctoral Dissertation, Univ. of Illinois, Urbana, IL, USA, 1997, 184 pp., Diss. Abstr. Int., B 1997, 57(12), 7630 AVAIL: Univ. Microfilms Int., Order No. DA9717324.
215. S. Liu, J. R. Haigis, M. B. DiTaranto, K. Kinsella, J. R. Markham, Q. Li, D. B. Fenner, P. R. Solomon, S. Farquharson, *et al.*, "Process monitoring and control of integrated circuit manufacturing using Fourier transform infrared spectroscopy, *Proc. SPIE-Int. Soc. Opt. Eng.* **2367** (Optical Sensors for Environmental and Chemical Process Monitoring): (1995) 171–182.
216. Q. Fan, L. M. Ng, In situ Fourier transform infrared-attenuated total reflection spectroscopy monitoring of polyaniline synthesis mechanism on the p-type silicon electrode, *J. Vac. Sci. Technol.*, **A14**(3, Pt. 2) (1996) 1326–1329.
217. F. Qin, E. E. Wolf, Spatially resolved infrared spectroscopy: a novel technique for in situ study of spatial surface coverage during CO oxidation on supported catalysts, *Catal. Lett.* **39**(1,2) (1996) 19–25.
218. N. Fujino, I. Karino, M. Ohmori, M. Yasutake, S. Wakiyama, Method and apparatus for analyzing minute foreign substances, and process for manufacturing semiconductor or LCD elements, Seiko Instruments Inc.; Mitsubishi Denki Kabushiki Kaisha; EP Patent No. 727659A2; Issue Date:08/21/1996; Filing Date:02/14/1996; 24 pp. And also as EP Patent No. 727659A3; Issue Date:03/04/1998; Filing Date:02/14/1996.
219. G. R. Howes, The application of Fourier transform infrared spectroscopy to the optimization of chemically-amplified photoresist processing, Doctoral Dissertation, Univ. of Wisconsin, Madison, WI, USA; 1996; 192 pp.; Diss. Abstr. Int., B 1996, 57(5), 3175.

220. J. S. Namkung, M. Hoke, R. S. Rogowski, S. Albin, Optical fiber FTIR evanescent wave absorption spectroscopy of natural aluminum corrosion," Department Physics, College William and Mary, Williamsburg, VA, 23185, USA *Proc. SPIE-Int. Soc. Opt. Eng.* **2883** (Optical Remote Sensing for Environmental and Process Monitoring) (1996) 655–669.
221. Q. Pan, C. Pu, E. S. Smotkin, In situ Fourier transform infrared-diffuse reflection spectroscopy of direct methanol fuel cell anodes and cathodes, *Dep. Chem. Environ. Eng., Illinois Inst. Technol.* **143**(10) (1996) 3053–3057.
222. J. Onoe, K. Takeuchi, In situ high-resolution infrared spectroscopy of a photopolymerized C60 film, *The Inst. of Physical and Chemical Research (RIKEN)*, Saitama, Japan, 351–01 *Phys. Rev. B:* Condens. Matter **54**(9) (1996) 6167–6171.
223. S. M. Stephens, R. A. Dluhy, In situ and ex-situ structural analysis of phospholipid-supported planar bilayers using infrared spectroscopy and atomic force microscopy, *Thin Solid Films* 284–285 (1996) 381–386.
224. Y. V. Pan, E. Z. Barrios, D. D. Denton, In situ infrared spectroscopy as a real time diagnostic for plasma polymer film deposition, *Appl. Phys. Lett.* **68**(24) (1996) 3386–3388.
225. J. A. Mielczarski, Z. Xu, J. M. Cases, Qualitative and Quantitative Evaluation of Heterogeneous Adsorbed Monolayers on Semiconductor Electrode by Infrared Reflection Spectroscopy, *J. Phys. Chem.* **100**(17) (1996) 7181–7184.
226. F. Ozanam, A. Djebri, J.-N. Chazalviel, The hydrogenated silicon surface in organic electrolytes probed through in situ IR spectroscopy in the ATR geometry, *Electrochim. Acta* **41**(5) (1996) 687–692.
227. A. E. Russell, L. Rubasingham, P. L. Hagans, T. H. Ballinger, Cell and optics design for in situ far-infrared spectroscopy of electrode surfaces, *Electrochim. Acta* **41**(5) (1996) 637–640.
228. S. Salim, In situ Fourier transform infrared spectroscopy of chemistry and growth in chemical vapor deposition, Doctoral Dissertation, Massachusetts Institute of Technology, Cambridge, MA, USA Diss. Abstr. Int., B. **56**(7) (1995) 3894.
229. Q. Fan, C. Pu, K. L. Ley, E. S. Smotkin, In situ FTIR-diffuse reflection spectroscopy of the anode surface in a direct methanol/oxygen fuel cell, *J. Electrochem. Soc.* **143**(2) (1996) L21–L23.
230. B. Beden, On the use of 'in situ' UV-visible and infrared spectroscopic techniques for studying corrosion products and corrosion inhibitors, *Mater. Sci. Forum,* **192–194** (Pt. 1, Electrochemical Methods in Corrosion Research V) (1995) 277–290.
231. I. T. Bae, M. Sandifer, Y. W. Lee, D. A. Tryk, C. N. Sukenik, D. A. Scherson, In Situ Fourier Transform Infrared Spectroscopy of Molecular Adsorbates at Electrode-Electrolyte Interfaces: A Comparison between Internal and External Reflection Modes, *Anal. Chem.* **67**(24) (1995) 4508–4513.
232. D. Olligs, U. Stimming, J. Stumper, On the mechanism of CO-electrooxidation on Pt-Ru alloys. An in situ infrared spectroscopic study, *New Mater. Fuel Cell Syst.* I, Proc. 1st Int. Symp., O. Savadogo, P. R. Roberge, T. N. Veziroglu, (Eds) (1995) 677–687.
233. T. Wadayama, Y. Maiwa, H. Shibata, A. Hatta, In situ IR spectroscopic study of the reaction of dimethylaluminum hydride with photochemically deposited amorphous silicon," J*pn. J. Appl. Phys.*, Part 2. **34**(6B) (1995) L779–L781.
234. B. J. Kip, E. A. T. Peters, J. Happel, T. Huth-Fehre, F. Kowol, Identifying recyclable carpet materials using a hand-held infrared spectrometer, Dsm N. V.; Institut Fuer Chemo- Und Biosensorik; Kip, Berend Jan; Peters, Edo Augustinus Titus; Happel, Jens; Huth-Fehre, Thomas; Kowol, Frank; WO Patent No. 9702481 A1; DATE: 01/23/97; 33 pp.

235. D. Lorente, N. Aleixos, J. Gómez-Sanchis, S. Cubero, O. L. García-Navarrete, J. Blasco, Recent Advances and Applications of Hyperspectral Imaging for Fruit and Vegetable Quality Assessment, *Food and Bioprocess Technology*, **5**(4) (2012) 1121–1142.
236. S. G. Kazarian and K. L. Andrew Chan, ATR-FTIR spectroscopic imaging: recent advances and applications to biological systems, *Analyst* **138** (2013) 1940–1951.
237. B. J. Eggleton, B. Luther-Davies, K. Richardson, Chalcogenide photonics, *Nature Photonics* **5** (2011) 141–148.
238. M. Trchová and J. Stejskal, Polyaniline: The infrared spectroscopy of conducting polymer nanotubes (IUPAC Technical Report), *Pure and Applied Chemistry* **83**(10) (2011) 1803–1817.
239. USP 29 (2005) General Chapter <197>, "Spectrophotometric Identification Tests. United States Pharmacopeia and National Formulary (USP 29-NF 24). Rockville, MD: United States Pharmacopeia Convention; 2979.
240. USP 29 (2005) General Chapter <851>, "Spectrophotometry and Light Scattering. United States Pharmacopeia and National Formulary (USP 29-NF 24). Rockville, MD: United States Pharmacopeia Convention; 2979.
241. International Organization for Standardization, ISO Central Secretariat, 1, ch. de la Voie-Creuse, CP 56-CH-1211 Geneva 20, Switzerland.
242. ISO Link for search using "Infrared Methods", 34 Methods (2014). http://www.iso.org/iso/home/search.htm?qt=infrared+method&sort=rel&type=simple&published=on
243. F. R. Van de Voort, J. Sedman, S. T. Sherazi, Improved determination of isolated trans isomers in edible oils by Fourier transform infrared spectroscopy using spectral reconstitution, *J. AOAC Int.* **90**(2) (2007) 446–51.
244. Association of Official Analytical Chemists (AOAC), Official Methods for Analysis. 18th edn. (AOAC International, 481 N. Frederick Avenue, Suite 500, Gaithersburg, MD, USA 2005).
245. American Society for Testing and Materials (ASTM) International search for "Infrared Spectroscopy", 182 results (2014). http://www.astm.org/search/fullsite-search.html?query=Mid Infrared.
246. ASTM Practices for IR Spectroscopy, Annual *Book* of *ASTM Standards*, Volume 03.06, 2014. ASTM International, 100 Barr Harbor Drive, PO Box C700, West Conshohocken, PA, 19428–2959 USA.

Chapter 8

Infrared Reference Material Spectra and Tables

Contents

8.1 Standard Reference Materials (SRMs) for Photometric Accuracy Measurements

The National Institute for Standards and Technology (NIST) previously known as the National Bureau of Standards (NBS) provides materials designed as reference sources to verify the transmittance and absorbance scale accuracy of spectrophotometers. The materials are termed SRMs or Standard Reference Materials. In the case of SRM 930D, a set of three individual glass filters are provided in cuvette-like black metal holders. In addition, one empty filter holder is provided for use as a reference sample.[1–3]

8.1.1 *The Meaning of Photometric Accuracy*

Definition:

Photometric accuracy is a formal term used to describe how well a measuring device (spectrophotometer or illuminance meter) is able to determine the total energy flux transmitted through (or reflected from) a Standard Reference Material (SRM). As stated in volume 14.01 of the American Society for Testing and Materials (ASTM) International[1], the measurement of the photometric accuracy for an instrument is obtained as follows:

> **20.1** "Select the appropriate Standard Reference Material and obtain ten successive readings of the apparent absorbance or transmittance at the specified wavelength. Average the ten readings. The photometric accuracy is the difference between the true absorbance or transmittance value and the average observed value."

It is further stated that the following apply to the reporting of photometric accuracy.

> **20.3** "Report the photometric accuracy in the following order: reference material, wavelength, true absorbance or transmittance, observed absorbance or transmittance plus or minus the standard deviation."

8.1.2 *Example of Reporting Photometric Accuracy:*

As an example the stated specifications for photometric accuracy for a test instrument might be written as:

Photometric Accuracy
±0.002 Au (for range of 0.0 to 0.5 Absorbance units)
±0.004 Au (for range of 0.5 to 1.0 Absorbance units)
±0.3% T
Measured with NIST 930D (or 930d) filters

Here is an evaluation of these specifications in greater detail. A photometric accuracy of "0.002 Au at 0 to 0.5 Au" indicates that after the instrument has been set to zero, a standard sample which transmits between 30.1%T (i.e., 0.5 Au) and 100%T (i.e., 0.0 Au) measures within ±0.002 Au as compared to the specified value for that reference standard material at any wavelength position. Tables 8.1 and 8.2 are used to convert from transmittance or reflectance scale to absorbance scale. This issue of how to explain a specification such as ±0.004 at 0.5 to 1.0 Au, when using the NIST 930D filter set has been brought forward in competitive sales situations between instrument manufacturers. The tolerance for each of the 930d filters is stated by NIST as ±0.5 percent relative. Thus, a 10%T filter (which corresponds to 0.1 T or 1.0 Au) neutral density 930D-type reference standard would be expected to be within the range of 9.95 to 10.05%T. These values are equivalent to 1.0022 Au to 0.9978 Au. The difference in Au is 1.0022 Au minus 0.9978 Au equals 0.0044 Au. The standard tolerance in Au is 1.0000 ± 0.0022 Au, rounded to 1.000 ± 0.002 Au.

Table 8.1 Relationship of %T and absorbance unit (Au) values.

T (as ppm)	%T	T	1/T	Au
10000000	1000	10	$0.1 = 10^{-1}$	−1.0
1000000	100	1	$1.0 = 10^{0}$	0.0
316230	31.6	0.316	3.16	0.5
100000	10.0	0.1	$10 = 10^{1}$	1.0
10000	1	0.01	$100 = 10^{2}$	2.0
1000	0.1	0.001	$1000 = 10^{3}$	3.0
100	0.01	0.0001	$10000 = 10^{4}$	4.0
10	0.001	0.00001	$100000 = 10^{5}$	5.0
1	0.0001	0.000001	$1000000 = 10^{6}$	6.0

Table 8.2 Expanded table showing the relationship of %T and %R to absorbance unit (Au) values.

Transmittance or reflectance	Percent T or R	Absorbance (Au)
1.000	100.0	0.000000
0.995	99.5	0.002177
0.990	99.0	0.004365
0.985	98.5	0.006564
0.980	98.0	0.008774
0.975	97.5	0.010995
0.970	97.0	0.013228
0.965	96.5	0.015473
0.960	96.0	0.017729
0.955	95.5	0.019997
0.950	95.0	0.022276
0.945	94.5	0.024568
0.940	94.0	0.026872
0.935	93.5	0.029188
0.930	93.0	0.031517
0.925	92.5	0.033858
0.920	92.0	0.036212
0.915	91.5	0.038579
0.910	91.0	0.040959
0.905	90.5	0.043351

(*Continued*)

Table 8.2 (*Continued*)

Transmittance or reflectance	Percent T or R	Absorbance (Au)
0.900	90.0	0.045757
0.895	89.5	0.048177
0.890	89.0	0.050610
0.885	88.5	0.053057
0.880	88.0	0.055517
0.875	87.5	0.057992
0.870	87.0	0.060481
0.865	86.5	0.062984
0.860	86.0	0.065502
0.855	85.5	0.068034
0.850	85.0	0.070581
0.845	84.5	0.073143
0.840	84.0	0.075721
0.835	83.5	0.078314
0.830	83.0	0.080922
0.825	82.5	0.083546
0.820	82.0	0.086186
0.815	81.5	0.088842
0.810	81.0	0.091515
0.805	80.5	0.094204
0.800	80.0	0.096910
0.795	79.5	0.099633
0.790	79.0	0.102373
0.785	78.5	0.105130
0.780	78.0	0.107905
0.775	77.5	0.110698
0.770	77.0	0.113509
0.765	76.5	0.116339
0.760	76.0	0.119186
0.755	75.5	0.122053
0.750	75.0	0.124939
0.745	74.5	0.127844
0.740	74.0	0.130768
0.735	73.5	0.133713
0.730	73.0	0.136677
0.725	72.5	0.139662
0.720	72.0	0.142668
0.715	71.5	0.145694
0.710	71.0	0.148742
0.705	70.5	0.151811
0.700	70.0	0.154902
0.695	69.5	0.158015
0.690	69.0	0.161151
0.685	68.5	0.164309

(*Continued*)

Table 8.2 (*Continued*)

Transmittance or reflectance	Percent T or R	Absorbance (Au)
0.680	68.0	0.167491
0.675	67.5	0.170696
0.670	67.0	0.173925
0.665	66.5	0.177178
0.660	66.0	0.180456
0.655	65.5	0.183759
0.650	65.0	0.187087
0.645	64.5	0.190440
0.640	64.0	0.193820
0.635	63.5	0.197226
0.630	63.0	0.200659
0.625	62.5	0.204120
0.620	62.0	0.207608
0.615	61.5	0.211125
0.610	61.0	0.214670
0.605	60.5	0.218245
0.600	60.0	0.221849
0.595	59.5	0.225483
0.590	59.0	0.229148
0.585	58.5	0.232844
0.580	58.0	0.236572
0.575	57.5	0.240332
0.570	57.0	0.244125
0.565	56.5	0.247952
0.560	56.0	0.251812
0.555	55.5	0.255707
0.550	55.0	0.259637
0.545	54.5	0.263603
0.540	54.0	0.267606
0.535	53.5	0.271646
0.530	53.0	0.275724
0.525	52.5	0.279841
0.520	52.0	0.283997
0.515	51.5	0.288193
0.510	51.0	0.292430
0.505	50.5	0.296709
0.500	50.0	0.301030
0.495	49.5	0.305395
0.490	49.0	0.309804
0.485	48.5	0.314258
0.480	48.0	0.318759
0.475	47.5	0.323306
0.470	47.0	0.327902
0.465	46.5	0.332547

(*Continued*)

Table 8.2 (*Continued*)

Transmittance or reflectance	Percent T or R	Absorbance (Au)
0.460	46.0	0.337242
0.455	45.5	0.341989
0.450	45.0	0.346787
0.445	44.5	0.351640
0.440	44.0	0.356547
0.435	43.5	0.361511
0.430	43.0	0.366532
0.425	42.5	0.371611
0.420	42.0	0.376751
0.415	41.5	0.381952
0.410	41.0	0.387216
0.405	40.5	0.392545
0.400	40.0	0.397940
0.395	39.5	0.403403
0.390	39.0	0.408935
0.385	38.5	0.414539
0.380	38.0	0.420216
0.375	37.5	0.425969
0.370	37.0	0.431798
0.365	36.5	0.437707
0.360	36.0	0.443697
0.355	35.5	0.449772
0.350	35.0	0.455932
0.345	34.5	0.462181
0.340	34.0	0.468521
0.335	33.5	0.474955
0.330	33.0	0.481486
0.325	32.5	0.488117
0.320	32.0	0.494850
0.315	31.5	0.501689
0.310	31.0	0.508638
0.305	30.5	0.515700
0.300	30.0	0.522879
0.295	29.5	0.530178
0.290	29.0	0.537602
0.285	28.5	0.545155
0.280	28.0	0.552842
0.275	27.5	0.560667
0.270	27.0	0.568636
0.265	26.5	0.576754
0.260	26.0	0.585027
0.255	25.5	0.593460
0.250	25.0	0.602060
0.245	24.5	0.610834

(*Continued*)

Table 8.2 (*Continued*)

Transmittance or reflectance	Percent T or R	Absorbance (Au)
0.240	24.0	0.619789
0.235	23.5	0.628932
0.230	23.0	0.638272
0.225	22.5	0.647817
0.220	22.0	0.657577
0.215	21.5	0.667562
0.210	21.0	0.677781
0.205	20.5	0.688246
0.200	20.0	0.698970
0.195	19.5	0.709965
0.190	19.0	0.721246
0.185	18.5	0.732828
0.180	18.0	0.744727
0.175	17.5	0.756962
0.170	17.0	0.769551
0.165	16.5	0.782516
0.160	16.0	0.795880
0.155	15.5	0.809668
0.150	15.0	0.823909
0.145	14.5	0.838632
0.140	14.0	0.853872
0.135	13.5	0.869666
0.130	13.0	0.886057
0.125	12.5	0.903090
0.120	12.0	0.920819
0.115	11.5	0.939302
0.110	11.0	0.958607
0.105	10.5	0.978811
0.100	10.0	1.000000
0.095	9.50	1.022276
0.090	9.00	1.045757
0.085	8.50	1.070581
0.080	8.00	1.096910
0.075	7.50	1.124939
0.070	7.00	1.154902
0.065	6.50	1.187087
0.060	6.00	1.221849
0.055	5.50	1.259637
0.050	5.00	1.301030
0.045	4.50	1.346787
0.040	4.00	1.397940
0.035	3.50	1.455932
0.030	3.00	1.522879
0.025	2.50	1.602060

(*Continued*)

Table 8.2 (*Continued*)

Transmittance or reflectance	Percent T or R	Absorbance (Au)
0.020	2.00	1.698970
0.015	1.50	1.823909
0.010	1.00	2.000000
0.009	0.90	2.045757
0.008	0.80	2.096910
0.007	0.70	2.154902
0.006	0.60	2.221849
0.005	0.50	2.301030
0.004	0.40	2.397940
0.003	0.30	2.522879
0.002000	0.2000	2.698970
0.001500	0.1500	2.823909
0.001000	0.1000	3.000000
0.000100	0.0100	4.000000
0.000010	0.0010	5.000000
0.000001	0.0001	6.000000

Conservatively for transmittance we state 10.00 ± 0.03%T; and related to 0.0 Au or 100%T, we specify 100.0 ± 0.3%T.

If the stated tolerance for the NIST 930D filter is ±0.002 Au at 1.0 Au, we are stating that our maximum deviation in photometric accuracy for our instrument at 1.0 Au is also ±0.002 Au. For the worst case measurement for the NIST filter, we would have the filter error of ±0.002 added to the instrument error of ±0.002, to result in a total maximum variation ±0.004 Au error (This is our stated specification for the example given).

And one can readily calculate the reflectance of transmittance of any measurement given the absorbance of that measurement as follows. For conversion of Absorbance to Transmittance we use:

$$A = -\log_{10}\left(\frac{I}{I_0}\right) = \log_{10}\left(\frac{I_0}{I}\right) \tag{8.1}$$

$$\frac{I}{I_0} = 10^{-A} \Rightarrow \frac{I_0}{I} = \frac{1}{10^{-A}} \tag{8.2}$$

$$\frac{I}{I_0} = 10^{-A} = T \tag{8.3}$$

$$\%\text{T} = \text{T} \times 100 \tag{8.4}$$

To convert Absorbance to Reflectance we use:

$$A = -\log_{10}\left(\frac{I}{I_0}\right) = \log_{10}\left(\frac{I_0}{I}\right) \tag{8.5}$$

$$\frac{I}{I_0} = 10^{-A} \Rightarrow \frac{I_0}{I} = \frac{1}{10^{-A}} \tag{8.6}$$

$$\frac{I}{I_0} = 10^{-A} = R \tag{8.7}$$

$$\%\mathrm{R} = \mathrm{R} \times 100 \tag{8.8}$$

8.2 NIST Wavenumber Standard Reference Material for Infrared

The available NIST standard for infrared measurements is named NIST SRM 1921b, IR Transmission Wavelength (Polystyrene Film), and comes as a card mounted

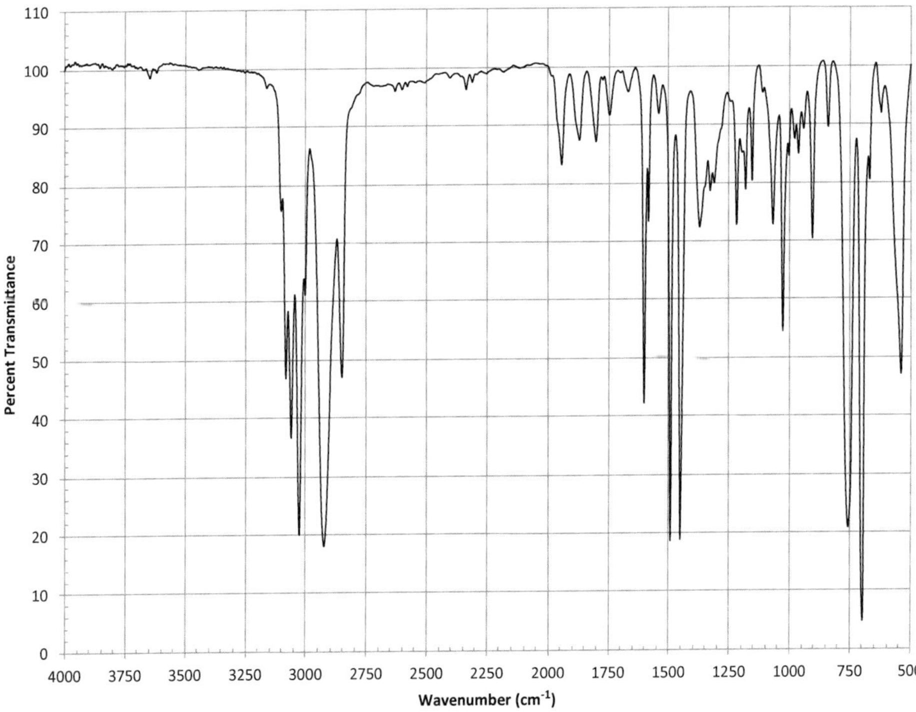

Fig. 8.1 Polystyrene film measured in transmittance with film holder and 4 cm^{-1} resolution. 64 co-added scans per spectrum, Nicolet Model 510 FTIR Spectrometer.

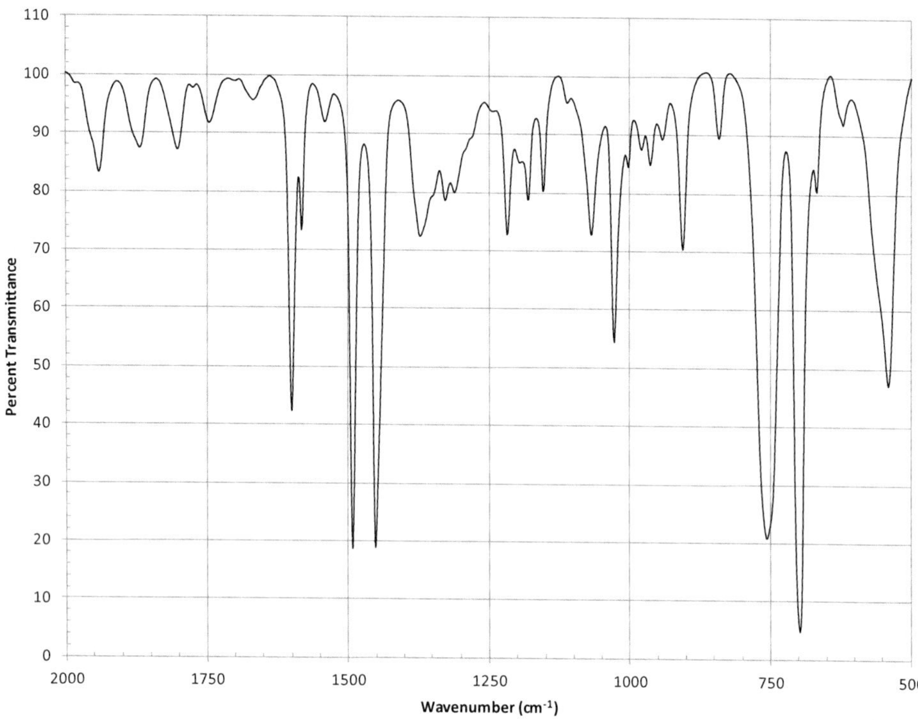

Fig. 8.2 Close-up of extended fingerprint region from Figure 8.1; Polystyrene film measured in transmittance with film holder and 4 cm^{-1} resolution. 64 co-added scans per spectrum, Nicolet Model 510 FTIR Spectrometer.

polymer film. This standard is useful for measurements from 3 to 18 microns, 3000 to 18000 nm, or 3333.3 to 555.6 cm^{-1} (wavenumbers).[4–7] See this link for more information: https://www-s.nist.gov/srmors/detail.cfm

Polystyrene is also measured in the near infrared and Volume 3, Chapter 8 describes various reflectance measurements of polystyrene in greater detail. The Figure 8.4 below shows 1.0 mm thick polystyrene measured in transmittance and converted to absorbance as $A = \log_{10}(1/T)$ for 2, 4, 8, 16, and 32 wavenumber resolution using an FTIR instrument over the near infrared range. The peak tables for each resolution are given in Table 8.4. So for infrared and near infrared polystyrene is an excellent and stable material for verification of the x-axis (i.e., wavelength or wavenumber) registration.

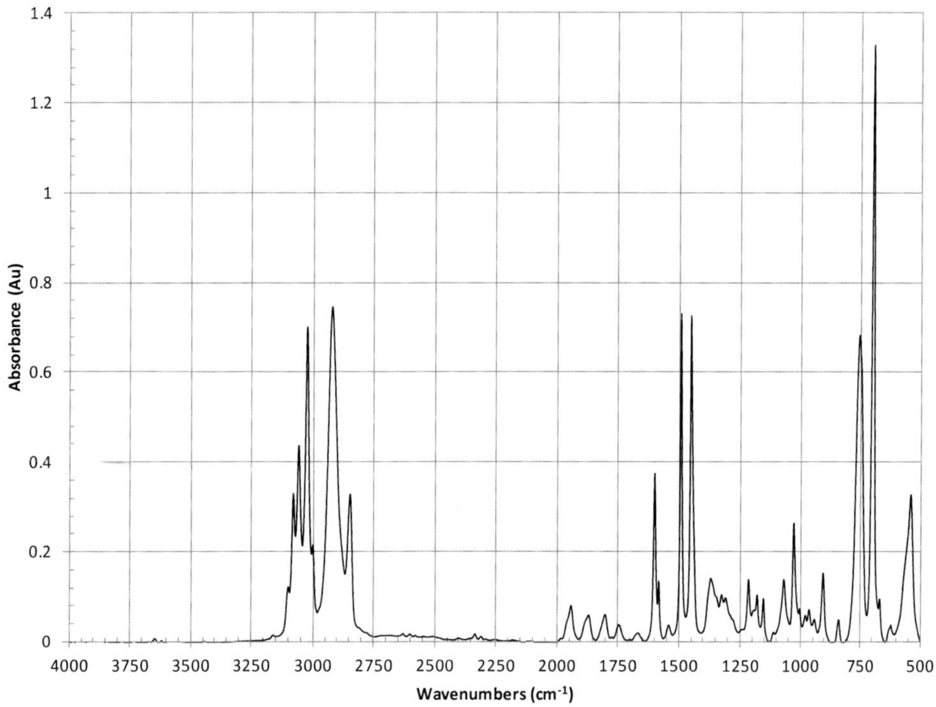

Fig. 8.3 Polystyrene film converted to absorbance from transmittance spectrum with film holder and $4\,cm^{-1}$ resolution. 64 co-added scans per spectrum, Nicolet Model 510 FTIR Spectrometer.

Table 8.3 Polystyrene Film Peak Table from Sample in Figures 8.1 and 8.2. Peak selection made using a five-point cubic spline, peak interpolation method on data with $4\,cm^{-1}$ resolution.[8]

Peak position (cm^{-1})	%T	NIST certified peak position*
3100.9744	75.6287	
3081.6898	46.8141	3082.19
3058.5482	36.6493	3060.03
3025.7642	19.9316	3026.42
3000.6942	61.1944	3001.40
2923.5555	17.9556	
2848.3453	47.0112	2850.13
1941.9659	83.2527	
1870.6127	87.4307	
1803.1163	87.1904	
1747.1908	91.7715	

(Continued)

Table 8.3 (*Continued*)

Peak Position (cm^{-1})	%T	NIST certified peak position*
1600.6273	42.2176	1601.35
1583.2711	73.4623	1583.13
1540.8449	91.9904	
1492.6332	18.6207	
1452.1354	18.9335	
1371.1398	72.4310	
1328.7135	78.6375	
1311.3573	79.9875	
1218.7909	72.8249	
1195.6493	85.1615	
1182.1501	78.8698	
1155.1515	80.3401	1154.64
1068.3705	72.8427	1069.20
1027.8727	54.4572	1028.35
1002.8027	84.5943	
979.6611	87.5156	
964.2333	84.9182	
943.0202	89.2676	
906.3793	70.3824	906.82
840.8115	89.5645	842.08
755.9589	20.7826	
698.1049	4.7028	
667.2495	80.4758	
620.9663	91.9022	
539.9707	47.0627	545.48

*NIST measurements are at made using 0.5 cm^{-1} resolution setting, with center of mass peak picking algorithm.

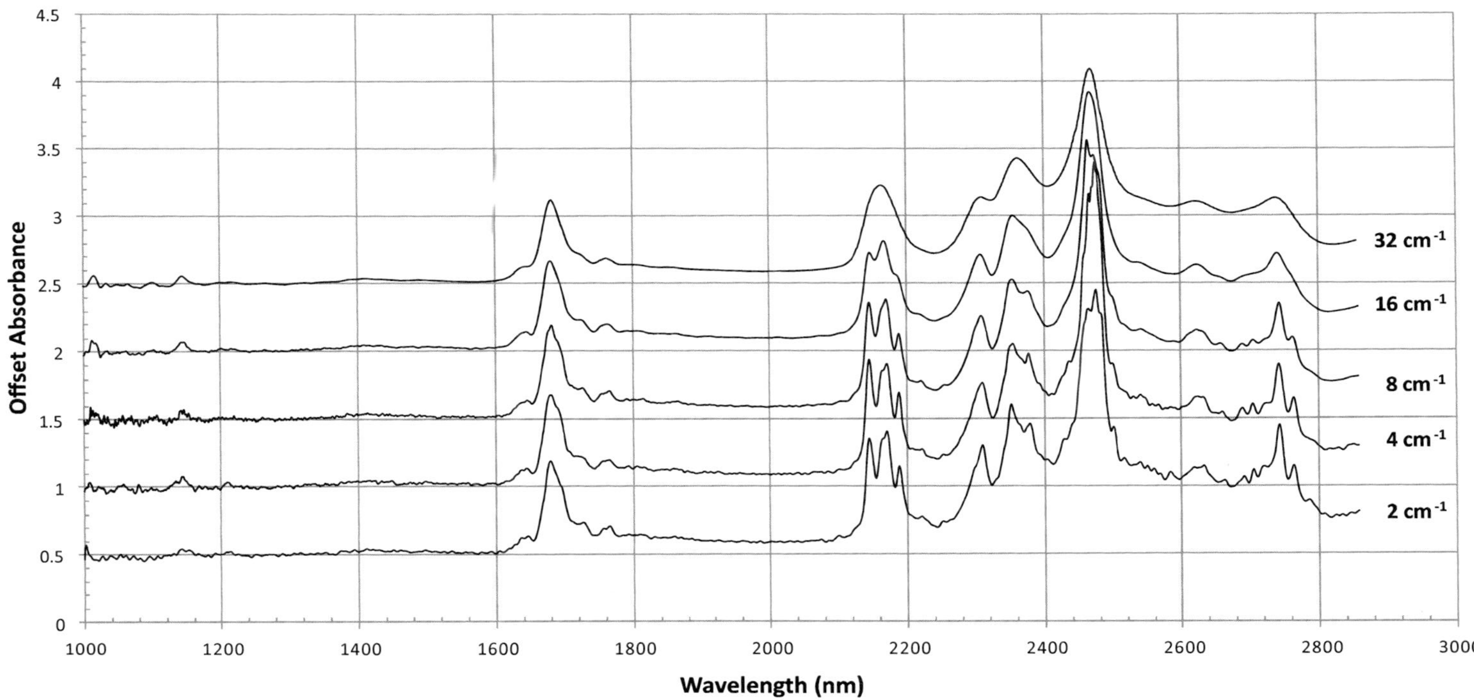

Fig. 8.4 Measurement of 1.0 mm thick polystyrene measured in transmittance and converted to absorbance as $A = \log_{10}(1/T)$ for 2, 4, 8, 16, and 32 wavenumber resolution using an FTIR instrument over the near infrared range. Savitzky-Golay smoothing, with 15 convolution points Quadratic-Cubic used for 2 and 4 wavenumber spectra; no smcothing for remaining spectra.

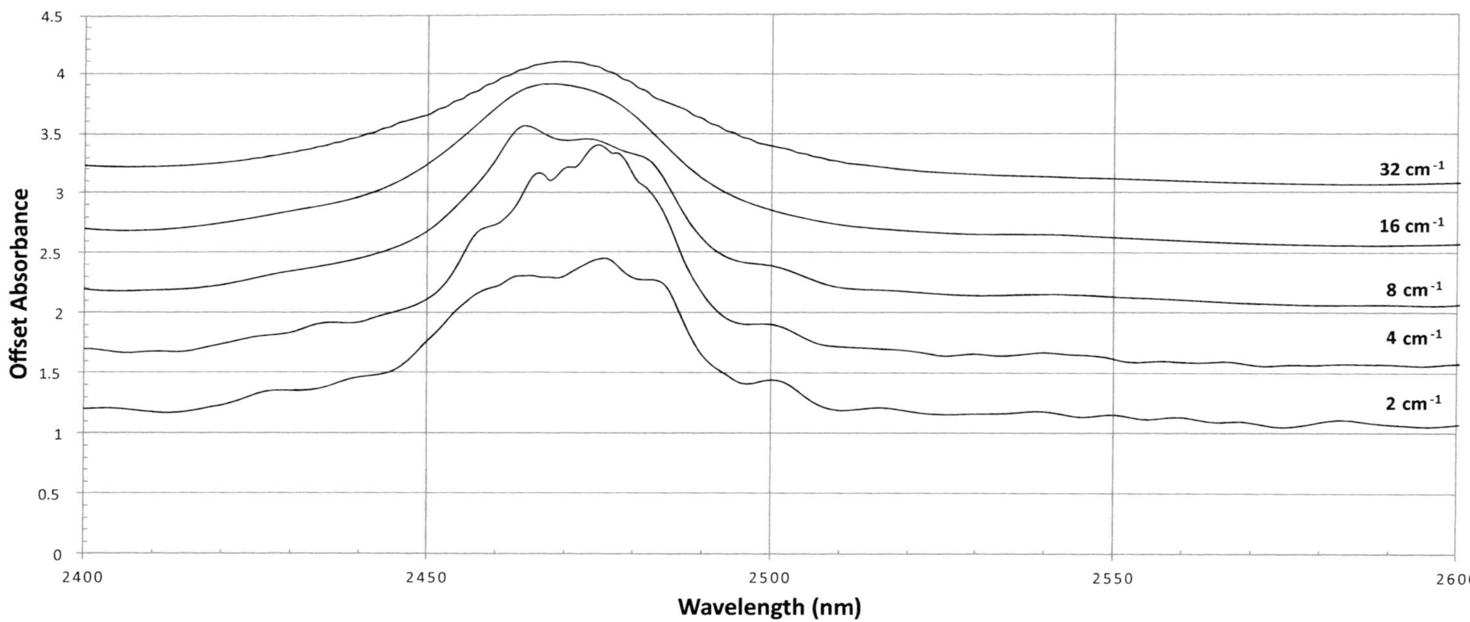

Fig. 8.5 Measurement of 1.0 mm thick polystyrene measured in transmittance and converted to absorbance as $A = \log_{10}(1/T)$ for 2, 4, 8, 16, and 32 wavenumber resolution using an FTIR instrument over the near infrared range. Savitzky-Golay smoothing, with 15 convolution points Quadratic-Cubic used for 2 and 4 wavenumber spectra; no smoothing for remaining spectra. Close-up of region from 2400 to 2600 nm.

Table 8.4 Peak positions for Polystyrene spectra at different resolution settings (from Figure 8.3), using a five-point cubic spline, peak interpolation method.[9]

$2\,cm^{-1}$	Peak height	$4\,cm^{-1}$	Peak height	$8\,cm^{-1}$	Peak height	$16\,cm^{-1}$	Peak height	$32\,cm^{-1}$	Peak height
2741.922	0.955	2741.922	0.903	2741.922	0.853	2740.833	0.720	2737.791	0.635
2617.731	0.642	2622.283	0.659	2621.633	0.654	2623.737	0.635	2622.482	0.611
2539.056	0.681	2539.706	0.674	2541.006	0.658	2536.565	0.652	NA	NA
2475.335	1.948	2474.684	2.396	2463.631	2.064	2468.909	1.917	2469.901	1.599
2351.794	1.108	2354.395	1.054	2353.745	1.029	2355.716	0.998	2362.745	0.933
2309.530	0.808	2309.530	0.765	2307.580	0.760	2308.227	0.712	2310.332	0.643
2169.734	0.912	2170.385	0.907	2169.734	0.882	2167.711	0.813	2163.575	0.732
2144.376	0.858	2144.376	0.937	2145.026	0.858	2146.894	0.727	NA	NA
1765.301	0.213	1764.651	0.202	1765.301	0.211	1763.729	0.198	1761.740	0.192
1678.823	0.693	1679.473	0.682	1681.424	0.695	1680.460	0.665	1682.537	0.624

8.3 Infrared Photometric Measurement Standards

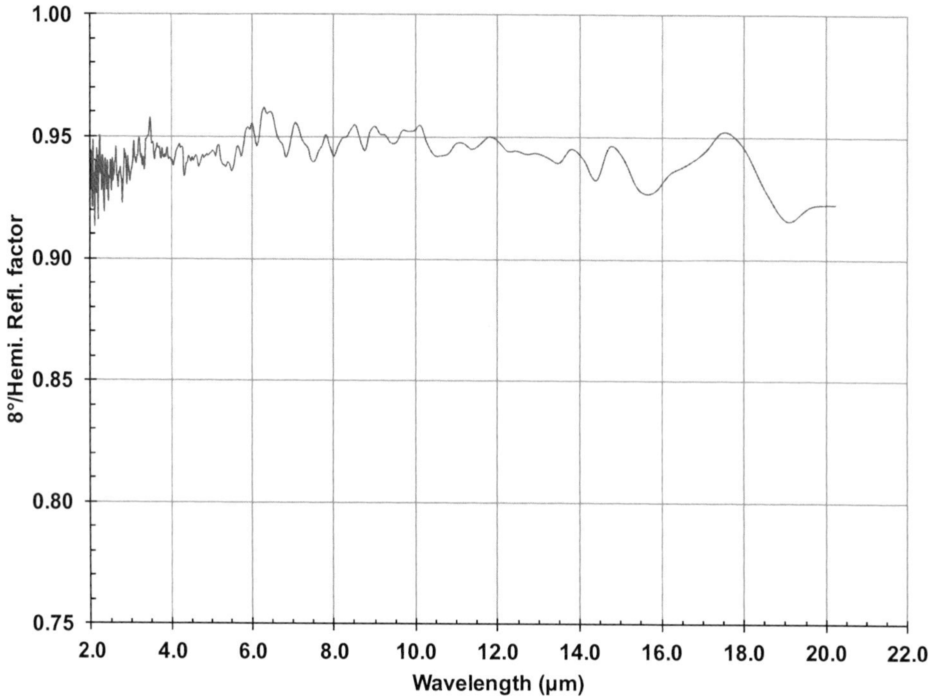

Fig. 8.6 Diffuse gold photometric reflectance standard measured using 8°/Hemispherical Spectral Reflectance Factor geometry with a calibrated FTIR instrument. Mean Reflectance over the spectral wavelength region (in microns) shown is 0.941. Measured photometric data is given in the Table below. (Data and spectrum used with permission from Avian Technologies, New London, NH).

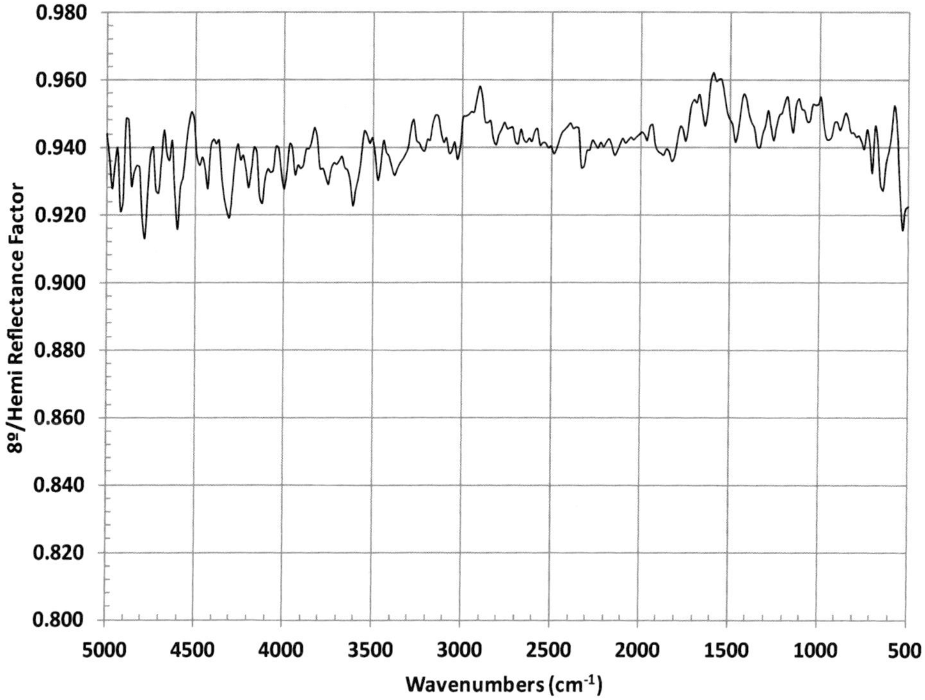

Fig. 8.7 Diffuse gold photometric reflectance standard measured using 8°/Hemispherical Spectral Reflectance Factor geometry with a calibrated FTIR instrument. Mean Reflectance over the spectral wavelength region (in microns) shown is 0.941. Measured photometric data is given in the Table below. (Data and spectrum used with permission from Avian Technologies, New London, NH).

Table 8.5 Diffuse Gold Reflectance Values in Wavenumber and Wavelength Space for specimen measured for Figures 8.6 and 8.7.

Wavenumber (cm^{-1})	Wavelength (μm)	Reflectance
5000	2.00	0.944
4000	2.50	0.928
3333	3.00	0.936
2857	3.50	0.947
2500	4.00	0.940
2222	4.50	0.941
2000	5.00	0.944
1818	5.50	0.937
1667	6.00	0.954
1538	6.50	0.957
1429	7.00	0.953

(Continued)

Table 8.5 (*Continued*)

Wavenumber (cm^{-1})	Wavelength (μm)	Reflectance
1333	7.50	0.940
1250	8.00	0.942
1176	8.50	0.954
1111	9.00	0.954
1053	9.50	0.947
1000	10.00	0.954
952	10.50	0.942
909	11.00	0.947
870	11.50	0.947
833	12.00	0.948
800	12.50	0.944
769	13.00	0.943
741	13.50	0.939
714	14.00	0.943
690	14.50	0.939
667	15.00	0.942
645	15.50	0.928
625	16.00	0.931
606	16.50	0.937
588	17.00	0.943
571	17.50	0.951
556	18.00	0.946
541	18.50	0.929
526	19.00	0.916
513	19.50	0.919
500	20.00	0.922

8.3.1 *The Meaning of the NIST Uncertainty Number for SRMs*

NIST uncertainty is a measure of uncertainty unlike the typical manufacturer report as either precision or accuracy. For example, NIST uncertainty has been calculated from Equation 8.9.[10]

$$U = \sqrt{A^2 + B^2 + C^2} \tag{8.9}$$

Where A = twice the largest standard deviation of measurement of multiple emission lines (measurands)versus the spectrometer measured values over a period of 31 days (or some designated period); B = twice the standard deviation of the uncertainty in the peak selection method used; and C is the maximum variation in the standard due to temperature changes over a specified range and conditions. (Note: The number 2 (twice) is referred to as the k value also known as the *coverage factor*, see reference.[11]

In testing SRM 1920a, the reported error for multiple wavelengths using emission lines for the NIST spectrometer was ±0.265 nm for all the data points measured over 31 days. The NIST actual reported uncertainty for SRM 1920a is equal to Equation 8.10.

$$U = \sqrt{0.57^2 + 0.41^2 + 0.15^2} = 0.75\,\text{nm} \tag{8.10}$$

This value was rounded up to ±1 nm considering the added uncertainty of the Cary spectrometer measurements and material stability over the lifetime of the SRM.[10]

A more detailed description of uncertainty computations and rationale with respect to wavelength SRMs is found in several references.[11–13] There are significant complexities to the question of uncertainty in reference materials and so the reader is referred directly to these three citations. One may also obtain more detailed information on the history of the international metrical committees responsible for defining the technical details for uncertainty associated with measurement results; the reader is referred to Appendices C and D in reference 11 for historical details.

8.4 Infrared Materials Data

The following tables and figures illustrate the properties of detectors, sources and optical materials updated to modern material specifications. These include: detector response curves (Tables 8.6 through 8.8 and Figures 8.8 through 8.10); Infrared Source emissivity (Table 8.9 and Figure 8.11); and optical material transmittance data (Tables 8.10 through 8.13 and Figures 8.12 through 8.16).[14–18]

8.4.1 *Infrared Detector Response Curves*

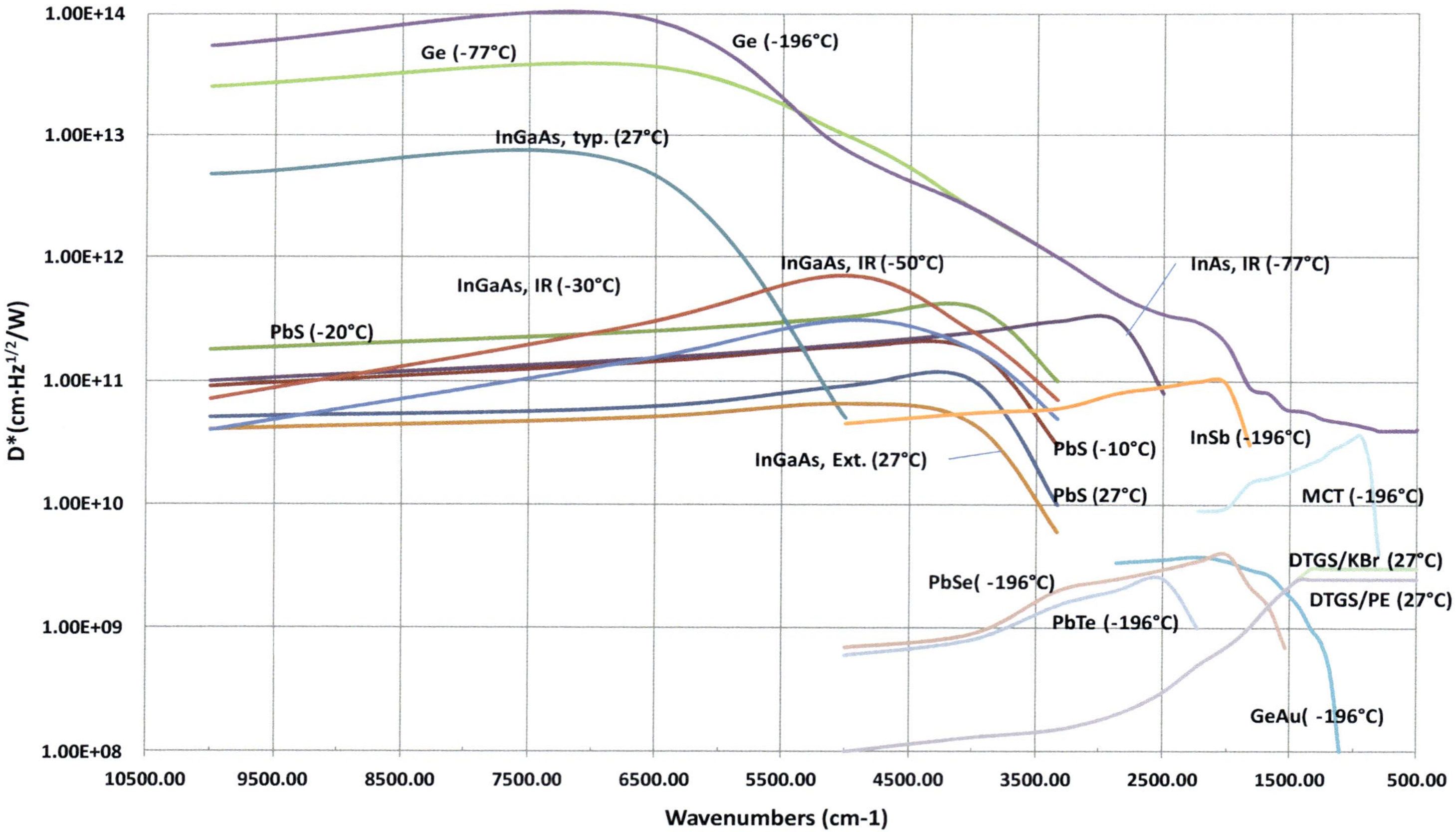

Fig. 8.8 Spectral Response (as $D^* = (cm \cdot Hz^{1/2})/W$) for infrared detectors (in wavenumbers (cm^{-1})).

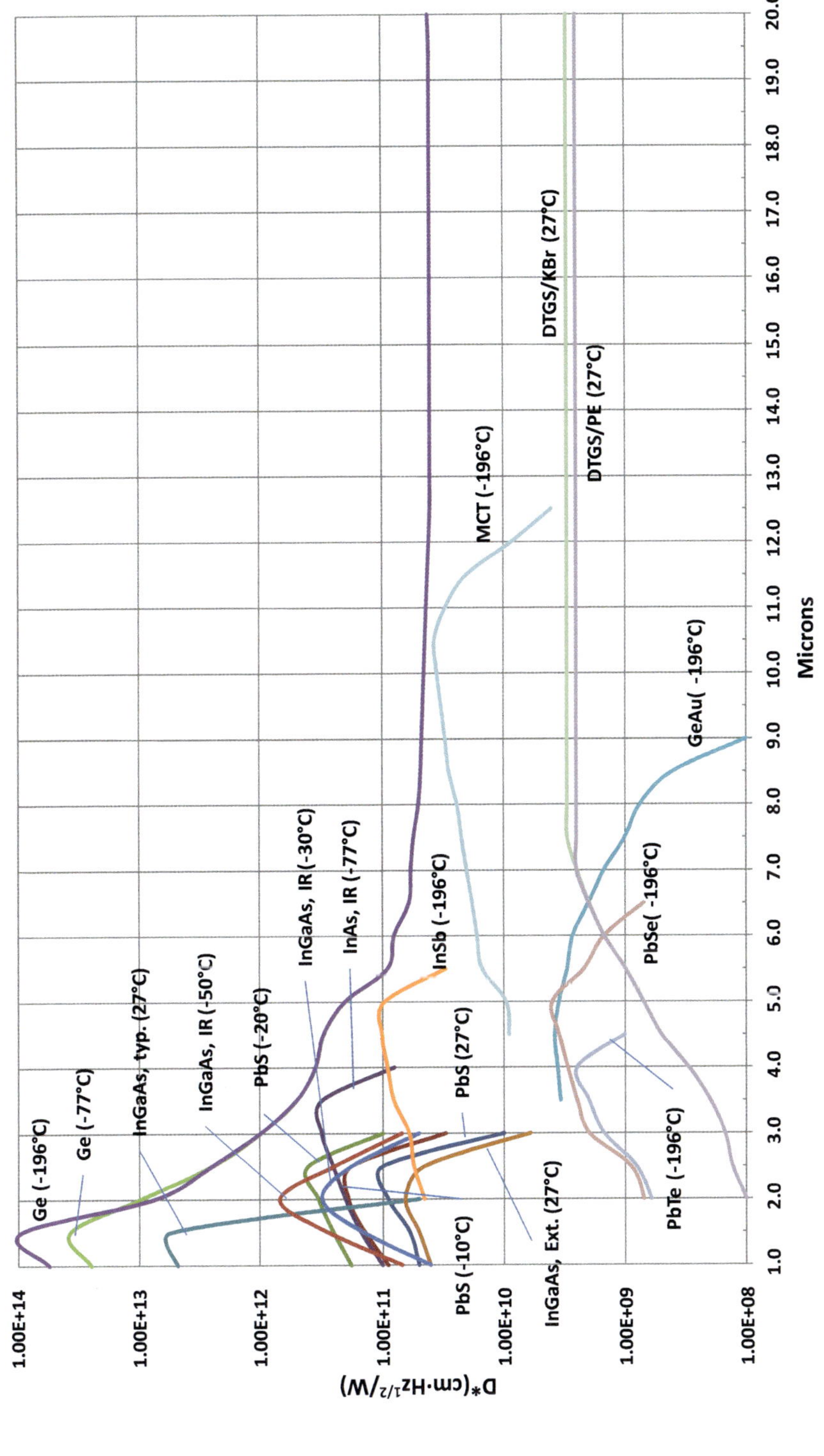

Fig. 8.9 Spectral Response (as $D^* = (cm \cdot Hz^{1/2})/W$) for infrared detectors (in microns).

Table 8.6 Spectral Response (as $D^* = (cm \cdot Hz^{1/2})/W$) for infrared detectors (in microns).

Microns	PbS (27°C)	PbS (−10°C)	PbS(−20°C)	InAs (−77°C)	InGaAs, typ (27°C)	InGaAs, ext (27°C)	InGaAs-IR (−30°C)	InGaAs-IR (−50°C)	Ge(−77°C)	Ge(−196°C)
1.0	5.00E+10	9.00E+10	1.80E+11	1.00E+11	4.80E+12	4.00E+10	4.00E+10	7.00E+10	2.50E+13	5.50E+13
1.5	6.00E+10	1.40E+11	2.50E+11	1.50E+11	5.50E+12	5.00E+10	1.45E+11	2.80E+11	3.75E+13	9.50E+13
2.0	9.00E+10	1.90E+11	3.30E+11	2.00E+11	5.00E+10	6.50E+10	3.10E+11	7.00E+11	1.00E+13	7.50E+12
2.5	1.00E+11	1.80E+11	4.00E+11	2.50E+11		4.50E+10	1.80E+11	2.50E+11	2.50E+12	2.50E+12
3.0	1.00E+10	3.00E+10	1.00E+11	3.10E+11		6.00E+09	5.00E+10	7.00E+10	1.00E+12	1.00E+12
3.5				3.10E+11					5.00E+11	5.00E+11
4.0				8.00E+10					3.50E+11	3.50E+11
4.5									3.00E+11	3.00E+11
5.0									2.00E+11	2.00E+11
5.5									9.00E+10	9.00E+10
6.0									8.00E+10	8.00E+10
6.5									6.00E+10	6.00E+10
7.0									5.80E+10	5.80E+10
7.5									5.50E+10	5.50E+10
8.0									5.00E+10	5.00E+10
8.5									4.80E+10	4.80E+10
9.0									4.70E+10	4.70E+10
9.5									4.60E+10	4.60E+10

(Continued)

Table 8.6 *(Continued)*

Microns	PbS (27°C)	PbS (−10°C)	PbS(−20°C)	InAs (−77°C)	InGaAs, typ (27°C)	InGaAs, ext (27°C)	InGaAs-IR (−30°C)	InGaAs-IR (−50°C)	Ge(−77°C)	Ge(−196°C)
10.0									4.50E+10	4.50E+10
10.5									4.40E+10	4.40E+10
11.0									4.30E+10	4.30E+10
11.5									4.20E+10	4.20E+10
12.0									4.10E+10	4.10E+10
12.5									4.00E+10	4.00E+10
13.0									4.00E+10	4.00E+10
13.5									4.00E+10	4.00E+10
14.0									4.00E+10	4.00E+10
14.5									4.00E+10	4.00E+10
15.0									4.00E+10	4.00E+10
15.5									4.00E+10	4.00E+10
16.0									4.00E+10	4.00E+10
16.5									4.00E+10	4.00E+10
17.0									4.00E+10	4.00E+10
17.5									4.00E+10	4.00E+10
18.0									4.00E+10	4.00E+10
18.5									4.00E+10	4.00E+10
19.0									4.00E+10	4.00E+10
19.5									4.00E+10	4.00E+10
20.0									4.10E+10	4.10E+10

Table 8.7 Spectral Response (as D* = (cm·Hz$^{1/2}$)/W) for infrared detectors (in microns).

microns	Ge:Au(−196°C)	InSb(−196°C)	PbTe (−196°C)	PbSe (−196°C)	DTGS/KBr (27°C)	DTGS/PE (27°C)	MCT(−196°C)
1.0							
1.5							
2.0		4.50E+10	6.00E+08	7.00E+08	1.00E+08	1.00E+08	
2.5		5.50E+10	8.00E+08	9.00E+08	1.30E+08	1.30E+08	
3.0		6.00E+10	1.50E+09	2.00E+09	1.50E+08	1.50E+08	
3.5	3.40E+09	8.00E+10	2.00E+09	2.50E+09	2.00E+08	2.00E+08	
4.0	3.60E+09	9.00E+10	2.50E+09	3.00E+09	3.00E+08	3.00E+08	
4.5	3.80E+09	1.00E+11	1.00E+09	3.50E+09	5.00E+08	5.00E+08	9.00E+09
5.0	3.50E+09	9.50E+10		4.00E+09	7.00E+08	7.00E+08	9.50E+09
5.5	3.00E+09	3.00E+10		2.20E+09	1.00E+09	1.00E+09	1.50E+10
6.0	2.70E+09			1.50E+09	1.50E+09	1.50E+09	1.65E+10
6.5	2.00E+09			7.00E+08	2.00E+09	2.00E+09	1.80E+10
7.0	1.50E+09				2.50E+09	2.50E+09	2.00E+10
7.5	1.00E+09				3.00E+09	2.50E+09	2.20E+10
8.0	7.50E+08				3.00E+09	2.50E+09	2.40E+10
8.5	4.00E+08				3.00E+09	2.50E+09	2.80E+10
9.0	1.00E+08				3.00E+09	2.50E+09	3.00E+10
9.5					3.00E+09	2.50E+09	3.25E+10

(Continued)

Table 8.7 (*Continued*)

microns	Ge:Au(−196°C)	InSb(−196°C)	PbTe (−196°C)	PbSe (−196°C)	DTGS/KBr (27°C)	DTGS/PE (27°C)	MCT(−196°C)
10.0					3.00E+09	2.50E+09	3.50E+10
10.5					3.00E+09	2.50E+09	3.70E+10
11.0					3.00E+09	2.50E+09	3.00E+10
11.5					3.00E+09	2.50E+09	2.00E+10
12.0					3.00E+09	2.50E+09	8.50E+09
12.5					3.00E+09	2.50E+09	4.00E+09
13.0					3.00E+09	2.50E+09	
13.5					3.00E+09	2.50E+09	
14.0					3.00E+09	2.50E+09	
14.5					3.00E+09	2.50E+09	
15.0					3.00E+09	2.50E+09	
15.5					3.00E+09	2.50E+09	
16.0					3.00E+09	2.50E+09	
16.5					3.00E+09	2.50E+09	
17.0					3.00E+09	2.50E+09	
17.5					3.00E+09	2.50E+09	
18.0					3.00E+09	2.50E+09	
18.5					3.00E+09	2.50E+09	
19.0					3.00E+09	2.50E+09	
19.5					3.00E+09	2.50E+09	
20.0					3.00E+09	2.50E+09	

Table 8.8 Infrared Detector Sensitivity Ranges (in microns).

Detector Sensitivity Ranges	Start	End	Difference
Silicon (Si)	0.3	1.1	0.8
Lead Sulfide (PbS)	1.1	3	1.9
Indium Arsenide (InAs)	1.7	5.7	4.0
Indium Gallium Arsenide (InGaAs), typical	0.9	1.7	0.8
Indium Gallium Arsenide (InGaAs), extended	0.65	2.4	1.8
Indium Gallium Arsenide (InGaAs), IR extended	0.65	2.7	2.1
Germanium (Ge:X)	2	40	38.0
Germanium Gold (Ge:Au)	2	9	7.0
Germanium Cadmium (Ge:Cd)	2	24	22.0
Lead Selenide (PbSe)	1.7	5.5	3.8
Germanium Zinc (Ge:Zn)	2	40	38.0
Indium Antimonide (InSb)	1.8	6.8	5.0
Lead Telluride (PbTe)	1.5	4.5	3.0
Lead Selenide (PbSe)	1.5	4.5	4.0
Deuterated Triglycine Sulfate (DTGS/KBr)	0.83	25	24.2
Deuterated Triglycine Sulfate (DTGS/PE)	10	120	110.0
Mercury Cadmium Telluride (MCT)	1	17	16.0
Triglycine Sulfate (TGS)	10	120	110.0
Pyroelectric Lithium Tantalate ($LiTaO_3$) (PLT)	1.5	30	28.5
Infrared	2.5	25	22.5
Raman	2.5	25	22.5

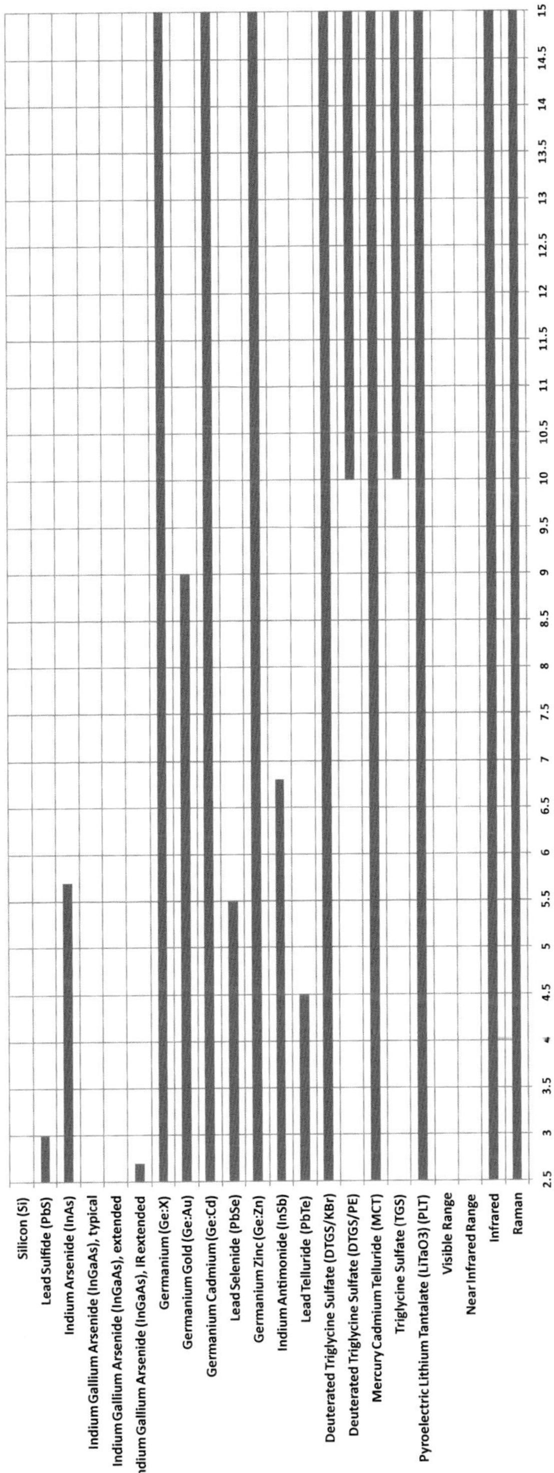

Fig. 8.10 Detector sensitivity ranges (in microns): Note the IR range is officially from 2.5 microns to 25 microns (i.e., 4000 to 400 cm^{-1}).

8.4.2 *Infrared Source Emissivity Ranges*

Table 8.9 Infrared Source Emissivity Ranges (in microns).

Source Emissivity Ranges	Start	End	Difference
Globar	1	100	99
Nernst glower	0.3	35	35
Carbon arc	0.5	100	100
Near Infrared Range	0.78	2.5	2
Infrared	2.5	25	23

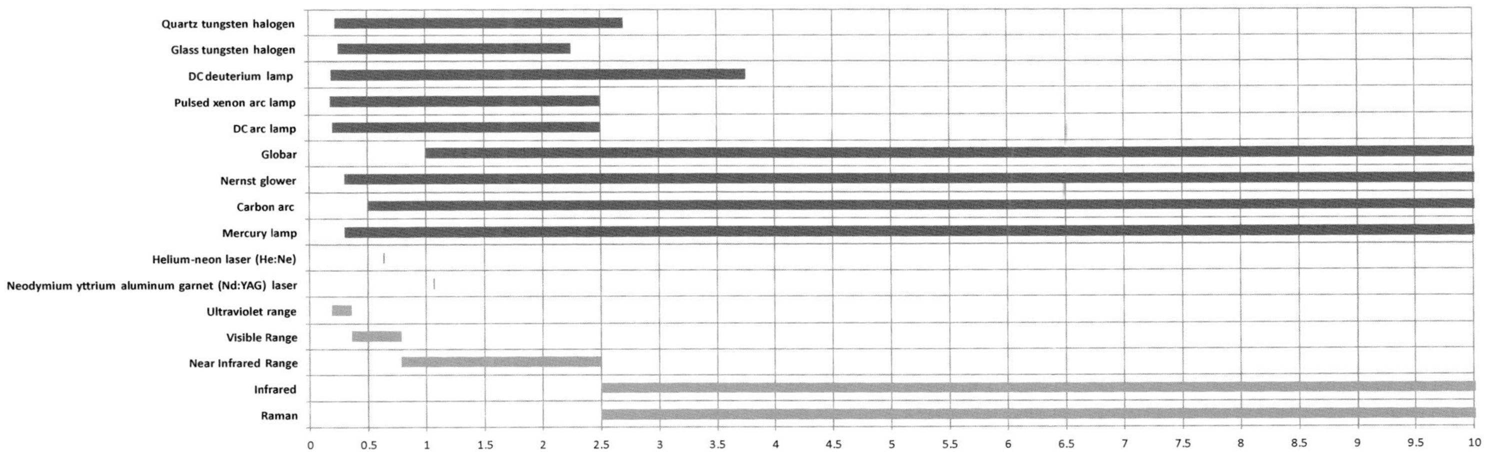

Fig. 8.11 Spectroscopy Emission Source energy ranges (in microns): Note the IR range is officially from 2.5 microns to 25 microns (i.e., 4000 to 400 cm^{-1}).

8.4.3 *Infrared Optical Material Transmittance Ranges*

Table 8.10 Infrared Optical Materials Transmittance Ranges (in microns).

Optical Material Spectral Ranges	Start	End	Difference
Sodium Chloride (NaCl)	0.25	16	6
Chalcogenide (NIR-IR measurements)	2.22	11.1	9
Cesium Iodide (CsI)	0.3	50	50
Potassium Bromide (KBr)	0.25	26	26
Potassium Chloride (KCl)	0.25	20	20
Thallium Bromide-Iodide (KRS-5)	0.6	40	39
Germanium (Ge)	1.1	30	29
Arsenic Sulfide (As_2S_3)	0.6	15	14
Magnesium Fluoride (MgF_2)	0.6	9.5	9
Zinc Selenide (ZnSe) (IRTRAN-4)	0.6	26	25
Barium Fluoride (BaF_2)	0.2	13	13
Zinc Sulfide (ZnS) (Cleartran or IRTRAN-2)	0.6	15	14
Calcium Fluoride (CaF_2) (IRTRAN-3)	0.2	9	9
Silver Bromide (AgBr)	0.5	35	35
Polyethylene, Polymer (High Density)	16	333	317
Infrared	2.5	25	23
Raman	2.5	25	23

Table 8.11 Infrared Optical Materials Transmittance Data (in T with respect to nm).

Wavelength (nm)	NaCl	CsI	KBr	KCl	KRS-5	Ge	MgF_2	ZnSe	BaF_2	ZnS	CaF_2
100	0.000	0.000	0.000	0.000	0.000	0.000	0.000	0.000	0.100	0.000	0.100
200	0.100	0.000	0.650	0.650	0.000	0.000	0.780	0.000	0.900	0.000	0.840
300	0.880	0.620	0.840	0.840	0.000	0.000	0.890	0.000	0.900	0.100	0.890
400	0.900	0.740	0.895	0.895	0.000	0.000	0.920	0.000	0.900	0.500	0.900
500	0.900	0.780	0.900	0.900	0.000	0.000	0.920	0.000	0.900	0.640	0.900
600	0.900	0.810	0.900	0.900	0.000	0.000	0.920	0.650	0.900	0.685	0.900
700	0.900	0.810	0.900	0.900	0.620	0.000	0.920	0.680	0.900	0.685	0.900
800	0.900	0.810	0.900	0.900	0.650	0.000	0.920	0.705	0.900	0.686	0.905
900	0.900	0.815	0.900	0.900	0.660	0.000	0.920	0.710	0.900	0.686	0.905
1000	0.900	0.815	0.900	0.900	0.670	0.000	0.920	0.715	0.900	0.687	0.910
1450	0.900	0.818	0.900	0.900	0.672	0.000	0.920	0.720	0.900	0.687	0.915
1500	0.900	0.818	0.900	0.900	0.673	0.000	0.920	0.722	0.900	0.695	0.915
1950	0.900	0.818	0.900	0.900	0.674	0.000	0.920	0.723	0.900	0.695	0.915
2000	0.900	0.820	0.900	0.900	0.675	0.300	0.920	0.725	0.910	0.700	0.920
2350	0.900	0.820	0.900	0.900	0.677	0.415	0.920	0.725	0.910	0.700	0.920

(Continued)

Table 8.11 (*Continued*)

Wavelength (nm)	NaCl	CsI	KBr	KCl	KRS-5	Ge	MgF_2	ZnSe	BaF_2	ZnS	CaF_2
3000	0.900	0.821	0.900	0.900	0.677	0.415	0.925	0.725	0.910	0.705	0.921
4000	0.900	0.822	0.900	0.900	0.677	0.415	0.930	0.725	0.910	0.705	0.923
5000	0.900	0.823	0.900	0.910	0.677	0.415	0.910	0.725	0.920	0.710	0.925
6000	0.905	0.824	0.900	0.900	0.677	0.415	0.750	0.730	0.920	0.710	0.905
7000	0.910	0.824	0.900	0.900	0.677	0.415	0.550	0.730	0.600	0.710	0.825
8000	0.910	0.824	0.900	0.900	0.678	0.415	0.300	0.730	0.500	0.710	0.740
9000	0.910	0.824	0.900	0.900	0.679	0.415	0.050	0.730	0.400	0.710	0.400
10000	0.910	0.825	0.900	0.900	0.680	0.405	0.000	0.735	0.300	0.670	0.100
15000	0.500	0.818	0.900	0.880	0.677	0.365	0.000	0.750	0.000	0.400	0.000
18000	0.100	0.815	0.900	0.800	0.675	0.340	0.000	0.650	0.000	0.000	0.000
20000	0.000	0.812	0.905	0.625	0.673	0.210	0.000	0.100	0.000	0.000	0.000
30000	0.000	0.800	0.100	0.000	0.670	0.000	0.000	0.000	0.000	0.000	0.000
40000	0.000	0.650	0.000	0.000	0.640	0.000	0.000	0.000	0.000	0.000	0.000
50000	0.000	0.100	0.000	0.000	0.480	0.000	0.000	0.000	0.000	0.000	0.000
60000	0.000	0.000	0.000	0.000	0.000	0.000	0.000	0.000	0.000	0.000	0.000

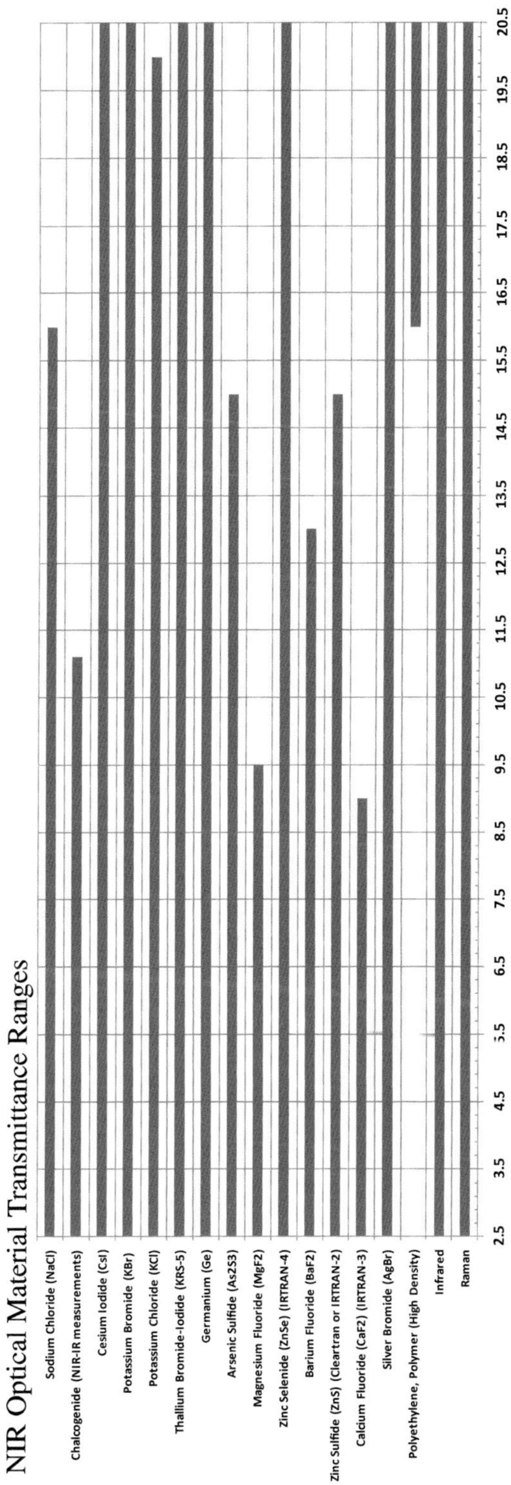

Fig. 8.12 Infrared Optical Materials Transmittance Ranges (in microns): Note the NIR range is officially from 780 nm to 2500 nm (0.78 microns to 2.5 microns).

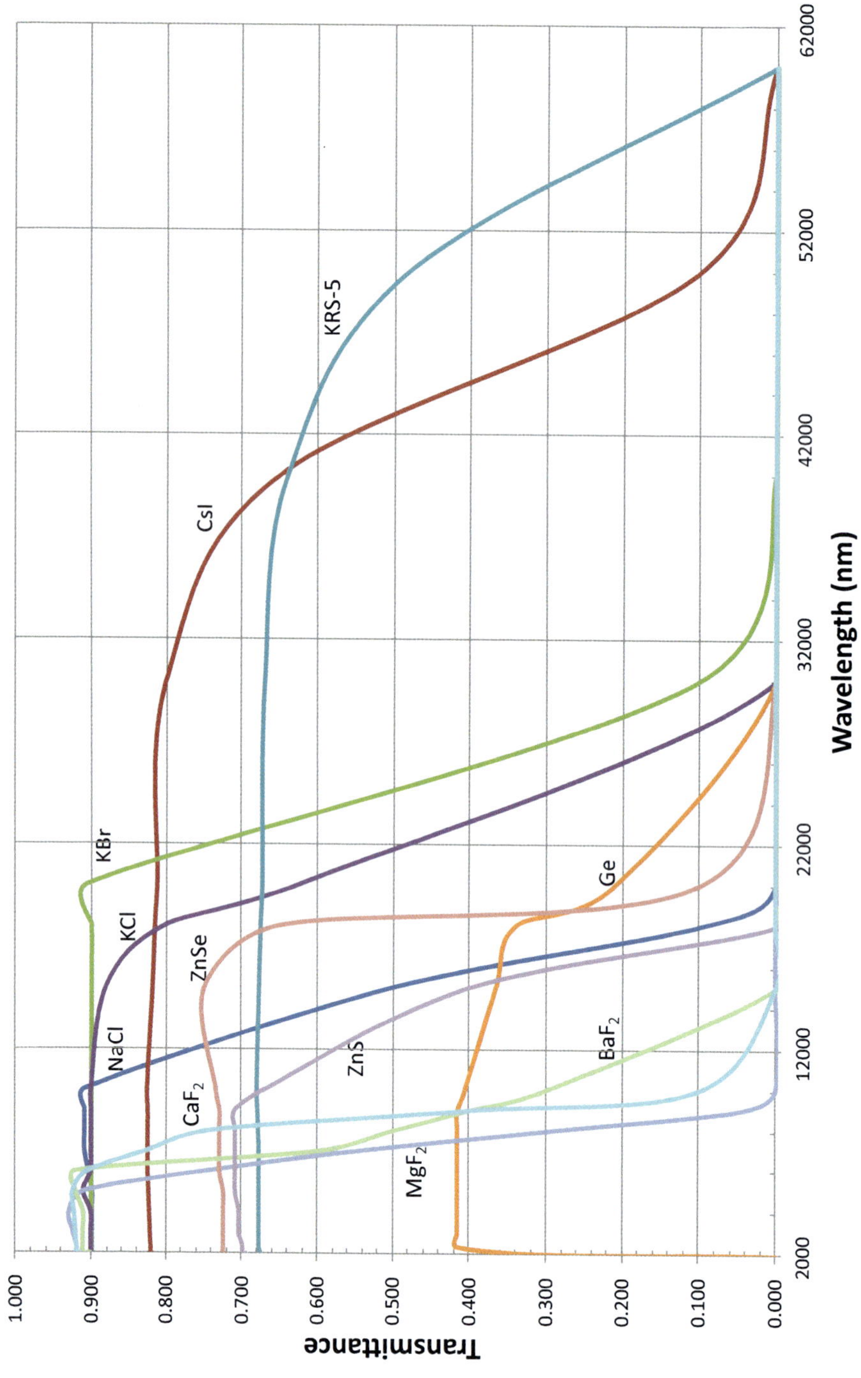

Fig. 8.13 Infrared Optical Materials Transmittance Data (in T with respect to nm).

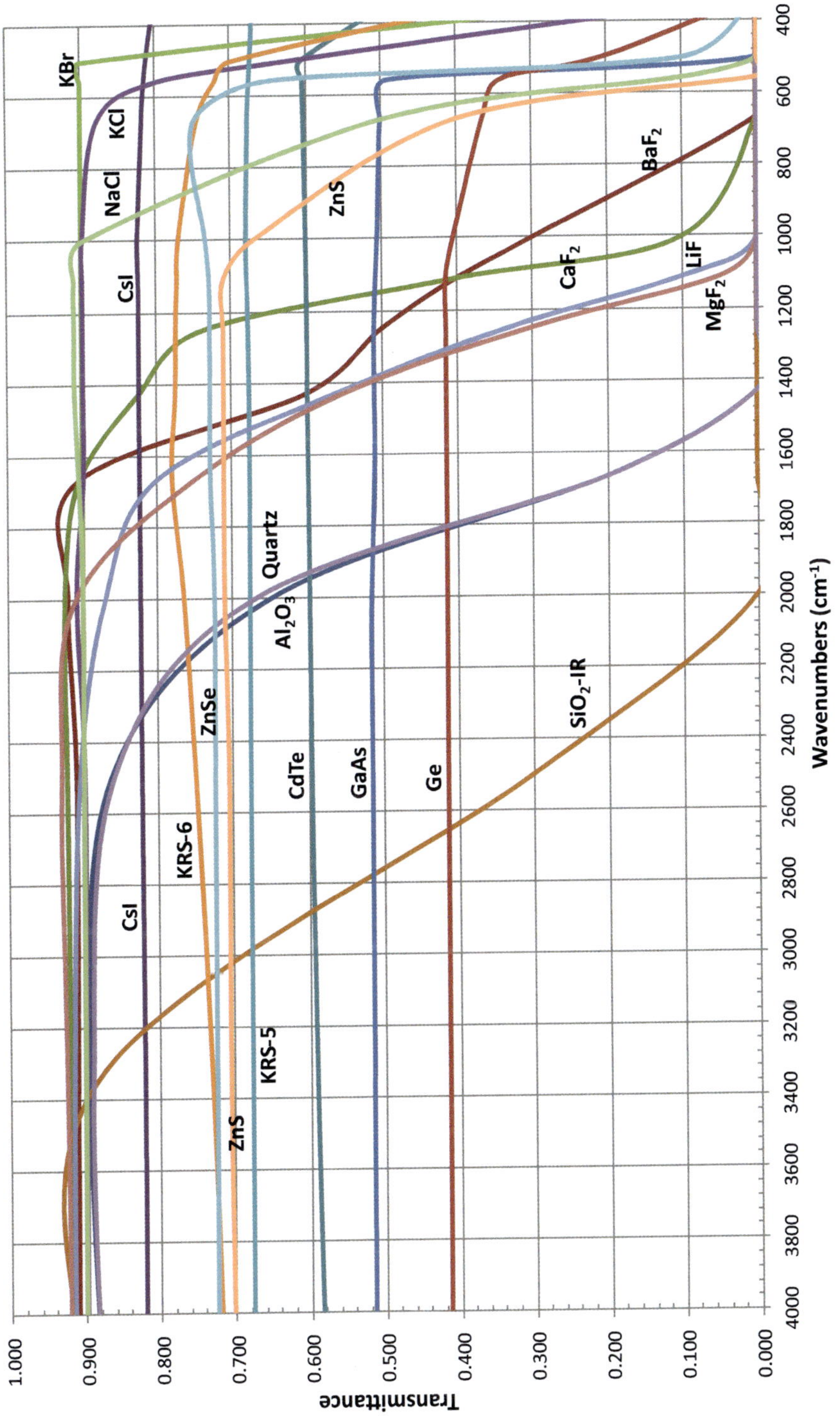

Fig. 8.14 Infrared Optical Materials Transmittance Data (in T with respect to wavenumbers, cm^{-1}).

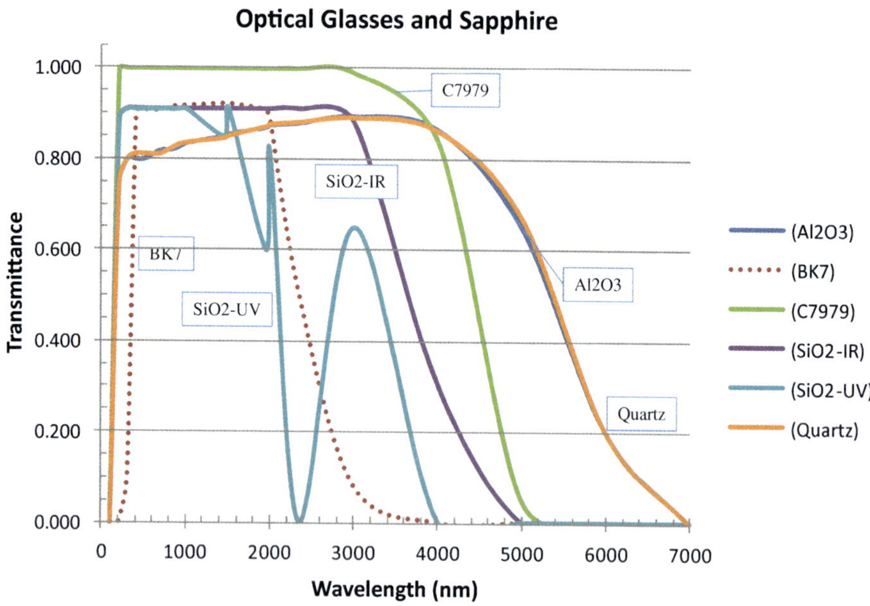

Fig. 8.15 Optical Glasses (i.e., BK7, Corning 7979, IR glass, UV glass, Quartz) and Sapphire (Al_2O_3) Transmittance Data versus Wavelength (in nanometers).

Table 8.12 Optical Glass and Sapphire Transmittance Data versus Wavelength (in nanometers).

Lambda (nm)	(Al2O3)	(BK7)	(C7979)	(SiO2-IR)	(SiO2-UV)	(Quartz)
100	0.000	0.000	0.0000	0.000	0.000	0.000
200	0.750	0.000	0.9970	0.890	0.890	0.750
300	0.800	0.100	0.9999	0.910	0.910	0.800
400	0.800	0.900	1.0000	0.910	0.910	0.810
500	0.800	0.905	1.0000	0.910	0.910	0.810
600	0.810	0.905	1.0000	0.910	0.910	0.810
700	0.820	0.905	1.0000	0.910	0.910	0.810
800	0.820	0.910	1.0000	0.910	0.910	0.820
900	0.825	0.915	1.0000	0.910	0.910	0.830
1000	0.835	0.915	1.0000	0.910	0.910	0.835
1450	0.850	0.920	1.0000	0.910	0.850	0.845
1500	0.855	0.920	1.0000	0.910	0.910	0.850
1950	0.870	0.905	0.9999	0.910	0.600	0.870
2000	0.875	0.880	0.9997	0.910	0.810	0.875
2350	0.880	0.500	1.0000	0.910	0.000	0.880
3000	0.895	0.080	0.9907	0.880	0.650	0.890
4000	0.865	0.000	0.8369	0.300	0.000	0.860
5000	0.650	0.000	0.0500	0.000	0.000	0.670
6000	0.200	0.000	0.0000	0.000	0.000	0.200
7000	0.000	0.000	0.0000	0.000	0.000	0.000

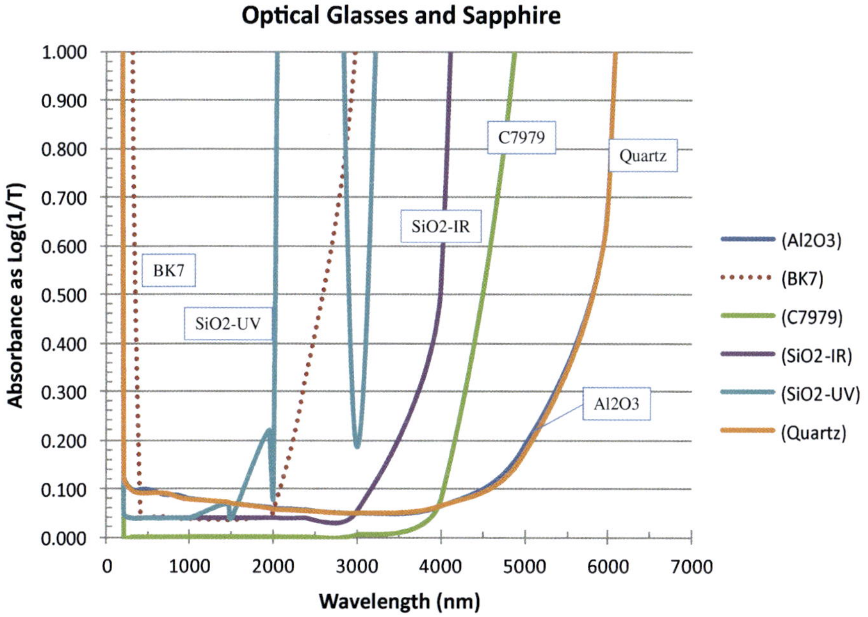

Fig. 8.16 Optical Glasses (i.e., BK7, Corning 7979, IR glass, UV glass, Quartz) and Sapphire (Al_2O_3) Absorbance Data as $Log_{10}(1/T)$ versus Wavelength (in nanometers).

Table 8.13 Optical glass and sapphire absorbance (Log(1/T)) data.

Lambda (nm)	(Al2O3)	(BK7)	(C7979)	(SiO2-IR)	(SiO2-UV)	(Quartz)
100	7.000	7.000	7.0000	7.000	7.000	7.000
200	0.125	7.000	0.0013	0.051	0.051	0.125
300	0.097	1.000	0.0000	0.041	0.041	0.097
400	0.097	0.046	0.0000	0.041	0.041	0.092
500	0.097	0.043	0.0000	0.041	0.041	0.092
600	0.092	0.043	0.0000	0.041	0.041	0.092
700	0.086	0.043	0.0000	0.041	0.041	0.092
800	0.086	0.041	0.0000	0.041	0.041	0.086
900	0.084	0.039	0.0000	0.041	0.041	0.081
1000	0.078	0.039	0.0000	0.041	0.041	0.078
1450	0.071	0.036	0.0000	0.041	0.071	0.073
1500	0.068	0.036	0.0000	0.041	0.041	0.071
1950	0.060	0.043	0.0000	0.041	0.222	0.060
2000	0.058	0.056	0.0001	0.041	0.092	0.058
2350	0.056	0.301	0.0000	0.041	7.000	0.056
3000	0.048	1.097	0.0041	0.056	0.187	0.051
4000	0.063	7.000	0.0773	0.523	7.000	0.066
5000	0.187	7.000	1.3010	7.000	7.000	0.174
6000	0.699	7.000	7.0000	7.000	7.000	0.699
7000	7.000	7.000	7.0000	7.000	7.000	7.000

References

1. ASTM (American Society for Testing and Materials) Method E275-01, "Standard Practice for Describing and Measuring Performance of Ultraviolet, Visible, and Near Infrared Spectrophotometers", ASTM International, 100 Barr Harbor Drive, PO Box C700, West Conshohocken, PA (2001–1014).
2. ASTM (American Society for Testing and Materials) Method E 131-06, "Standard Terminology Relating to Molecular Spectroscopy", ASTM International, 100 Barr Harbor Drive, PO Box C700, West Conshohocken, PA (2006).
3. NIST SRM website: URL at http://www.nist.gov/srm/index3column.cfm
4. NIST SRM URL at https://www-s.nist.gov/srmors/detail.cfm
5. D. Gupta, L. Wang, L. M. Hanssen, J.J. Hsia, and R.U. Datla, Polysytrene Films for Calibrating the Wavelength Scale of Infrared Spectrophotomers - SRM 1921, NIST Special Publication 260–122, National Institute of Standards and Technology, Gaithersburg, MD (April 1995), 25 pages.
6. Standard Reference Material (SRM)®1921 Certificate, Infrared Transmission Wavelength Standard, National Institute of Standards and Technology, Gaithersburg, MD 20899, 6 pages (7 April 1995).
7. Standard Reference Material (SRM)®1921b Certificate, "Infrared Transmission Wavelength Standard", National Institute of Standards and Technology, Gaithersburg, MD 20899, 7 pages, Certificate Issue Date: 19 February 2014.
8. W.H. Press, S.A. Teukolsky, W.T. Vetterling, B.P. Flannery, *Numerical Recipes: The Art of Scientific Computing,* 3rd edn. (Cambridge University Press, 2007, 120–124).
9. eFTIR Software, http://www.essentialftir.com/ (Algorithm as implemented into eFTIR software peak picking algorithm.)
10. V.R. Weidner, P.Y. Barnes, and K.L. Eckerle, A Wavelength Standard for the Near Infrared Based on the Reflectance Of Rare-Earth Oxides, *J. Research NBS* **91**(5) (1986) 243–253.
11. B.N. Taylor and C.E. Kuyatt, Guidelines for Evaluating and Expressing the Uncertainty of NIST Measurement Results, National Institute of Standards and Technology (NIST), Technical Note 1297, 20 pages (1994).
12. Standard Reference Material (SRM)®2036 Certificate, Near Infrared Wavelength/Wavenumber Reflection Standard, National Institute of Standards and Technology, Gaithersburg, MD 20899, 7 pages (24 January 2013).
13. S.J. Choquette, D.L. Duewer, L.M. Hanssen, E.A. Early, Standard reference material 2036 near infrared reflection wavelength standard, *Appl. Spectrosc.* **59**(4) (2005) 496–504.
14. http://www.hamamatsu.com/resources/pdf/ssd/infrared_techinfo_e.pdf
15. A. Rogalski, Infrared Detectors at the Beginning of the Next Millenium, *Opto–Electron. Rev.* **9**(2) (2001) 173–187.
16. A. Rogalski, History of infrared detectors, *Opto–Electron. Rev.* **20**(3) (2012) 279–308.
17. E.L. Dereniak and G.D. Boreman, *Infrared detectors and systems* (Wiley, New York, NY, 1996).
18. D.L. Shumaker and J.S. Accetta (Eds), *The Infrared and Electro Optical Systems Handbook* (8 Volume Set) Hardcover (Press Monographs , SPIE-International Society for Optical Engine (January 1993).

Chapter 9

Glossary of Infrared Spectroscopy Terms

Contents

MIR — Rudimentary Glossary

(Note: More details for each topic are found within Volume 3 (Chapter 9), the index, the glossary, and the various other volumes of this text set).[1–3]

ABSORBANCE — Measure of concentration of material present: Negative log (base 10) of Transmittance or Reflectance $[-\log_{10} T]$ or $[-\log_{10} R]$ of product of extinction coefficient/absorptivity (ε), pathlength (b), and concentration (c), written as $A = \varepsilon bc$.

ABSORPTIVITY — Probability of light absorbing at a particular wavenumber for a specific analyte under specific conditions, e.g., pH, solvent, and temperature. Thus, a specific amount of material at specified conditions will absorb a specific fraction of the light striking it. Absorptivity is signified by an epsilon (ε) or a small "a".

ANALYTE — The particular material or quality to be determined in an analytical analysis.

APODIZATION OR APODIZATION FUNCTION — A function used to remove the "ripple" effect on either side of spectral peaks or bands following the application of the Fourier transform to an interferogram. These ripples are minimized by applying a triangular function or Happ-Genzel (smoothed triangular function) to the spectral data to produce smaller ripples and slightly broadened peak widths.

BEER'S LAW — Relationship between the amount of light absorbed by an analyte and its concentration (c), pathlength (b), and absorptivity (ε), written as $A = \varepsilon bc$.

Note that ε is the molar absorptivity (referred to as molar extinction coefficient by earlier physicists) in units of Liter · Mole^{-1} · cm^{-1}; c is the concentration of molecules in the spectrometer beam in units of Mole · Liter^{-1} (Note: this is a volume fraction unit not weight fraction); and pathlength (i.e., the thickness of the sample in units of cm). The absorptivity for any specific molecule type is calculated by careful measurements of the absorbance of a sample by applying the relationship.

BOXCAR APODIZATION — Is the apodization function applied to FT transformed interferogram spectra that maximizes both peak height and resolution (band halfwidth) as compared to other apodization functions. It is the apodization function most often used for gas phase infrared spectrophotometry. This is a square waveform apodization function as contrasted to triangular or Happ-Genzel (smoothed triangular function) apodization functions.

CALIBRATION — Determining the optical response (absorbance) of some analytical method to known amounts of pure analyte. This may be a univariate or multivariate method for near infrared spectroscopy.

CALIBRATION CURVE — The results of a calibration when graphed, usually Cartesian coordinates, e.g., concentration (in moles/liter or grams/liter) versus absorbance.

CENTERBURST — An interferogram has a centerburst, also referred to as the ZPD or 'zero-path-difference' location. When the path of the fixed and moving mirror are identical in the interferometer, the interferogram is at maximum interference; referred to as the centerburst.

CONCENTRATION — The amount of analyte in a given volume of solution, e.g., moles per liter.

CUVET (A.K.A. CUVETTE) — Transparent receptacle in which sample solutions are introduced into the light path of spectrometers. These are usually cylindrical or rectangular for clear liquid transmittance measurements.

DETECTOR — Device used to detect the intensity of the radiation reflected from, or transmitted through, a sample. There is detailed information for near infrared detectors within this set of volumes.

DOUBLE-SIDED INTERFEROGRAM — An interferogram has a 'center-burst', also referred to as the ZPD or 'zero-path-difference' location. When the path of the fixed and moving mirror are identical in the interferometer, the interferogram is at maximum interference. The centerburst may be placed anywhere along the recorded data for the interferogram. When it is placed in the center of the interferogram it is termed a double-sided interferogram (also known as an asymmetric interferogram).

EXTINCTION, AND EXTINCTION COEFFICIENT — Terms sometimes substituted for absorptivity (ε), respectively.

FREQUENCY — The number of times per unit that the magnitude of an electromagnetic wave goes from maximum to minimum then back to maximum amplitude. The unit for the number of waves per second is Hertz (Hz).

INTERFEROGRAM — A wave pattern created using an interferometer where waves are superimposed and later interpreted to compute a pattern of amplitude versus frequency for a measurement of light interaction with materials. This interferometer pattern is interpreted using a Fourier transform to compute an infrared spectrum.

LIMIT OF DETECTION — Lowest amount of analyte that may be seen above noise level of instrument for those measurement conditions.

LINEAR RESPONSE — Straight line response between concentration and measured signal. In context, this means that for a doubling of the analyte concentration, the detectable and measureable signal will double. This allows a linear calibration model or equation to be used to determine the concentration of analyte through a range of absorbance levels.

LINEARITY — The response of an instrument changing in a predictable and linear fashion with increasing analyte concentration.

NOISE — Any signal generated by the detector that is not directly responding to the light transmitted at the required wavenumber.

PATHLENGTH — The distance the light passes through the sample in its holder. In practical terms, the inside dimension of the sample holding device, or in reflection measurement mode, the length of the light passing through the sample holding device.

REFERENCE — A material with known characteristics, such as peaks of identified wavenumber positions, or photometric values of reflected energy in Absorbance units (Au) at a specific wavenumber.

REFERENCE STANDARD — Reference material wherein the photometric properties are documented or certified. This reference standard is used to calibrate the basic performance of a spectrometer. Government issued reference standards are often termed SRMs or Standard Reference Materials.

RETARDATION — Is defined as the optical path difference between the two beams in the interferometer. Higher retardation gives better resolution, increases the number of data points, and results in larger computer files.

SCANNING — The process where the wavenumber range of the system is viewed in serial order, usually from highest to lowest wavenumber i.e., highest frequency. This usually occurs when the grating is rotated about its axis, or an interferometer is sequenced for scanning an interferogram and converting to a spectrum.

SENSITIVITY — See "Limits of Detection".

SIGNAL — The output of the detector due to its response to energy emerging from the sample in the sample holder. It consists of an analog signal converted to a digital signal and converted to units such as counts, reflectance, transmittance, or absorbance; depending upon the measurement geometry and computational methods.

SIGNAL-TO-NOISE RATIO — The numerical ratio of the total signal I/I_0 to the noise of the instrument as root mean squared (RMS) deviation of repeated signal measurements. The RMS is the standard deviation of the photometric measurements at each wavenumber.

SINGLE-SIDED INTERFEROGRAM — An interferogram has a 'center-burst', also referred to as the ZPD or 'zero-path-difference' location. When the path of the fixed and moving mirror are identical in the interferometer, the interferogram is at maximum interference. The centerburst may be placed anywhere along the recorded data for the interferogram. When it is placed near the start of the interferogram it is termed a single-sided interferogram (also known as a symmetric interferogram).

SOURCE — Also known as the "Globar". This is the origin of the light used in the spectrometer, an infrared heat source is most often used for infrared spectrometers.

SPECTRUM — Series of wavenumbers of radiation, belonging to a specific portion of the electromagnetic spectrum, e.g., the near infrared spectrum, where the absorbance of a sample measurement is recorded with respect to wavenumber.

WAVENUMBER (CM^{-1}) — For infrared spectroscopy, the wavenumber represents the number of waves (cycles) occurring per centimeter designated as cm^{-1}. It is equivalent to 1×10^7/wavelength in nanometers, e.g., 2500 nm is equivalent to $4000\,cm^{-1}$.

ZERO FILLING — Is a process of adding a number of data points consisting of amplitude values located between corresponding interferogram data points to increase the number of data points. This action is completed prior to applying the FFT to the interferogram in order to produce a spectrum. Zero-filling adds interpolated points that lie between the real interferogram data points. The resultant data is smoother in appearance and more suitable for application of the FFT algorithm (proper number of data points). The appearance and photometric accuracy of the spectrum is improved, but there is no improvement in the linewidth resolution of the spectrum.

ZPD — An interferogram has a centerburst, also referred to as the ZPD or 'zero-path-difference' location. When the path of the fixed and moving mirror are identical within the interferometer, the interferogram is at maximum interference; there is zero path difference between the light traveling through the two channels of the Michelson interferometer.

References

1. J. M. Chalmers, P. R. Griffiths, *Handbook of Vibrational Spectroscopy* 1st edn. (John Wiley & Sons, 2002).
2. Roy Cain, personal communications, *A User's Guide to Fourier Transform Infrared Spectrometers*, (ASTM, March 2006). Roy was an active leader in ASTM activities and was a recipient of the ASTM Fellow, Award of Merit in (1999).
3. ASTM (American Society for Testing and Materials) Method ASTM E1421 — 99(2009), "Standard Practice for Describing and Measuring Performance of Fourier Transform Mid-Infrared (FT-MIR) Spectrometers: Level Zero and Level One Tests", ASTM International, 100 Barr Harbor Drive, PO Box C700, West Conshohocken, PA, 19428-2959 USA (2009).

Volume 4 Index

E

F

G

H

I

J

K

L

M

N

O

P

Q

R

S

T

U

V

W